Analysis of Turbulence in Fusion Plasmas

Online at: https://doi.org/10.1088/978-0-7503-4856-0

IOP Series in Plasma Physics

Series Editor

Richard Dendy

Culham Centre for Fusion Energy and the University of Warwick, UK

About the series

The IOP Plasma Physics ebook series aims at comprehensive coverage of the physics and applications of natural and laboratory plasmas, across all temperature regimes. Books in the series range from graduate and upper-level undergraduate textbooks, to research monographs and reviews.

The conceptual areas of plasma physics addressed in the series include:

- Equilibrium, stability, and control.
- Waves: fundamental properties, emission, and absorption.
- Nonlinear phenomena and turbulence.
- Transport theory and phenomenology.
- Laser–plasma interactions.
- Non-thermal and suprathermal particle populations.
- Beams and non-neutral plasmas.
- High energy density physics.
- Plasma–solid interactions, dusty, complex, and non-ideal plasmas.
- Diagnostic measurements and techniques for data analysis.

The fields of application include:

- Nuclear fusion through magnetic and inertial confinement.
- Solar–terrestrial and astrophysical plasma environments and phenomena.
- Advanced radiation sources.
- Materials processing and functionalisation.
- Propulsion, combustion, and bulk materials management.
- Interaction of plasma with living matter and liquids.
- Biological, medical, and environmental systems.
- Low-temperature plasmas, glow discharges, and vacuum arcs.
- Plasma chemistry and reaction mechanisms.
- Plasma production by novel means.

A full list of titles published in this series can be found here: https://iopscience.iop.org/bookListInfo/iop-plasma-physics-series.

Analysis of Turbulence in Fusion Plasmas

Boudewijn Ph van Milligen
National Fusion Laboratory, CIEMAT, Madrid, Spain

Raul Sanchez
Department of Physics, Universidad Carlos III de Madrid, Leganés, Madrid, Spain

IOP Publishing, Bristol, UK

ISBN 978-0-7503-4856-0 (ebook)
ISBN 978-0-7503-4854-6 (print)
ISBN 978-0-7503-4857-7 (myPrint)
ISBN 978-0-7503-4855-3 (mobi)

DOI 10.1088/978-0-7503-4856-0

Version: 20221201

IOP ebooks

British Library Cataloguing-in-Publication Data: A catalogue record for this book is available from the British Library.

Published by IOP Publishing, wholly owned by The Institute of Physics, London

IOP Publishing, No.2 The Distillery, Glassfields, Avon Street, Bristol, BS2 0GR, UK

US Office: IOP Publishing, Inc., 190 North Independence Mall West, Suite 601, Philadelphia, PA 19106, USA

Cover image: Transfer Entropy graph for the TJ-II stellarator. Image credit: Boudewijn Van Milligen.

Contents

Preface

Progress in science comes when experiments contradict theory.

—Richard P Feynman

When reviewing the literature on turbulence in fusion plasmas, one observes that while several books on the theory of turbulence are available [1–3] and also on general aspects of turbulent or anomalous transport [4], there are only few books focusing specifically on data analysis techniques for turbulence. We believe the reason for this is easy to understand: fusion plasmas constitute a hostile environment for measurement, making the measurement of turbulent quantities a significant challenge. Nevertheless, progress in the understanding of turbulence and its significance for the achievement of fusion as a power source is only possible when theoretical predictions are verified or falsified, and this requires measurement and analysis.

This book attempts to fill this gap by focusing not on the theory of turbulence—treated in profusion elsewhere—but on the analysis of turbulence measurements. It should be noted that this book provides a strictly personal view of the field and is not all-encompassing. Most of the material presented is based on personal work or work performed in close collaboration with others. Thus, this book does not intend to be a review of the field, and we extend our apologies to colleagues who may feel their contributions are underrepresented.

We have attempted to introduce the concepts used in the analysis of turbulence data in a gradual manner, providing explanations and emphasizing the underlying motivation for the approaches we followed. The book is organized in a manner that reflects our personal journey of discovery of the complexity of plasmas, gradually revealing ever more detail and understanding of the behaviour of these extreme systems.

The dynamics of present-day research require one to publish regularly, which has led us to write a considerable number of specialized, rather terse papers on this subject, usually with little space for explanation and context. This brings us to another reason for writing this book: we felt it was necessary to attempt to provide some insight into how these different pieces of the puzzle, published individually, fit together into a global view of turbulence in fusion plasmas. Many of the described methods and techniques are also used in, or could be applied to, fields other than fusion plasma research.

The reader may wonder why we sometimes used rather old figures (from 20 years in the past, or more) to illustrate concepts and results. This is inherent in the scientific tendency to only publish novel results—if a result reproduces what has already been published, it is often not shown in more recent publications. Thus, although in many cases we could have produced similar graphs from more recent measurements, we have preferred to show published results whenever possible, with the advantage that the interested reader can obtain more details from the corresponding publication.

Organization of this book

Chapter 1 provides a general introduction to the study of turbulence in fusion plasmas, highlighting the many issues and remarkable phenomena occurring in these complex systems. Our interest in turbulence is motivated by the need to understand transport in fusion plasmas, in which turbulence is found to play an essential role.

Chapter 2 describes an initial characterization of turbulence, starting with standard analysis tools such as probability distributions, correlations, and Fourier spectra, and progressing to some more advanced techniques to detect structures in plasmas, such as the biorthogonal decomposition and (wavelet) bicoherence. These analyses already reveal some of the complexity that is the focus of this book.

Our general understanding of transport is based on the collective effects resulting from particles moving individually according to random (Brownian) motion. Chapter 3 reviews some aspects of probability theory (the theory of random variables). It is shown that the theory allows for a much broader range of behaviours than are usually considered by 'standard' transport theories, and that these 'alternative' descriptions fit some aspects of turbulence extraordinarily well. The associated analysis techniques are suited for the study of single turbulent time series, and allow quantifying complex behaviour, self-similarity, and long-range memory effects in a systematic way.

Chapter 4 then studies the *interactions* between turbulent variables using a technique originating in information theory, namely the transfer entropy. It is used both to study the causal relationship between fluctuating variables and to analyze the propagation of perturbations through a turbulent system. In other words, this chapter addresses the essential multivariate and/or spatio-temporal nature of turbulence.

Chapter 5 focuses on a specific yet important aspect of turbulence: intermittence or 'burstiness'. This property is analyzed using a technique from chaos theory. It is shown how the intermittence provides a new window on turbulent properties in the plasma interior.

Finally, chapter 6 summarizes the insights obtained in the course of the book and discusses some ideas for expanding and applying this work and future perspectives.

Acknowledgements

Both of the authors were introduced to the study of turbulence and anomalous transport by Benjamin Carreras, who has been a crucial influence in our careers, guiding our nascent understanding and always pushing us onward to do more. Among other notable achievements, Ben was among the first to recognize the importance of zonal flows in fusion plasmas [5, 6].

Modern science is not an individual endeavour and obviously much, if not all, of the work and results presented in this book are the fruits of collaborations with many people. In this respect, we should mention the TJ-II Team [7] at the National Fusion Laboratory of CIEMAT (Madrid, Spain), including Carlos Hidalgo, M Ángeles Pedrosa, Antonio López Fraguas, Teresa Estrada, Enrique Ascasíbar, Eduardo de la Cal, Edi Sánchez, Álvaro Cappa, Isabel García Cortés,

Kieran McCarthy, José Luis de Pablos, Arturo Alonso, Emilia Solano, Elena de la Luna, Ignacio Pastor, Marian Ochando, and many others. Likewise, at the University of Carlos III in Leganés (Madrid, Spain), we would like to highlight collaborations with Luis García, Victor Tribaldos, José Ramón Martín-Solís, José Miguel Reynolds, José Angel Mier (now at University of Cantabria), Jacobo Varela, and others.

We have also collaborated with many people from across the world, among whom are Radu Balescu† from the Université Libre de Bruxelles (Belgium) [2, 8], David Newman from the University of Alaska at Fairbanks (USA), Diego del Castillo from ORNL (Oak Ridge, TN, USA), Paul Terry from the University of Wisconsin at Madison, the Wendelstein 7-X Team at Greifswald (Germany) [9] including Matthias Hirsch and Andreas Dinklage, Gustavo Grenfell, and Gregor Birkenmeier from the Max Planck Institute (Garching, Germany), Anantanarayanan (Chippy) Thyagaraja from UKAEA (UK), the JET Contributors [10] at Culham (UK), the HIBP Team—Alexander Melnikov and his team from the Kurchatov institute (Moscow, Russia) and the team from the Institute of Plasma Physics (Kharkov, Ukraine)—, Carlos Silva from IST (Lisbon, Portugal), and Paul Bons from Universität Tübingen (Germany).

We would like to thank all these people, and many others whom we have not mentioned explicitly, as this book could not have become a reality without their direct or indirect collaboration in this collective research effort spanning many decades and involving hundreds of people. We offer our apologies in advance to anyone who should have been included here and was inadvertently left out.

Last but not least, we should mention the unwavering support and patience of our families. Boudewijn would like to thank his wife, Deysi Coromoto, and his daughter María for providing a joyful and interesting life and his mother Emma and sisters Marjolein and Ingrid in The Netherlands for their steadfast encouragement with this project. Raul would also like to thank his wife, Estefanía Cuevas, his mother Carmina, his brother Ruben, and his wife Sonia, and his two wonderful nephews, Ivan and David, for keeping his sanity in check while doing science and teaching.

References

[1] Yoshizawa A, Itoh K and Itoh S I 2003 *Plasma and Fluid Turbulence: Theory and Modelling* 1st edn (Boca Raton, FL: CRC Press)

[2] Balescu R 2005 *Aspects of Anomalous Transport in Plasmas* (Bristol: Institute of Physics))

[3] Scott B D 2021 *Turbulence and Instabilities in Magnetised Plasmas: Fluid Drift Turbulence* vol 1 (Bristol: IOP Publishing)

[4] Horton W 2017 *Turbulent Transport in Magnetized Plasmas* 2nd edn (Singapore: World Scientific)

[5] Diamond P H, Liang Y-M, Carreras B A and Terry P W 1994 Self-regulating shear flow turbulence: a paradigm for the L to H transition *Phys. Rev. Lett.* **72** 2565

[6] Ware A S, Terry P W, Diamond P H and Carreras B A 1996 Transport reduction via shear flow modification of the cross phase *Plasma Phys. Control. Fusion* **38** 1343

[7] Hidalgo C *et al* 2022 Overview of the TJ-II stellarator research programme towards model validation in fusion plasmas *Nucl. Fusion* **62** 042025

[8] Klages Rainer, Radons Günter and Sokolov Igor M (ed) 2008 *Anomalous Transport: Foundations and Applications* (New York: Wiley)
[9] Beidler C D *et al* 2021 Demonstration of reduced neoclassical energy transport in Wendelstein 7-X *Nature* **596** 221–6
[10] Joffrin E *et al* 2019 Overview of the JET preparation for deuterium-tritium operation with the ITER like-wall *Nucl. Fusion* **59** 112021

Author biographies

Boudewijn van Milligen

Boudewijn van Milligen studied physics at the University of Utrecht in The Netherlands, where he also obtained his PhD thesis. This thesis was based on work performed at the FOM Institute for Plasma Physics, then located in Rijnhuizen, and JET in the UK, among others, thus starting a life-long career in which international collaborations have always been predominant. The thesis work, performed under the motivating guidance of Niek Lopes Cardozo, aroused his interest in transport studies and unusual transport phenomena.

Following a six-month holiday break in Venezuela and elsewhere, he started work at the CIEMAT institute in Madrid, Spain, with financing from the European Commission. After an initial phase in which he was involved in the data acquisition systems for the TJ-I tokamak and the TJ-IU stellarator, he refocused his attention on data analysis. Under the guidance of Carlos Hidalgo and Benjamin Carreras, this turned out to be an extremely fruitful area. Over the years, he has explored data from many fusion devices around the world, using a wide range of data analysis methods, with the purpose of clarifying the physics underlying transport and turbulence, as reported in this book.

Raul Sanchez

Raul Sanchez studied theoretical physics at the Universidad Complutense de Madrid, where he also obtained his PhD thesis. In this work, performed under the direction of Professor Luis Garcia at Universidad Carlos III de Madrid (UC3M), he developed a fast and accurate solver to calculate ideal MHD ballooning stability in arbitrary magnetic configurations. This was the basis of the widely used COBRA code, which he later developed while already working as a post-doc at Oak Ridge National Laboratory (ORNL) in the USA. He remained associated with ORNL for almost a decade until finally relocating back to Madrid as a physics professor at UC3M. During his time at ORNL he co-developed the ideal MHD equilibrium code SIESTA with Steve Hirshman, his advisor at the time and one of the several people that showed him the joy of working in plasma physics.

In parallel with his work in magnetohydrodynamics, he also developed a strong interest in plasma turbulence after meeting Benjamin Carreras and David Newman at ORNL. This collaboration, that remains active and fruitful to this day, introduced him to the complexity of turbulence and the turbulent nature of complex systems. With the help of many collaborations and friends, this has been a topic that has kept him busy and entertained many days and many nights.

IOP Publishing

Boudewijn Ph van Milligen and Raul Sanchez

Chapter 1

Introduction

In the first quarter of the twenty-first century humanity is facing a number of challenges of enormous scope, impact, and complexity, such as climate change and the degradation of the biosphere, overpopulation, food and water scarcity, health crises and pandemics, disinformation, corruption and bad government, inequalities between regions and population classes leading to tensions, migration, and war, etc, to mention just a few. Many of these problems are related to each other. Some of these issues could be handled better, in principle, if we had access to a plentiful, cheap, and non-contaminating source of energy. Evidently, a source of energy free from greenhouse gas emissions would be enormously beneficial for the control of climate change, and for improving living conditions in crowded cities. In addition, one could think about desalinating sea water, in order to increase the availability of drinking water, reduce disease transmitted through contaminated water, and allow cultivating areas that are currently arid and unproductive.

1.1 Energy policy options

Most people would agree that it is beneficial to reduce the use of fossil fuels, to mitigate climate change, reduce contamination, and avoid exhausting precious resources—bearing in mind, for example, that oil is not only used as an energy source, but also to manufacture plastics and many other materials. When deciding what 'energy mix'[1] to adopt in the near and far future, it is important to consider many factors. One important issue is the flexibility of energy sources to adapt the supply to the fluctuating demand of electricity, regardless of external circumstances. Flexible energy sources are vital to keep power grids operating at times of high load or when other energy sources fail. Fossil fuel power plants are flexible, but have well-known ecological issues. Biofuels could substitute fossil fuels, but they compete with food production. Hydroelectric plants are also flexible, up to a point, but their

[1] Energy mix: a collection of energy sources to cover the energy needs of society.

availability depends on the local orology, climate, and ecological considerations. Geothermal energy can be very reliable, but is unlikely to be able to meet a significant fraction of the energy demand. Wind and solar energy are not very flexible, as they depend on highly variable external factors that cannot be controlled (the weather)—and the same can be said of many other 'alternative' energy sources (alternative with regard to fossil fuels). If methods are developed to allow massive energy storage, the need for flexible energy sources would be much reduced; but such methods are currently not sufficiently abundant, although hydrogen fuel could potentially take this role [1]. The robustness of the energy supply can also be increased by interconnecting power grids, up to a point. To summarize, the future energy mix should probably consist of a combination of the power sources mentioned and others, but it is not entirely clear what power source can take over the role of fossil fuels as the main flexible power supply. Fusion reactors would fall into this category.

When deciding the energy mix, other factors in addition to the climate, the environment, and flexibility should also be taken into account. These additional considerations are of a more societal nature and include the social acceptance of each option, opportunities in terms of job creation and the economy, the possible creation of additional social inequalities (for example when an option is only available to the rich), etc. Furthermore, one should take into account strategic considerations regarding the potential threats associated with nuclear proliferation or the political stability of regions from which raw materials are obtained. In any case, it appears certain that the decisions that will be taken in the coming years or decades regarding the energy mix will be very complex and have a major impact on society.

1.2 Fusion

Fusion is the energy source that fuels stars, that burn uninterruptedly for billions of years, consuming the most plentiful element of the Universe—hydrogen—and converting it into other elements and energy. It is natural to wonder whether we could tap into this energy source somehow. In a way, some 'alternative' energy sources already do this: they convert solar energy into electricity, either directly or indirectly. For example, photovoltaic panels convert the solar photon flux directly into electricity, wind turbines make use of the air currents associated with weather systems that are driven indirectly by the solar power influx, hydroelectric power plants use the gravitational energy stored in water that was lifted by evaporation driven by solar heat, and so on. One could argue that the same argument would apply to fossil fuels, whose energy content also originates, ultimately, from solar energy; but the difference is, of course, that fossil fuels release solar power that was stored aeons ago.

Research into fusion as a power source is different in the sense that it is an attempt to reproduce the fusion reactions occurring in the Sun directly in the laboratory [2]. Fusion refers to the reaction in which the nuclei of light elements combine to produce a new, slightly heavier element, such as the two hydrogen isotopes that combine to yield helium in figure 1.1. The binding energy of the resulting nucleus is slightly higher than the binding energy of the original two nuclei, and the difference

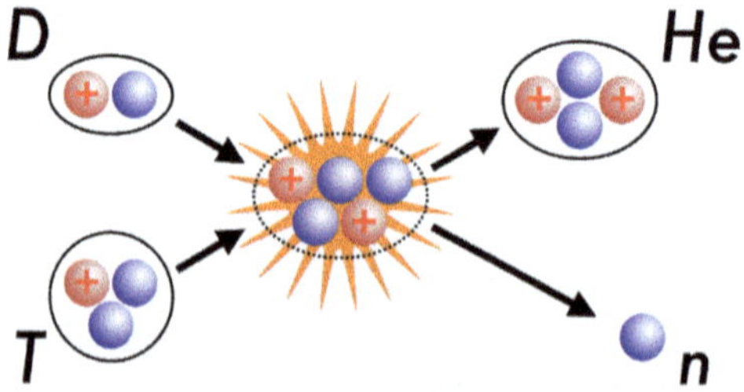

Figure 1.1. Cartoon of a D–T fusion reaction.

is released in the form of the kinetic energy of the fusion products. Nuclear binding energies are associated mainly with the strong nuclear force, one of the fundamental forces of the Universe. As a consequence, nuclear binding energies are orders of magnitude larger than the chemical binding energies associated with the much weaker electromagnetic force, so that the amount of energy released in each fusion reaction is very large in comparison to the chemical reactions occurring when burning fossil fuels.

However, bringing together two light nuclei is not easy because at a relatively large distance, they repel each other due to the fact that the nuclei have a positive electric charge, as indicated by the '+' symbols in figure 1.1. This effect is called Coulomb repulsion, a manifestation of the electromagnetic force. Although it is much weaker than the strong force, it has a much longer range, so that it dominates at relatively large distances. In the Sun, this 'Coulomb barrier' is overcome by the huge pressure in the Sun's core due to its enormous mass, which is compressed by its own gravitational force. On Earth, we cannot apply this method, so it is necessary to overcome the Coulomb barrier by other means, and this means investing energy. The first deliberate fusion reactions on Earth were produced by accelerating deuterons towards a deuteron target [3]. It turns out to be relatively easy to produce fusion reactions. For a moderate amount of money, one can acquire a fusion device called a 'fusor' in which ions are accelerated using electrodes, or even build one at home. However, this type of device is highly inefficient and has never been able to show any net energy gain.

If fusion reactors can be built successfully, there are several important potential benefits: the basic fuel is abundant[2], the energy production per reaction is enormous[3], the reaction product is non-radioactive helium, which is not a greenhouse gas and not toxic, and a fusion reactor does not, in principle, constitute a nuclear hazard (neither with regard to meltdowns nor proliferation). That being said, there are of course also some drawbacks, such as the activation of the device as a consequence of the neutron bombardment that is produced by the D–T reaction (figure 1.1), in which the fusion of deuterium (D) and tritium (T) produces helium

[2] The fuels for fusion are isotopes of hydrogen, such as deuterium that can be obtained from sea water, and tritium that will need to be bred in the reactor, at least in first-generation reactors based on the D–T reaction.

[3] The energy that may be obtained from 1 kg of deuterium and tritium is equivalent to the energy that can be obtained from burning about ten million liters of gasoline!

(He) and a neutron. Although other fusion reactions are possible, they are more difficult to achieve, so alternative aneutronic reactor designs that would suffer much less from this drawback remain firmly in the conceptual stage, for now. Also, current experimental reactor designs are very large, complex and expensive, although one may hope that future designs based on new technological advances, e.g. regarding magnet coils, could lead to improvements.

Several routes are being explored to achieve efficient fusion on Earth. Currently, the most promising route appears to be magnetic confinement, which is the focus of this book. Nevertheless, we should certainly not omit mentioning that alternative methods are being explored, some financed publicly and some privately. In fact, the last few years have seen a veritable explosion of fusion start-ups [4]. However, most of these efforts are still far from proving their viability and the main competitor to magnetic fusion remains inertial fusion, based on a method in which a pellet of hydrogen fuel is compressed and heated with high power lasers [5, 6].

1.3 Magnetic confinement

The basic idea of fusion is to heat hydrogen fuel to such high temperatures that the associated violent thermal motion is energetic enough to overcome the mentioned Coulomb barrier. Once this barrier is overcome, the two nuclei will fuse, releasing a large amount of energy. The required temperatures are well above the ionization energy of the hydrogen fuel atoms, so that the fuel mixture will be in a fully ionized state, called a plasma (figure 1.2).

Of course, it is not enough to overcome the Coulomb barrier only occasionally. For efficient power generation, the fusion energy production rate should exceed the

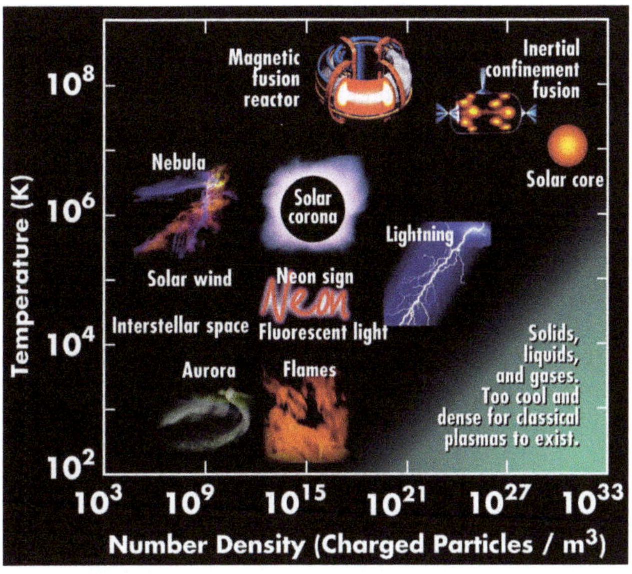

Figure 1.2. The plasma state. (Source: The Contemporary Physics Education Project.) © CPEP https://www.cpepphysics.org/

applied power input needed to maintain the hot plasma state. This requirement translates into a relatively simple criterion for the properties of the confinement system, known as the 'Lawson criterion', which is stated as follows [7]:

$$n\tau_E T \geqslant 3 \times 10^{21}\ \text{keV s m}^{-3}. \tag{1.1}$$

According to this criterion, efficient energy production requires confining the fuel at a particle density n (m^{-3}) for a time τ_E(s) and heating it to a temperature T (expressed in keV)[4], such that the above triple product exceeds the specified number. In view of other constraints[5], it turns out that this condition can only be achieved realistically for extremely high temperatures of between 100 and 1000 million degrees K, significantly above the temperature in the Sun's core (around 10 million degrees K). Since our best materials only resist melting or disintegration up to a few thousand degrees, such a plasma cannot be confined in a material vessel. Fortunately, charged particles are subject to electromagnetic forces and can therefore be confined using magnetic fields.

Charged particles moving in a magnetic field $\mathbf{B}$ experience the Lorentz force $\mathbf{F} = q(\mathbf{E} + \mathbf{v} \times \mathbf{B})$. If the magnetic field is strong enough, particles will gyrate around the magnetic field lines in tight helical orbits, so that they become effectively 'tied' to the field line. By bending the magnetic field lines back on themselves, a region could in principle be created in which the particles are confined; since they follow the field lines, they will not leave the mentioned region, to first approximation. This straightforward idea gives rise to the toroidal (donut-shaped) design of magnetic confinement devices. The unavoidable curvature of the field lines and non-uniformity of the field strength induce particle drifts that push the particles away from the field lines. Good confinement therefore requires a careful design of the magnetic trap to compensate for these drifts.

To balance the pressure gradient (∇p) due to the confined plasma, the toroidal[6] magnetic field needs to be supplemented by a poloidal[7] field component. Generally speaking, the poloidal magnetic field can be produced in two ways. In devices called tokamaks, it is mainly produced by means of a net toroidal current flowing in the plasma. In this case, the device design can be essentially axisymmetric[8], so that the whole system can be considered to be two-dimensional, which is an important simplification. An example of such a device is the JET tokamak (see appendix A.5). In devices called stellarators, the poloidal field is mainly produced by currents flowing in external coils, which is incompatible with axisymmetry. An example is the W7-X stellarator (see appendix A.4).

The resulting balance between the outward pressure gradient due to the confined plasma and the inward magnetic force, $\nabla p = \mathbf{j} \times \mathbf{B}$, where $\mathbf{j}$ represents the currents flowing in the plasma, is called magneto-hydrodynamic (MHD) equilibrium.

---

[4] 1 keV is equivalent to about 10 million °C or K.

[5] The quantum-mechanical collision cross section [2].

[6] Toroidal: the direction that goes the long way around the torus (coordinate ϕ).

[7] Poloidal: the direction that goes the short way around the torus (coordinate θ).

[8] Axisymmetry: symmetry with regard to toroidal rotations.

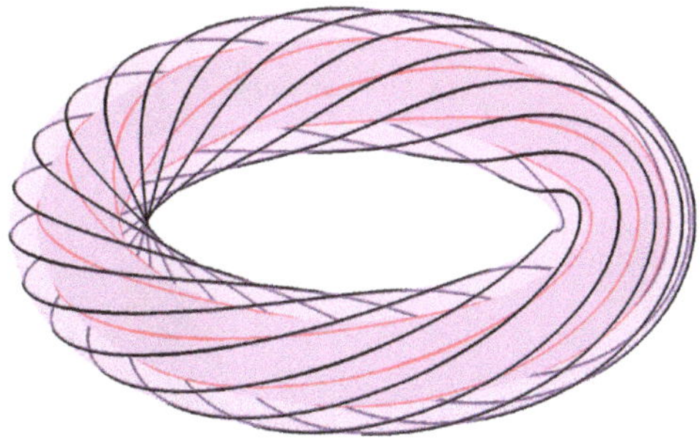

Figure 1.3. Sketch of a toroidal plasma (purple) and two field lines (red and black) on two corresponding flux surfaces. The rotational transform is the winding ratio between poloidal and toroidal turns.

A simplified sketch of the resulting magnetic field is shown in figure 1.3. The figure shows two flux surfaces with embedded field lines. A flux surface is a surface S such that the magnetic field is always directed parallel to the surface ($\mathbf{B} \cdot \mathbf{n} = 0$, where $\mathbf{n}$ is the normal to the surface S). Flux surfaces play a major role in the confinement and transport of magnetically confined plasmas, as particles can move with great freedom along the magnetic field lines (within the flux surfaces), but are strongly restrained in their motion perpendicular to the field lines (perpendicular to the flux surfaces)—in particular, in a well-designed magnetic trap, particle drifts perpendicular to the flux surfaces will be almost vanishingly small. The sketch shows that each field line within the confined region winds both the long and short way around the magnetic flux surface in which it is embedded. The rotational transform is defined as the mean winding ratio between poloidal and toroidal turns of a field line, $\iota\!\!\!-\, = \langle d\theta/d\phi \rangle$.[9] Thus, the black field line, on its corresponding flux surface, has a given rotational transform. The red field line, on a flux surface with a somewhat smaller minor radius, has a slightly different (lower, in this case) rotational transform. By plotting the rotational transform as a function of minor radius r, one obtains the rotational transform profile $\iota\!\!\!-(r)$.

The rotational transform plays an important role in magnetic confinement and stability: when $\iota\!\!\!-$ takes on a rational value (n/m), the field line on the corresponding flux surface closes in on itself after a finite number of turns, making it easier to displace it radially and giving rise to the birth of MHD modes or instabilities when the local plasma conditions are suitable. On the other hand, a flux surface with an irrational $\iota\!\!\!-$ value has a higher stability against such displacements, in principle. If the MHD modes grow sufficiently, they may give rise to topological bifurcations called 'magnetic islands'. An example of such a magnetic island is shown in figure 1.4. The radial deformation of an originally thin rational surface with rotational transform $\iota\!\!\!- = n/m = 3/4$ has produced a helical flux tube with a periodicity matching the field lines of the surface. The poloidal (vertical) cross section shown reveals clearly the $m = 4$ nature of the island through its four-fold symmetry. A similar toroidal (horizontal) cut (not shown) would reveal its $n = 3$

[9] Often, the safety factor q is used, which is the inverse of the rotational transform, $q = 1/\iota\!\!\!-$.

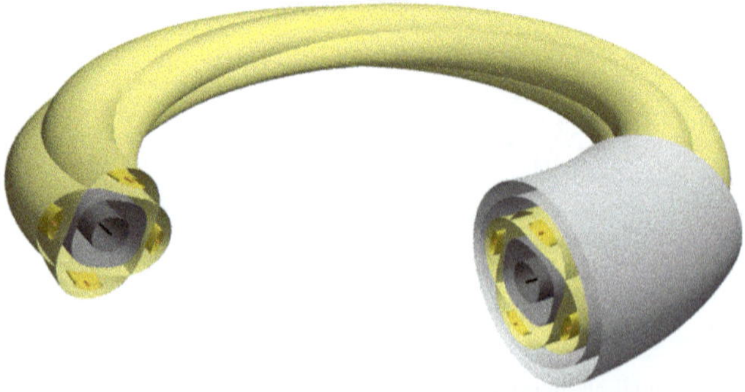

Figure 1.4. The flux surfaces of a magnetic island in a circular plasma configuration. The island (yellow) constitutes a helical flux tube matching the rotational transform of the corresponding flux surface. (Reproduced from [11] © Matthias Hölzl.)

character. The growth of the island leads to the deformation of the flux surfaces immediately outside the island, which however remain nested. Secondary nested flux surfaces appear inside the island, topologically separate from the main plasma. Field lines inside the island never connect to the region outside the island and vice versa. The two disconnected regions are separated by a flux surface called a 'separatrix'. As a consequence, conditions (density, temperature) inside the island may sometimes differ strongly from conditions outside [8]. Such islands have an impact on confinement and transport in fusion plasmas: they may constitute a barrier to radial transport, or they may constitute a kind of radial 'short circuit' through fast transport occurring inside the island, depending on the circumstances [9, 10].

Once the magnetic configuration is decided, the confined plasma needs to be heated and fuelled to achieve the required conditions of density and temperature. The simplest technique is to drive a toroidal current inside the plasma using an external transformer; dissipation due to the finite resistivity of the plasma then leads to heating. However, this technique becomes inefficient as the plasma heats up, as plasma resistivity drops with temperature. Consequently, other techniques are needed. To name a few: the injection of electromagnetic waves at specific absorption frequencies (electron and ion cyclotron resonant heating), the injection of energetic neutral particles (neutral beam injection) that transfer their kinetic energy to the plasma via collisions and contribute ions and electrons after ionization, or the injection of pellets (cryogenic hydrogen or other materials) that are ionized and contribute to the plasma density [12].

1.4 Fusion plasmas: a special class of systems

The goal of heating and fuelling the confined plasma is to create the required conditions for fusion reactions to occur. Some of the injected energy and mass is lost to the surroundings of the plasma, for example through the emission of radiation, or through particles whose orbits exit the plasma and hit the vacuum vessel, thereby transferring mass, momentum, and energy. Thus, fusion plasmas are essential 'open' rather than closed systems.

While extreme conditions are created in the plasma core (e.g. temperatures of many millions of Kelvin), the surrounding laboratory is kept at essentially ordinary ambient conditions (room temperature, sea level pressure, etc) for practical purposes. Thus, fusion plasmas are special in the sense that they are characterized by some of the steepest gradients of any systems in the Universe, in which the temperature drops from the required value for fusion to occur to room temperature in a distance of the order of a metre. Therefore, fusion plasmas are necessarily far from thermodynamic equilibrium. This is an important fact to keep in mind when studying fusion plasmas, as many theoretical approaches are only really valid for systems in thermodynamic equilibrium, and so should be used with considerable caution.

The steep gradients provide 'free energy' that allow the growth of all kinds of instabilities and, consequently, fusion plasmas exhibit an important level of turbulence[10]. This turbulence is not some kind of irrelevant 'noise' that sits on top of an otherwise stable background, but rather it may have a determining role in establishing the global state of the plasma. As noted in the seminal work by Ilya Prigogine[11], open systems far from equilibrium may *self-organize* and spontaneously generate 'dissipative' structures that require energy to sustain them. According to Prigogine, 'in non-equilibrium systems, fluctuations determine the global outcome' [14].

All this makes fusion plasmas rather interesting subjects of study, and it should not be surprising that they exhibit unusual phenomena. One could view turbulence as Nature's frantic attempt to flatten the gradients and try to return to a state of thermodynamic equilibrium. As a result, turbulence tends to work against our efforts to create fusion conditions in the laboratory—with a few exceptions, as we will see. In order to achieve a power source based on magnetically confined fusion plasmas, it is therefore essential to understand how turbulence works and how it can be controlled.

Plasmas are pervaded by many types of instabilities and waves [15]. The current understanding is that two types of instability are the main contributors to radial turbulent transport in fusion devices: drift waves and MHD modes.

Drift waves are collective oscillations driven by a pressure gradient perpendicular to the magnetic field. The motion of electrons and ions is very different: due to their very light mass, electrons move preferentially along field lines to restore quasi-neutrality, whereas the relatively heavy ions mainly experience drifts perpendicular to the field lines. Drift waves can be destabilized by effects interfering with the free ion and/or electron motion caused by, e.g. collisions, particle trapping, or particle drifts. Many different types of drift wave instabilities are distinguished, according to the driving mechanism. The main instabilities that are considered in the framework of turbulent transport are electron temperature gradient (ETG) and ion temperature

[10] Note that it is not easy to provide a concise definition of what turbulence is. For our purposes, a turbulent flow (a) is unpredictable in detail, (b) leads to increased mixing or transport, and (c) involves a broad range of scales [13].

[11] Nobel Prize in Chemistry (1977) 'for his contributions to non-equilibrium thermodynamics, particularly the theory of dissipative structures'.

gradient (ITG) driven instabilities, and the trapped electron mode (TEM) instability. These instabilities are thought to constitute the dominant mechanism for the transport of particles, energy, and momentum across magnetic field lines [16]. A vast amount of literature is devoted to studying their properties and quantifying the turbulent radial fluxes produced by these instabilities in various scenarios.

The validation of the predictions from gyrokinetic numerical models incorporating these instabilities (see the next section) against measurements is currently a very active field of study [17]. In the saturated state, the turbulent fluctuations often lead to the formation of large scale, radially extended structures (called 'streamers'), as well as radially varying poloidal and toroidal plasma flow patterns (called 'zonal flows') that interact nonlinearly, with a significant impact on the final turbulent flux—see figure 1.5. We note that the theory and simulation of these modes nearly always assumes a fixed magnetic geometry, i.e. magnetic fluctuations are not taken into account.

MHD modes, on the other hand, are instabilities that involve magnetic fluctuations in an essential way. These modes are likewise driven by the pressure gradient, particularly in regions of the magnetic configuration with so-called 'bad magnetic curvature'. Often, these instabilities are considered subdominant based on a simple argument: namely, that the relative amplitude of measured magnetic field fluctuations is typically found to be small[12], $\tilde{B}/B \sim 10^{-4}$, except in the case of major islands. This argument is, however, debatable due to the fact that transport is much faster along than across field lines, so that particles moving along the field lines are very sensitive to small radial magnetic field fluctuations [19]. From an experimental viewpoint, we note that these instabilities are closely associated with the formation of large scale structures (filaments), whose existence is quite evident, particularly in the plasma edge region. Furthermore, as we will discuss in more detail in chapter 4, magnetic field fluctuations may have a long-range, non-local impact via the radial coupling between modes. In addition, these instabilities likely facilitate the

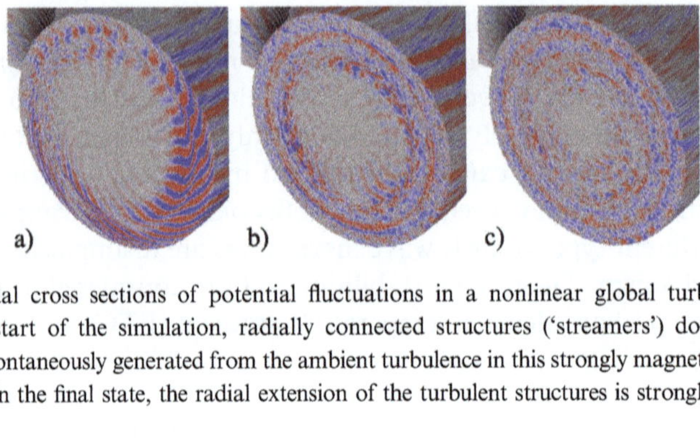

Figure 1.5. Poloidal cross sections of potential fluctuations in a nonlinear global turbulence simulation. (a) Close to the start of the simulation, radially connected structures ('streamers') dominate the system. (b) Zonal flows, spontaneously generated from the ambient turbulence in this strongly magnetised system, rupture the streamers. (c) In the final state, the radial extension of the turbulent structures is strongly reduced [18].

[12] In reversed field pinches, the relative magnetic field fluctuation amplitude is typically very much larger than the cited number.

formation of zonal flows and minor transport barriers, which may have an important indirect impact on plasma transport.

1.5 Transport in fusion plasmas

The efficiency of achieving the conditions required for fusion depends on the balance between the applied drive (fuelling and heating) and the losses. The power injected into the plasma is lost via radiation, convection, and conduction. Losses by radiation are more or less understood (e.g. Bremsstrahlung), and these losses can be controlled by keeping the plasma pure and fully ionized. Here, we will focus on convection and conduction, collectively denoted 'transport'. Assuming thermodynamic equilibrium, a rather doubtful assumption in view of the above, transport is expected to be dominated by diffusion.

The principle underlying diffusion is well understood: it is caused by the random motion of individual particles [20], namely the well-known Brownian motion, by which the mean square displacement of any particle from a given initial position increases in proportion to the elapsed time[13]. This microscopic, random movement of particles manifests itself collectively, on the macroscopic level, as diffusion. Interestingly, whereas the random motion of the microscopic particles does not have a preferred direction, the resulting macroscopic diffusion does. To see this, consider two neighbouring regions A and B, with different particle densities n_A and n_B (see figure 1.6). Brownian motion will lead to the spreading of particles from A into B and vice versa. Assuming the rate of spreading is the same for both regions, the net particle flux across the interface between A and B will then depend only on the density difference between the regions. Generalizing to a gradually varying density $n(x)$, this implies that the particle flux must depend on the density gradient.

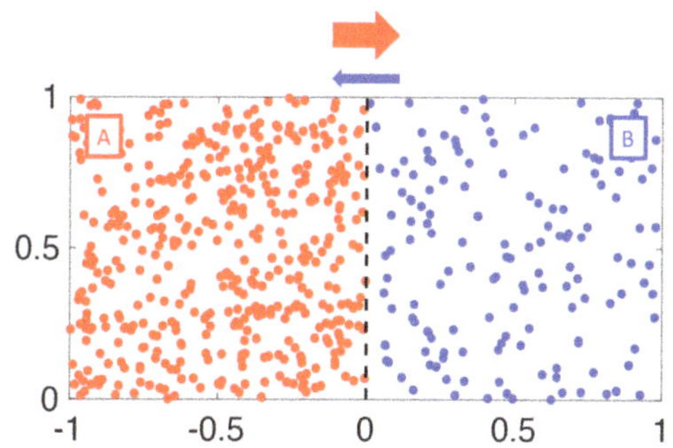

Figure 1.6. Cartoon showing two regions, A and B, with different particle densities, leading to a net diffusive flux across the interface (dashed line) from the high density to the low density region.

[13] The collective effect of particle spreading due to Brownian motion is described by the 'heat kernel', a normalized Gaussian distribution whose width increases in proportion to $\sqrt{t}$ (equation (3.14)).

Indeed, Fick's law states that the particle flux, Γ, is proportional to the negative density gradient[14]:

$$\Gamma = -D\nabla n. \tag{1.2}$$

The proportionality constant D is the diffusion coefficient, with a value that is determined only by the characteristics of the microscopic motion. This equation can be supplemented by the conservation law of the particle density n:

$$\frac{\partial n}{\partial t} = -\nabla \cdot \Gamma. \tag{1.3}$$

By combining these equations, one obtains the transport equation for the particle density:

$$\frac{\partial n}{\partial t} = \nabla \cdot (D\nabla n). \tag{1.4}$$

To actually model transport in a given system one would need to add a particle source and specify boundary conditions.

This transport equation is perhaps too simple for many practical applications, but it serves us here to illustrate that the above derivation skips over several important hidden assumptions. While the conservation equation is exact, Fick's law is phenomenological and only valid in limited cases (a homogeneous medium—with fixed D—, no external forces acting on the particles). The fact that the equations are written down in the form of differential equations implies that the random particle motion underlying the diffusion is assumed to consist of infinitesimal steps, i.e. the average particle step size Δx (between 'collisions') is much smaller than any relevant length scale, such as the system size or the gradient length $(\nabla n/n)^{-1}$, so the theory is 'local'. This locality is also reflected in the fact that the flux Γ at location x is assumed to depend only on D and ∇n at that same location x. By imposing locality, one is implicitly denying the relevance of radially extended structures or events for transport. It is also implicitly assumed that the process is 'Markovian', i.e. the next step of a particle is exclusively determined by its current situation (in a probabilistic sense). However, in stochastic magnetic field regions, persistent turbulent eddies, or transport barriers, this assumption may be violated (due to trapping effects), cases in which the history of the particle trajectory becomes important (see sections 3.4 and 4.6). Finally, the whole construct is based on 'standard' Brownian motion, but there are in fact other options consistent with the central limit theorem ('fractional' Brownian motion), which may arise naturally in strongly driven non-equilibrium systems, as we will discuss at some length in chapter 3. The main message of this paragraph is that the diffusive model is not set in stone, and addressing the complexities of transport in fusion plasmas may require one to keep an open mind. A secondary message is that one should never forget that 'transport' is the collective manifestation of the random motion

[14] If D depends on the medium (through the local temperature, say), the Fokker–Planck equation may provide a better approach than Fick's law [21], and one would write $\Gamma = -\nabla(Dn)$.

of individual particles; hence the importance of the underlying random processes for transport.

With all these caveats, the diffusive approach can be applied to fusion plasmas. Of course, the situation is somewhat different than the simple picture of Brownian motion sketched above: in hot, strongly magnetized plasmas, the charged particles describe tight helical orbits around the field lines. Nevertheless, the particles experience random 'Coulomb' collisions that cause their gyrocentres to shift from one field line to the next, and the cumulative effect of many such random displacements then gives rise to diffusion. Assuming a static and *uniform* background magnetic field that remains unaffected by the motion of the charged particles, transport equations can be written down for the particle density, particle energy (equivalent to temperature), etc. This approach leads to the 'classical' transport theory [22].

This straightforward theory was rapidly found to be unsatisfactory, underestimating the actual transport measured in magnetically confined plasmas by several orders of magnitude [15]. The geometry of the confining intense magnetic field significantly affects the orbits of the charged particles, giving rise, among others, to the existence of particles trapped in local magnetic well regions. The complex motion of the charged particles in the presence of steep gradients also induces electric fields that lead to particle drifts. All this has a strong impact on the resulting effective transport that cannot be ignored. Subsequently, a sophisticated theory was developed that takes the full impact of the (fixed) magnetic geometry into account, leading to corrected and rather complex expressions for transport. It is known as the 'neoclassical' transport theory [23].

The neoclassical theory is powerful and capable of clarifying many aspects of transport in fusion plasmas. Nevertheless, it still predicts radial fluxes that are often very much lower than what is actually found in experiments. To understand this, one should keep in mind that this is a theory based on several strong assumptions: it assumes a state of thermodynamic equilibrium, it is expressed in terms of differential equations (implicitly assuming small particle excursions, making it a 'local' theory), and, more importantly, neoclassical transport does not consider turbulence, since it ignores the nonlinear interaction between the confining field—assumed to be static—and the motion of the confined particles. Nevertheless, the transport resulting from the latter interaction often accounts for a large fraction of the total transport, called the 'anomalous' or 'turbulent' transport.

So, how can one properly account for turbulence? From the above argument, one clearly should include the evolution of the electric and magnetic fields self-consistently with the particle motion. Since the kinetic equations for the motion of the constituent particles of a plasma are known, as well as the Maxwell equations describing the evolution of the fields, the full plasma dynamics can in principle be simulated using the computer. In practice, this is a problem that not even the largest supercomputer available today can address. It is necessary to simplify the equations, using a minimal set of approximations, to make the calculations tractable. The standard procedure is to average the equations over the gyromotion of the charged particles, thus eliminating the fastest temporal scale of the problem. The resulting 'gyrokinetic' simulations provide the best numerical approximation to the turbulence problem to date [24]. However, these Gyrokinetic simulations are very expensive in terms of computer memory and time, so they are rarely used directly to model the gradually evolving situation of an experimental plasma. Instead, one typically uses Gyrokinetic simulations implemented in simplified geometries, with a magnetic field that is once more taken to be static, to obtain what is described as an 'effective diffusivity' for turbulence, calculated from the ratio between a prescribed gradient and the resulting turbulent flux. This effective diffusivity is then added to the Neoclassical predictions in the form of an additional diffusive term.

There are several problems with this approach, the main one being that the feedback between the turbulence and the background gradients is not taken into account, based on the assumption that there is a clear separation of timescales between the fast evolution of turbulence and the slow evolution of gradients (profiles) due to overall transport. We will see that this separation of scales may not always hold. A second issue is that the modification of the magnetic geometry due to turbulent fluctuations is ignored. On a deeper level, one may also question whether transport due to turbulence can really be described well by (local) diffusive equations, in view of the fact that experimental analyses show that turbulence exhibits many non-local effects: to name a few, intermittent events ('bursts' of the radial fluxes), long-range correlations arising between remote diagnostic systems, and non-Gaussian fluctuation distributions [25, 26]. Paradigmatic examples of long-range correlations are large scale structures emerging spontaneously from the ambient turbulence: e.g. zonal flows, responsible for the important L to H confinement transition, and 'avalanches' or 'streamers'. These important phenomena clearly reflect the complexity of fusion plasmas, as will be discussed in detail in subsequent chapters.

There are further problems, which might collectively be referred to as the 'boundary conditions'. Namely, the above approaches to transport in fusion plasmas are mainly applicable to the interior of the plasma. This plasma region is contained within a 'separatrix' that separates the interior region, where field lines close in upon themselves, from the scrape-off layer, a region where field lines connect to the surrounding material structure, such as the vacuum vessel. Transport in the latter region can be described by specific models that consider the material geometry, the physics of partially ionized gas, and the interaction of particles with the wall. The outcome of such models must then be interfaced with transport models for the interior region. This book will not dwell on the latter issues, but rather focus on the understanding of turbulence, mostly in the interior region.

1.6 How important is turbulence?

Before delving into a large effort to understand turbulence and its effect on the plasma, one would like to have some idea of how important it really is [27, 28]. To do so, one would like to compare the transport (of particles, heat, etc) between the experiment (where turbulence is present) and a hypothetical situation that is identical with one exception, namely the absence of turbulence. The difference is sometimes referred to as 'anomalous transport'[15], which is a bit of a misnomer, as if to deny the origin of the transport enhancement (namely, turbulence).

The fact that it is hard to model even non-turbulent transport, due to the mentioned 'boundary conditions' and other factors, makes this into a complicated task. Admitting the difficulty of comparing the absolute values of global transport coefficients between experiment and model, one is led to explore other options.

One important way to test transport hypotheses is to perform perturbative experiments [29]. In such experiments, instead of focusing on the absolute value of transport coefficients, one introduces a perturbation (such as a change in the heating source) and observes the reaction of the local plasma state. From this response, one can estimate the relevant transport coefficients [30] (essentially given by the ratio between a given flux and the corresponding driving gradient) which can be compared to local estimates from neoclassical transport, obtained from calculations based on the measurement of local profiles. The advantage of this approach is that the effects can be studied locally, so that the boundary conditions (hopefully) play only a minor role.

Another purely experimental method is measuring the net fluxes caused by temporally fluctuating fields explicitly, and comparing the result to the total flux from power balance estimates[16] [31]. In these studies, one typically finds, in the edge plasma where such measurements are feasible, that the fluctuating (turbulent) flux accounts for a large fraction of the total flux.

Under specific circumstances, turbulence can be largely suppressed at specific radial locations and times. These circumstances are known as 'H-mode' barriers [32] and internal transport barriers [33, 34]. Here, turbulence suppression is a consequence of the formation of local sheared radial electric fields, generated by the plasma itself [35]. The resulting plasma state, however, differs from the plasma state prior to the formation of the transport barriers not only in that turbulence is suppressed, but many profiles (plasma potential, temperature, and density) are also changed significantly, so that a clean comparison between states with/without turbulence is difficult. Something similar happens with plasmas in which a radial electric field is imposed by external means: turbulence can be suppressed, but profiles change [36]. On the other hand, it definitely seems to be the case that turbulence suppression is correlated with improved confinement, both at the edge and well inside the plasma, so that it seems reasonable to attribute low confinement to turbulent transport.

The level of turbulent transport can nowadays be predicted with a high degree of confidence using gyrokinetic simulations [24], even though not all situations can be handled (e.g. internal transport barriers). In any case, these calculations confirm that turbulence plays a major role in the understanding of transport in fusion devices.

---

[15] Balescu makes a distinction between 'anomalous' transport, a transport component due to turbulence that nevertheless behaves diffusively, and 'strange' transport, a component of transport that does not behave according to the standard diffusive equations [25].

[16] The total outward (heat or particle flux) at a given radius can be estimated from the net total power or number of particles deposited inside that radius, in steady state.

Summarizing, as far as we know, in the edge plasma, where gradients tend to be steep and fluctuations tend to be large, turbulent transport dominates overall transport. In the core region, where gradients tend to be small and fluctuations are relatively small, turbulent transport appears to be generally subdominant. For the intermediate region, no general statements can be made, but turbulence is expected to play a major role. In any case, the edge region has a very important weight in the overall confinement, which may justify the conclusion that 'turbulence matters for fusion'.

1.7 Unusual transport phenomena

The preceding section focused on the *value* of the transport coefficients. However, that is not the whole story. This section will describe some observations of remarkable phenomena that are difficult to explain on the basis of strictly local, diffusive models or using the cited approach in which turbulence is fully decoupled from the background transport and only considered as an additional term.

1.7.1 Power degradation and size scaling

In the absence of a complete transport theory that can predict the performance of a fusion device, one has turned to scaling laws. A scaling law is an engineering tool to predict the value of a system parameter as a function of some other significant variables in the absence of detailed knowledge of the underlying physics.

One of the main figures of merit of magnetically confined plasmas and one of the factors of the Lawson criterion for fusion is the energy confinement time, τ_E, which measures the efficiency of energy confinement in the device. It is defined as

$$\tau_E = \frac{W}{P - dW/dt}, \tag{1.5}$$

where W is the global plasma energy content and P the applied heating power.

This quantity is studied as a function of engineering parameters[17]:

$$\tau_E = CI^{\alpha_I}B^{\alpha_B}\bar{n}^{\alpha_n}P^{\alpha_P}R^{\alpha_R}\kappa^{\alpha_\kappa}\epsilon^{\alpha_\epsilon}S_{\rm cr}^{\alpha_S}M^{\alpha_M}, \tag{1.6}$$

where C is a constant, I (MA) is the plasma current, B (T) is the toroidal magnetic field, $\bar{n}$ (10^{19} m^{-3}) is the central line averaged density, P (MW) is the absorbed power, R (m) is the major radius, κ is the elongation, ϵ is the inverse aspect ratio, $S_{\rm cr}$ is the cross-sectional area (proportional to a^2, where a is the minor radius), and M is the hydrogen isotope mass. The parameters α are the corresponding scaling exponents, which are determined from a fitting procedure [37].

Decades of study in a huge range of devices, including different designs (tokamaks [38] and stellarators [39]) and machine sizes, have shown a remarkable consistency of the exponents.

[17] Using dimensional engineering parameters implies that the constant C must have a dimension that depends on the scaling exponents. This can be avoided through the use of appropriately normalized, dimensionless parameters.

One surprising result is the fact that the power scaling parameter systematically has the value $\alpha_P = -0.6 \pm 0.1$. In other words, as the heating power P is raised, the confinement quality as measured by τ_E degrades. This implies, necessarily, that the transport coefficients increase as the power is raised and gradients become steeper. It is worth noting that this result is largely independent from the type of toroidal confinement device, its size, the working gas, or the fuelling or heating method used, suggesting this is a feature common to all types of magnetically confined plasmas.

Various explanations have been suggested for this scaling behaviour. One idea is based on the assumption that heat transport tends to be of the 'gyro-Bohm' type, in which the radial diffusion is due to radial particle excursions of size $\rho_i = m_i|v_i|/eB$ and timescales of the order of the ion diamagnetic drift time. This would lead to an effective thermal diffusivity $\chi_{GB} \propto T^{1.5}/(aB^2)$, where T is the temperature, a the minor radius and B the magnetic field strength. It is evident that χ_{GB} increases with T, leading to a decrease of confinement time with T and hence heating. The final effect of this assumption on the scaling of τ_E can be assessed by making two further assumptions, namely: (i) the energy content of the plasma, W, can be estimated from $W \propto a^2 R \bar{n}_e T \propto P\tau_E$ (with this equation, T can be removed in favour of P) and (ii) $\tau_E \propto a^2/\chi_{GB}$. Combining results, one finds $\tau_E \propto P^{-0.6}$, quite close to the experimental result [40]. It should be noted that the apparent success of this straightforward argument to produce the correct exponent for the scaling with P is offset by its failure to produce the correct scaling for other parameters (such as B). Therefore, and also in view of the fact that plasmas do not always behave according to the gyro-Bohm expectation (particularly outside the core region [41]), this argument is not conclusive.

A second possible explanation is that an increase of heating power P will lead to an increase of (temperature and density) gradients, and thus an increase of the 'free energy' available to instabilities and turbulence. Steeper gradients will subsequently activate or enhance instabilities that increase radial transport [42], implying that turbulence is behind this scaling behaviour. We know that different circumstances (gradients) may trigger different types of turbulence, but apparently this does not matter to the scaling result, suggesting that the plasma somehow self-organizes towards a preferred end state, activating and deactivating turbulence as needed.

Another remarkable point is the scaling with system size. Assuming transport is dominated by diffusion (standard diffusion, non-fractional the sense of [43]), one expects τ_E to increase linearly with the cross-sectional area (or the square of the linear system size, L, see [44] for an argument). In actual fact, the scaling systematically falls short of this expectancy: building a bigger machine helps, but not as much as one would think. A possible explanation of this behaviour could be that transport is not strictly local, but somehow 'feels' the presence of the boundary, via long-range events (avalanches or streamers).

1.7.2 Profile consistency and stiffness

In the nineteen eighties it was observed that, often, a significant change in input electron cyclotron heating power P_{ECRH} only led to a moderate change of the resulting electron temperature profiles $T_e(r)$. The observation that profiles (of temperature, density, and pressure) often tend to adopt roughly the same shape, regardless of the applied heating and fuelling profiles, was dubbed profile consistency (or profile resilience) [45]. The resulting profiles, usually having a very similar shape, are known as canonical profiles [46, 47]. Like power degradation, the phenomenon has been observed over an extraordinarily broad range of devices and experimental conditions [48], although this does not mean that profiles always adopt the 'canonical' shape.

The explanation of this observation is found in the flux–gradient relation. In a standard diffusive model, the heat flux is linearly proportional to the temperature gradient, and similarly for other fields. In this case, no 'profile consistency' would occur, as profiles would evolve gradually as the sources are modified. Figure 1.7 provides a sketch of a strongly nonlinear flux–gradient relation. For small gradients, the flux increases linearly with the gradient, as one would expect for diffusion. But for larger gradients, the flux increases rather steeply over a small gradient interval, and returns to a more reduced slope for even higher gradient values. The solution of the transport problem, which consists of balancing sources and sinks by adjusting the flux would then lead to a very similar gradient for a broad range of fluxes (the blue area in the cartoon). Thus, in this range of fluxes, the profile (equal to the integral of the gradient) would look very similar, and one would thus obtain 'profile consistency'. In other words, the gradient would tend to be clamped to the range where the flux–gradient relation is steep over the relevant (broad) range of fluxes. This clamping of profiles is often called 'profile stiffness'.

What is the physical mechanism underlying this nonlinear flux–gradient relation? Several possible explanations have been put forward, but the main candidate appears to be the 'critical gradient' hypothesis [49]. This hypothesis assumes that there exists a critical gradient value, above which an instability is triggered, thus activating a new (fast) transport channel. It would explain the enhanced flux when the critical gradient value is exceeded, thus leading to a nonlinear flux–gradient relation as described above. This type of behaviour can be justified on the basis of the calculated growth rates of specific instabilities, changing from negative to positive at the critical gradient.

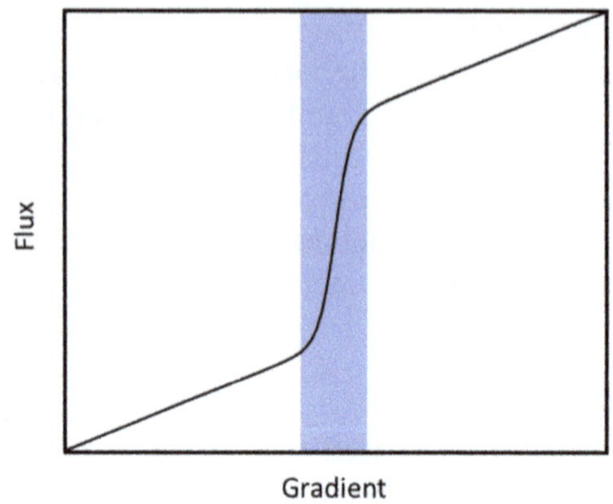

Figure 1.7. Cartoon of a 'critical gradient' type flux–gradient relation.

Therefore, turbulence again seems to be the essential ingredient to explain this phenomenon. It should be noted that the explanation is based on a feedback between profiles and turbulence, which poses an extraordinary challenge to turbulence simulations, as long and short temporal and spatial scales must be resolved simultaneously and evolved in time to reproduce the phenomenon.

1.7.3 Rapid transport phenomena and uphill transport

Changes in sources or sinks lead to adjustments of the transport fluxes and consequently the profiles. In 'ordinary' diffusive-type transport models, the timescales of these large scale profile adjustments are typically of the order of the confinement time. However, experiments reveal that sometimes phenomena occur on much shorter timescales.

A paradigmatic example was reported at the TEXT tokamak [50]. The edge of the ECRH heated plasma was cooled through the injection of an impurity. Figure 1.8 shows the experimental time traces of the electron temperature, T_e. The graph shows that the core T_e responds almost immediately to the edge cooling, i.e. within 0.5 ms, much faster than could be explained by diffusive processes. This observation led this group to postulate an ad hoc model in which the local transport coefficient is modified globally on this rapid timescale (reflecting a 'non-local' transport mechanism in which a modification at the plasma edge has an immediate impact on the

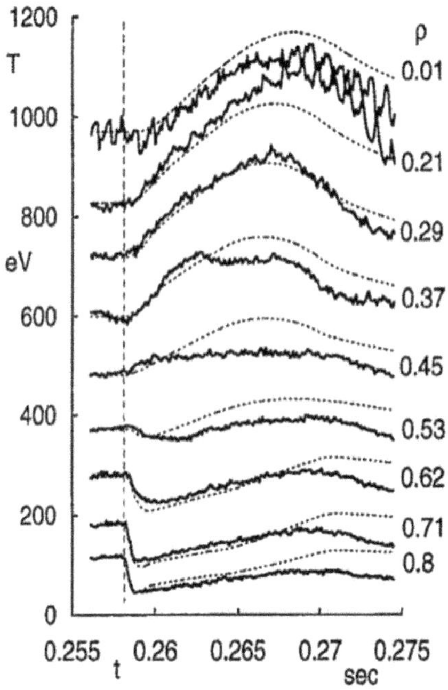

Figure 1.8. Electron temperature from ECE from a number of radii across the plasma, following strong edge cooling induced by impurity injection. (Reproduced with permission from [50]. Copyright (1992) by the American Physical Society.)

transport coefficient near the core). The resulting simulated plasma response (shown in the figure as dashed lines) approximately matched the observations, but the physical justification of this model remained unclear.

So-called 'uphill' transport (i.e. transport characterized by fluxes that go against the direction of the negative thermodynamic gradient) was observed in many experiments. Figure 1.9 shows an example in which strong off-axis heating leads to on-axis T_e profile peaking [51]. Standard diffusive models would require unrealistically intense inward convection to explain such behaviour, that would flow against the gradient ('uphill').

These remarkable phenomena are related and can be explained, at least qualitatively, on the basis of transport models with a strongly nonlinear flux–gradient relationship in combination with a supercritical superdiffusive transport channel [52], both of which have their roots in turbulence, as argued above. The nonlinear flux–gradient relation would produce the profile stiffness effect mentioned above, whereas the superdiffusive transport component would allow the unheated core region to 'fill up' [25].

1.7.4 Zonal flows

Although turbulence in fusion plasmas occurs in three-dimensional physical space, the presence of a strong magnetic field has the effect of reducing the effective dimension to near two. Unlike three-dimensional turbulence, two-dimensional turbulence is characterized by inverse energy cascades, implying the spontaneous formation of large scale structures from small scale turbulence. While noting that

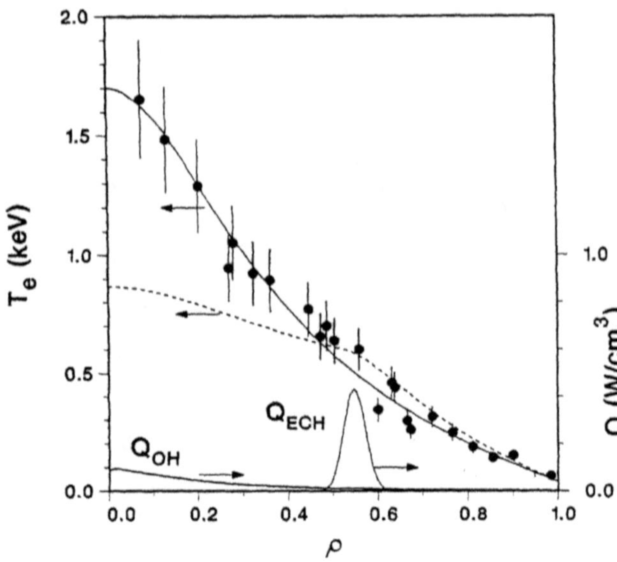

Figure 1.9. Electron temperature profile (dots) and power deposition from Ohmic heating (Q_{OH}) and ECR heating (Q_{ECH}). The dashed curve is a prediction. (Reproduced with permission from [51]. Copyright (1992) by the American Physical Society.)

this is an extreme simplification of the issue, it is still true that the presence of a strong magnetic field can lead to the spontaneous formation of large scale structures from the ambient turbulence, known as zonal flows [35]. One of the physical mechanisms for zonal flow production is Reynolds stress [53]. This effect is similar, in some respects, to the formation of 'bands' in the atmosphere of the planet Jupiter, where the dimension-reducing role of the magnetic field is taken over by the planet's fast rotation.

The radial variation of the zonal flow velocity ('sheared flow') generated by this effect then leads to turbulence suppression, as the radially varying flow speed tears apart turbulent eddies, thus reducing their size. In this way, a feedback loop is created: turbulence growth–zonal flow generation–sheared flow enhancement–turbulence suppression [54]. This is an example of self-organization with the amazing result that turbulence, driven by steep gradients, produces an effect that … reduces turbulence. The end state of the turbulence and the profiles then includes a transport barrier with steep gradients, associated with the sheared flow (with names such as H-mode barrier or internal transport barrier). Needless to say that the discovery of these spontaneous transport barriers was welcomed by the fusion community as a magnificent tool to improve the efficiency of future fusion reactors!

1.7.5 The impact of rational surfaces on transport

It has long been suspected that the topology of the confining magnetic field has an impact on plasma transport. Specifically, we are referring to the fact that minor transport barriers are often formed that are associated with low-order rational surfaces, which will be an issue that we will come back to often in this book. A legitimate question is why, if this effect is supposedly important, studies on this issue do not abound in the literature on transport in fusion devices—although some notable studies are available, as discussed below. A localized transport barrier (i.e. a local reduction of heat flux) implies a local change of *slope* of the temperature profile. Given the general turbulent state of the plasma and the prevailing measurement resolution and errors, such rather localized changes of slope are usually not easy to detect. As a result, many transport models completely ignore their possible existence and do not contemplate any effects that explicitly depend on the rational values of the rotational transform.

In 1997 an astounding report appeared that combined measurements using a very high resolution Thomson scattering diagnostic with a cleverly designed experiment in which the microwave power deposition position was scanned through the plasma [55]. Figure 1.10 shows one of the most surprising results from that report, reflecting clear evidence that a succession of minor transport barriers existed in the RTP tokamak. Several years later, a heuristic transport model was proposed to explain these results, the so-called 'q-comb' model, in which relatively sharp reductions of the electron heat transport coefficient were associated with the main low-order rational surfaces [56]. This simple hypothesis was sufficient to reproduce the observed behaviour.

Apart from the mentioned minor transport barriers, plasmas occasionally, and under special circumstances, can develop so-called internal transport barriers (ITBs) [33].

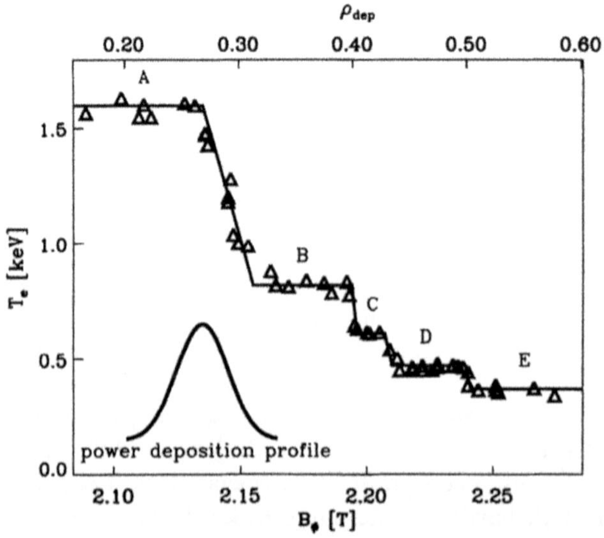

Figure 1.10. Stepwise response of the central electron temperature $T_e(0)$ as the deposition location of ECRH power is scanned: evidence of minor transport barriers. (Reproduced with permission from [55]. © IOP Publishing. All rights reserved.)

ITB arise only transiently, but are much stronger than the 'minor transport barriers' that are assumed to be semi-permanent. In tokamaks, strong ITBs can be established by creating a core reversed magnetic shear region. The location of such ITBs appears to be correlated with integral values of the (minimum of the) safety factor, q [57]. The impact of ITBs on heat transport has been studied in some detail at, e.g. Alcator C-Mod [58] and JET [59, 60], showing that the heat diffusivity drops strongly in the ITB region. ITBs have also been obtained and studied in stellarators [61] and, here too, a relationship with the magnetic configuration is suggested. The existence of ITBs is widely acknowledged and supported by experimental evidence on many machines.

It seems undeniable, therefore, that the magnetic topology has a non-trivial impact on transport. Minor transport barriers and ITBs are likely established through a similar mechanism. In the neighbourhood of low-order rational surfaces, it may be easier for sheared flow regions to form as a consequence of Reynolds stress forces associated with MHD turbulence—a type of turbulence that is sensitive to the rotational transform. This is not to say that *all* transport barriers are necessarily of this type; there may be regions where other types of turbulence dominate and where sheared flow regions can form without immediate relation to the local rotational transform. At this time, insufficient data are available to provide a clear statement about this issue.

1.8 Summary

The study of turbulence is intellectually and computationally challenging. In this chapter, we have tried to show, without going into too much detail, that fusion plasmas are rather extreme systems, far from thermodynamic equilibrium, subject to strong turbulence and riddled with remarkable phenomena. Estimates of the relative importance of turbulent transport as compared to regular (diffusive, neoclassical) transport, based on both theory and experiment, systematically indicate that turbulence is important and often even constitutes the dominant component of radial (heat and particle) transport. In addition, some of the mentioned unusual[18] phenomena can be ascribed unambiguously to turbulence, either through the non-linearity of the flux–gradient relation, or through its non-locality. Some other phenomena are likely associated with turbulence (e.g. the formation of minor transport barriers near rational surfaces via zonal flow formation, driven by turbulence), but remain to be clarified further. It is therefore clear that the study of turbulence in fusion plasmas is a rich subject with many open questions that require urgent answers.

One important observation is that the often assumed separation between scales may not hold: turbulence is not confined to small radial scales and short times, but its impact can be felt on the large scales of the system size and the confinement time. This impact may occur via large scale structures that form spontaneously from small scale turbulence (an example being the zonal flow), or via the mutual modification of 'background' profiles and turbulent transport in a feedback loop. This observation has a major impact on the understanding of this phenomenon, and implies that it cannot be simply dismissed as an additional contribution to diffusion (i.e. a local and instantaneous mechanism). A significant effort has therefore been devoted to the detection and analysis of the long-range (spatial and temporal) effects of turbulence, as will be discussed in this book.

There are also practical motives for undertaking the study of turbulence. An improved understanding of turbulence may lead to its control or mitigation, with potentially enormous economic benefits. In addition, the incorporation of the full complexity associated with turbulence in predictive models may allow optimizing plasma configurations, leading to improved fusion device designs [62].

The remainder of this book is mainly dedicated to efforts to improve our understanding of turbulence through data analysis and modelling.

References

[1] Zhang F, Zhao P, Niu M and Maddy J 2016 The survey of key technologies in hydrogen energy storage *Int. J. Hydrog. Energy* **41** 14535–52

[2] Freidberg J P 2007 *Plasma Physics and Fusion Energy* (Cambridge: Cambridge University Press)

[3] Oliphant M L E, Harteck P and Rutherford E 1932 Transmutation effects observed with heavy hydrogen *Proc. R. Soc. Lond.* A **144** 692

[4] Waldrop M M 2014 Plasma physics: the fusion upstarts *Nature* **511** 398

[5] Hurricane O A *et al* 2014 Fuel gain exceeding unity in an inertially confined fusion implosion *Nature* **506** 343–8

[6] Craxton R S *et al* 2015 Direct-drive inertial confinement fusion: a review *Phys. Plasmas* **22** 110501

[18] 'Unusual' in the sense of 'remarkable, extraordinary'; but not 'unusual' in the sense of 'rare'.

[7] Lawson J D 1957 Some criteria for a power producing thermonuclear reactor *Proc. Phys. Soc.* B **70** 6

[8] Gill R D, Edwards A W, Pasini D and Weller A 1992 Snake-like density perturbations in JET *Nucl. Fusion* **32** 723–35

[9] van Milligen B P, van Lammeren A C A P, Lopes Cardozo N J, Schüller F C and Verreck M 1993 Gradients of electron temperature and density across $m = 2$ magnetic islands in RTP *Nucl. Fusion* **33** 1119

[10] Waelbroeck F L 2009 Theory and observations of magnetic islands *Nucl. Fusion* **49** 104025

[11] Hölzl M 2010 Diffusive heat transport across magnetic islands and stochastic layers in tokamaks *PhD Thesis* Technische Universität München, Garching, Germany

[12] Okazaki T 2021 *Fusion Reactor Design* (New York: Wiley)

[13] Lesieur M 2008 *Turbulence in Fluids* 4th edn (Berlin: Springer)

[14] Prigogine I 1984 *Order Out of Chaos: Man's New Dialogue with Nature* (New York: Bantam)

[15] Chen F F 1984 *Introduction to Plasma Physics and Controlled Fusion: Plasma Physics* vol 1 2nd edn (New York: Plenum)

[16] Horton W, Kim J-H, Asp E, Hoang T, Watanabe T-H and Sugama H 2008 Drift wave turbulence *AIP Conf. Proc.* **1013** 1–19

[17] White A E 2019 Validation of nonlinear gyrokinetic transport models using turbulence measurements *J. Plasma Phys.* **85** 925850102

[18] Sanchez R, Newman D E, Leboeuf J N, Decyk V and Carreras B A 2008 Nature of transport across sheared zonal flows in electrostatic ion-temperature-gradient gyrokinetic plasma turbulence *Phys. Rev. Lett.* **101** 205002

[19] Callen J D 1977 Drift-wave turbulence effects on magnetic structure and plasma transport in tokamaks *Phys. Rev. Lett.* **39** 1540–3

[20] Einstein A 1905 Über die von der molekulartheoretischen Theorie der Wärme geforderte Bewegung von in ruhenden Flüssigkeiten suspendierten Teilchen *Ann. Phys., Lpz.* **17** 549

[21] van Milligen B Ph, Bons P D, Carreras B A and Sánchez R 2005 On the applicability of Fick's law to diffusion in inhomogeneous systems *Eur. J. Phys.* **26** 913

[22] Braginskii S I 1965 *Reviews of Plasma Physics* vol 1 (New York: Consultants Bureau) p 205

[23] Hinton F L and Hazeltine R D 1976 Theory of plasma transport in toroidal confinement systems *Rev. Mod. Phys.* **48** 239

[24] Garbet X, Idomura Y, Villard L and Watanabe T H 2010 Gyrokinetic simulations of turbulent transport *Nucl. Fusion* **50** 043002

[25] Balescu R 2005 *Aspects of Anomalous Transport in Plasmas* (Bristol: Institute of Physics)

[26] Naulin V 2007 Turbulent transport and the plasma edge *J. Nucl. Mater.* **363-365** 24–31

[27] Stroth U, Geist T, Koponen J P T and Hartfuss H-J 1999 Evidence for convective inward particle transport in a stellarator *Phys. Rev. Lett.* **82** 928

[28] Beidler C D *et al* 2021 Demonstration of reduced neoclassical energy transport in Wendelstein 7-X *Nature* **596** 221–6

[29] Mantica P and Ryter F 2006 Perturbative studies of turbulent transport in fusion plasmas *Compt. Rend. Phys.* **7** 634

[30] van Berkel M, Zwart H J, Tamura N, Hogeweij G M D, Inagaki S, de Baar M R and Ida K 2014 Explicit approximations to estimate the perturbative diffusivity in the presence of convectivity and damping. I. Semi-infinite slab approximations *Phys. Plasmas* **21** 112507

[31] Wootton A J, Carreras B A, Matsumoto H, McGuire K, Peebles W A, Ritz C P, Terry P W and Zweben S J 1990 Fluctuations and anomalous transport in tokamaks *Phys. Fluids* B **2** 2879

[32] Wagner F *et al* 1982 Regime of improved confinement and high beta in neutral-beam-heated divertor discharges of the ASDEX tokamak *Phys. Rev. Lett.* **49** 1408
[33] Wolf R C 2003 Internal transport barriers in tokamak plasmas *Plasma Phys. Control. Fusion* **45** R1
[34] Itoh K, Toda S, Fujisawa A, Itoh S-I, Yagi M, Fukuyama A, Diamond P H and Ida K 2007 Physics of internal transport barrier of toroidal helical plasmas *Phys. Plasmas* **14** 020702
[35] Diamond P H, Itoh S-I, Itoh K and Hahm T S 2005 Zonal flows in plasma—a review *Plasma Phys. Control. Fusion* **47** R35
[36] Grenfell G 2019 Measurement and control of turbulence spreading in the scrape-off layer of TJ-II *Nucl. Fusion* **59** 016018
[37] Kardaun O J W F 2005 *Classical Methods of Statistics: with Applications in Fusion-oriented Plasma Physics* (Berlin: Springer Science)
[38] ITER Physics Expert Group on Confinement and Transport 1999 Chapter 2: Plasma confinement and transport *Nucl. Fusion* **39** 2175
[39] Dinklage A *et al* 2007 Physical model assessment of the energy confinement time scaling in stellarators *Nucl. Fusion* **47** 1265
[40] Miyazawa J *et al* 2010 Bessel function model of the electron pressure profile normalized by gyro-Bohm type parameter dependence *J. Plasma Fusion Res.* **9** 094–9
[41] Wilks T M, Wolfe S, Hughes J W, Hubbard A E, Greenwald M, Cao N, Rice J E and Reinke M 2019 Experimental energy confinement time scaling with dimensionless parameters in C-Mod I-mode plasmas *Nucl. Fusion* **59** 126023
[42] Horton W, Hu B, Dong J Q and Zhu P 2003 Turbulent electron thermal transport in tokamaks *New J. Phys.* **5** 14
[43] Sánchez R, Carreras B A, Newman D E and Lynch V E 2006 Renormalization of tracer turbulence leading to fractional differential equations *Phys. Rev.* E **74** 016305
[44] van Milligen B Ph, Sánchez R and Carreras B A 2004 Probabilistic finite-size transport models for fusion: anomalous transport and scaling laws *Phys. Plasmas* **11** 2272
[45] Coppi B 1980 Nonclassical transport and the principle of profile consistency *Comments Plasma Phys. Control. Fusion* **5** 261
[46] Dnestrovskij Y N, Yu A, Dnestrovskij, Lysenko S E and Cherkasov S V 2002 Canonical profiles in tokamak plasmas with an arbitrary cross section *Plasma Phys. Rep.* **28** 887–99
[47] Yu N, Dnestrovskij K A and Razumova A J H 2006 Self-consistency of pressure profiles in tokamaks *Nucl. Fusion* **46** 953
[48] Ryter F *et al* 2001 Experimental studies of electron transport *Plasma Phys. Control. Fusion* **43** A323
[49] Garbet X *et al* 2004 Profile stiffness and global confinement *Plasma Phys. Control. Fusion* **46** 1351
[50] Gentle K W *et al* 1995 An experimental counter-example to the local transport paradigm *Phys. Plasmas* **2** 2292
[51] Luce T C, Petty C C and de Haas J C M 1992 Inward energy transport in tokamak plasmas *Phys. Rev. Lett.* **68** 52
[52] van Milligen B Ph, Carreras B A and Sánchez R 2004 Uphill transport and the probabilistic transport model *Phys. Plasmas* **11** 3787
[53] Korsholm S B, Michelsen P K, Naulin V, Rasmussen J J, Garcia L, Carreras B A and Lynch V E 2001 Reynolds stress and shear flow generation *Plasma Phys. Control. Fusion* **43** 1377

[54] Diamond P H, Liang Y-M, Carreras B A and Terry P W 1994 Self-regulating shear flow turbulence: a paradigm for the L to H transition *Phys. Rev. Lett.* **72** 2565
[55] Lopes Cardozo N J *et al* 1997 Electron thermal transport in RTP: filaments, barriers and bifurcations *Plasma Phys. Control. Fusion* **39** B303
[56] Schilham A M R, Hogeweij G M D and Lopes Cardozo N J 2001 Electron thermal transport barriers in RTP: experiment and modelling *Plasma Phys. Control. Fusion* **43** 1699
[57] Joffrin E *et al* 2003 Internal transport barrier triggering by rational magnetic flux surfaces in tokamaks *Nucl. Fusion* **43** 1167
[58] Wukitch S J *et al* 2002 Double transport barrier experiments on Alcator C-Mod *Phys. Plasmas* **9** 2149
[59] Mantica P *et al* 2002 Perturbative transport experiments in JET low or reverse magnetic shear plasmas *Plasma Phys. Control. Fusion* **44** 2185
[60] Marinoni A *et al* 2006 Analysis and modelling of power modulation experiments in JET plasmas with internal transport barriers *Plasma Phys. Control. Fusion* **48** 1469
[61] Fujisawa A 2002 Transport barriers and bifurcation characteristics in stellarators *Plasma Phys. Control. Fusion* **44** A1
[62] Highcock E G, Mandell N R, Barnes M and Dorland W 2018 Optimisation of confinement in a fusion reactor using a nonlinear turbulence model *J. Plasma Phys.* **84** 905840208

IOP Publishing

Analysis of Turbulence in Fusion Plasmas

Boudewijn Ph van Milligen and Raul Sanchez

Chapter 2

Characterization of turbulence in plasmas

To improve our understanding of turbulence in fusion plasmas, one may consider two complementary approaches. One approach is based on theory and computer simulations. The theory describing the behaviour of a fusion plasma is essentially known: it is based on Maxwell's equations and the equations of conservation of mass, momentum, and energy. The difficulty of the problem therefore does not reside in the underlying theory or equations, but in the collective nature of the behaviour of many particles moving according to the mentioned equations and interacting via electromagnetic fields, modified by the presence and motion of the particles themselves. In order to model turbulence, approximations must be made to obtain a reduced set of equations that allows practical simulation in a computer, and several alternative approaches are possible, each with advantages and drawbacks. Great progress has been made in recent years, partly due to the availability of ever increasing computational power [1].

The other approach is direct measurement, which is absolutely necessary to verify the validity of numerical predictions based on reduced equations. Unfortunately, fusion plasmas constitute an extremely hostile environment, so that measurements of turbulent quantities have not been as abundant as one would have liked. In the core plasma, it is usually not possible to insert physical probes due to the damage that such probes would suffer and due to the perturbation of the plasma itself that would ensue. Many available turbulent data, taken by probes, are therefore limited to the edge of the plasma and the scrape-off layer (SOL) region immediately outside the confined plasma region. This peripheral location obviously sheds some doubts on the relevance of experimental findings for core turbulence, as the edge plasma is affected by processes that usually play only a minor role in the core, such as the interaction with neutral particles, ionization, and recombination.

Core turbulence can be explored by detecting electromagnetic or particle emissions from the plasma ('passive' diagnostics) or by injecting particles or electromagnetic (microwave, laser) beams and observing secondary particles or

doi:10.1088/978-0-7503-4856-0ch2

emissions exiting the plasma ('active' diagnostics). This type of diagnostics typically has a high degree of complexity and requires a significant effort to interpret the data. For example, detected radiation is usually integrated over the line of sight of the detector, so that recovery of the local emission requires combining information from various detectors looking at the same point from various angles (Abel inversion techniques or coherence techniques). The precise reconstruction of the location of the interaction or observation volume in relation to the surrounding vacuum vessel is often not easy, and the determination of the magnetic co-ordinates of the observation volume, necessary for comparison with other diagnostics and theory, may be rather convoluted and riddled with uncertainty. In addition, a fusion experiment is a hostile electromagnetic environment in which many diagnostics require extraordinary measures to suppress ambient noise sources, and even so, signal to noise ratios are often less than optimal. Consequently, each measurement system has its own specific procedure to obtain the information of interest from the actual measurement data, which includes the conversion of the raw data into calibrated physical quantities with their corresponding uncertainties [2].

The analysis of the obtained data will depend in a large degree on their nature, spatiotemporal resolution and extent or duration, and signal to noise ratio. For example, data from turbulence simulations often provide very detailed (even three-dimensional) spatial information, something that rarely is the case for experimental data. The coherent structures that are usually present in such simulations can then be detected by, e.g., applying topological methods to tracers [3] or using Lagrangian descriptors [4]. Experimental data generally do not provide access to the full spatial flow field, rendering such approaches impractical. On the other hand, the temporal duration of the measurements often far exceeds the equivalent duration of numerical turbulence simulations, due to CPU time limitations. These differences in the spatiotemporal extent of data from simulations and experiment are not helpful when trying to make comparisons. In view of the predominantly temporal nature of experimental data, the analysis approaches discussed in the following will focus mainly on the analysis of time series, the kind of experimental data that are most abundant in this field of research.

Let us assume we obtain a measurement or simulation result that we would like to use to study turbulence. The first issue to be considered is the temporal resolution needed to resolve turbulence. The turbulent timescale is estimated as the turbulent autocorrelation time and is typically of the order of 10–100 μs in fusion plasmas (see section 2.2). Hence, measurements should be performed with a sampling rate of at least about 100 kHz and preferably significantly faster—always assuming that the measurement system itself does not restrict the measurable frequencies to even lower values.

The minimum spatial resolution is set by the typical size of turbulent eddies. Gyrokinetic studies typically yield a size of a few to ten times the ion Larmor radius, $\rho_i = m_i v_\perp / |q_i| B$. Experimental determinations of the turbulent correlation length from the spatial decay of the cross correlation typically yield sizes of the order of about 1 cm, in rough agreement with the theoretical result; so the spatial resolution should preferably be finer than this value.

Assuming one can perform suitable measurements of relevant quantities, with a reasonable signal to noise level, one obtains a discretely sampled measurement

($x_i = x(t_i)$, $i = 1, \ldots, N$), or a set of such measurements at various locations z_j ($x_{ij} = x(z_j, t_i)$, $j = 1, \ldots, K$, $i = 1, \ldots, N$), to be processed further in order to extract information of interest.

2.1 Probability distribution functions and statistical moments

The first step towards quantification involves determining the probability distribution function (PDF) $p(x)$. Typically, this is done by binning the data, i.e. subdividing the range [min(x), max(x)] in a number of M bins and counting the number n_m of x-values that fall in bin m; the probability corresponding to this bin would be $p_m = n_m/(N \cdot \Delta x_m)$, where Δx_m is the size of the corresponding bin. Thus, the sum of the probabilities over all M bins, considering the size of each bin, is $\sum_m p_m \Delta x_m = 1$, since $\sum_m n_m = N$.

This 'binning' procedure generally works very well for the densely populated central part of the PDF. However, one is often interested in the behaviour of the 'tail' of the PDF: for example, one could ask whether the function $p(x)$ decays exponentially or more slowly (like a power law, $p(x) \propto |x|^{-\alpha}$) for large values of $|x|$. This question is often difficult to answer, as the 'tail region' is much less densely populated and prone to large fluctuations. Each value $p_m(x)$ has an uncertainty in x equal to about $\Delta x_m/2$, and a relative error $\epsilon(p_m)/p_m \simeq n_m^{-1/2}$—noting that this estimate is very crude and may be incorrect in general; for example, in the case of non-Gaussian statistics. Nevertheless, it shows that the underpopulated bins in the tail regions may be very unreliable, and as a result it is surprisingly hard to determine the shape of the tails of the PDF. To partly remedy this situation, the bin sizes Δx_m, which are often taken equal for convenience, may be adjusted. In the case of heavy-tailed (i.e. slowly decaying) distributions, it may be convenient to define larger bins in the less populated 'tail' regions, in order to fill each bin m with a reasonable number of counts n_m—while noting that this procedure increases the uncertainty in x corresponding to such bins.

A convenient and more robust alternative to the PDF is provided by the cumulative distribution function (CDF), which does not require a binning procedure and therefore does not suffer from a loss of precision due to the finite bin size. It is constructed simply by sorting the data points $x(i)$ according to the ascending x-value. The sorted array is $x'(i)$, such that $x'(i+1) \geqslant x'(i)$; the CDF is then $P(x') = p(x \leqslant x') = i/N$. In theory, the corresponding PDF is given by $p(x) = dP(x)/dx$, although it is not a practical proposition to estimate $p(x)$ directly from the numerical derivative of $P(x)$. The tails of the distribution are often best visualized on a double logarithmic plot. Considering that x may take on positive and negative values[1] and $0 < P(x) \leqslant 1$ is monotonically increasing, one could therefore plot $\mathcal{H}(x) - \text{sgn}(x) \cdot P(x)$ versus $|x|$ on a double logarithmic scale to visualize the tails of the distribution. Here, sgn(x) is the sign of x and $\mathcal{H}(x)$ is the Heaviside function ($\mathcal{H}(x) = 1$ when $x \geqslant 0$ and $\mathcal{H}(x) = 0$ otherwise). If one is able to determine the decay rate of $P(x)$ for large $|x|$, the decay rate of $p(x)$ follows from $p(x) = dP(x)/dx$.

[1] If this is not the case, one can always subtract the mean of x from the data.

To illustrate these ideas, we have generated some random data with a distribution close to the Cauchy distribution[2] with $\gamma = 1$, consisting of 100 000 data points. Figure 2.1 shows an example of the determination of the decay rate of the tail of the distribution using the two cited methods. The top figure shows the PDF, constructed using logarithmic binning (i.e. linear binning of $\log|x|$) with 200 bins. The bottom figure shows the CDF, plotted as suggested above. The red dashed lines are power law fits (i.e. linear fits to the logarithmic data) over the fit range $3 \leqslant |x| \leqslant 800$. The chosen fit range determines the obtained index; one would like to choose $|x|$ as large as possible to determine the asymptotic value of the index and to select a broad range to reduce the fit error, but one is limited by the non-asymptotic part of the distribution for small $|x|$ and by data scatter for large $|x|$. The two indices are consistent within the indicated fit errors—considering that one should subtract 1 from the index of $P(x)$ in order to compare it to that of $p(x)$. The scatter in the tail region of the PDF is much larger than that of the CDF, due to the inaccuracies introduced by the binning procedure, and consequently, the error of the fit to $P(x)$ is much smaller than the error of the fit to $p(x)$. The CDF appears to outperform most (or all) probability density estimates—including those produced by more or less

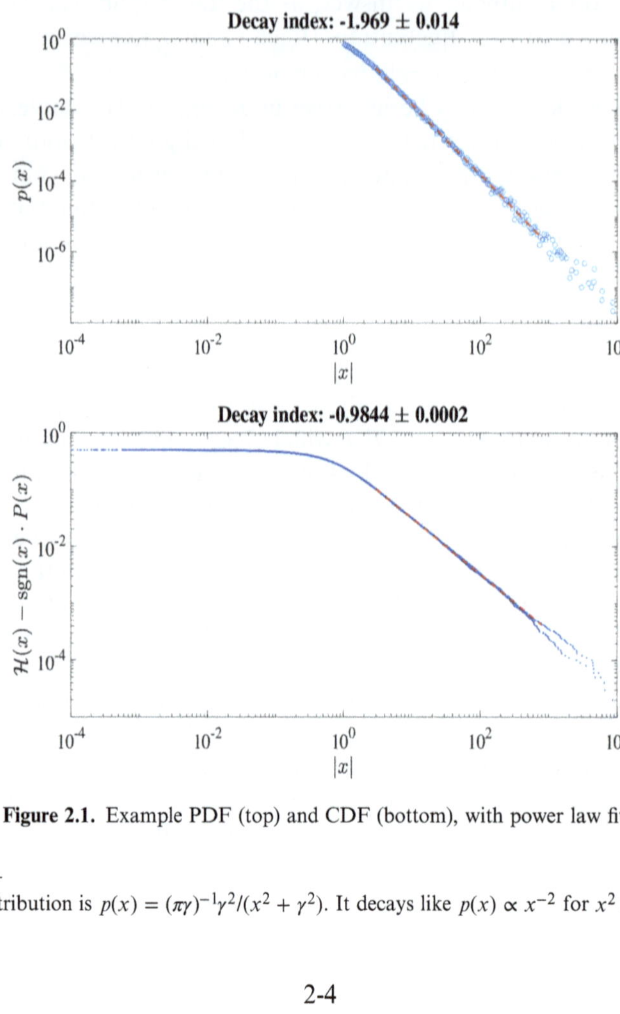

Figure 2.1. Example PDF (top) and CDF (bottom), with power law fits.

[2] The Cauchy distribution is $p(x) = (\pi\gamma)^{-1}\gamma^2/(x^2 + \gamma^2)$. It decays like $p(x) \propto x^{-2}$ for $x^2 \gg \gamma^2$.

sophisticated kernel density estimator methods [5]—with regard to the unbiased determination of the tail behaviour.

Assuming the PDF decays faster than algebraically, its shape can be captured by its (centred) moments:

$$M_1 = \langle x \rangle = \int xp(x)dx, \tag{2.1}$$

$$M_j = \int (x - \langle x \rangle)^j p(x)dx \qquad (j \geqslant 2). \tag{2.2}$$

As the construction of the PDF implies information loss due to the binning procedure, it is more convenient to determine the moments directly from the data, e.g.

$$M_1 = \langle x \rangle = \frac{1}{N}\sum_{i=1}^{N} x_i \tag{2.3}$$

$$M_j = \frac{1}{N}\sum_{i=1}^{N}(x_i - \langle x \rangle)^j \qquad (j \geqslant 2). \tag{2.4}$$

The first moment M_1 is the sample mean, $\langle x \rangle$, and the second one, M_2, is the square of the root-mean square (RMS) value or standard deviation, $\sigma(x) = \sqrt{M_2}$. Higher-order moments are usually normalized

$$\mu_j = M_j / \sigma(x)^j \qquad (j \geqslant 3). \tag{2.5}$$

It is customary to call the third normalized moment the skewness, $S = \mu_3$, and the fourth the kurtosis, $K = \mu_4$. Mathematically, $K \geqslant S^2 + 1$, and often a quadratic relation $K(S)$ is observed experimentally [6]. The 'flatness factor' or excess kurtosis is $F = K - 3$. The skewness S is a measure of the asymmetry of the 'tails' of the probability distribution of x, $p(x)$, while K provides information about the 'weight' of the tails. These two values are used to determine whether $p(x)$ is close to a Gaussian distribution (which has $S = 0$ and $K = 3$) or not. Note that some authors use the terms kurtosis and flatness indistinctly; which one is meant should always be clear from the value assigned to a Gaussian distribution.

The Gaussian distribution, sometimes also called the 'normal' distribution, plays a dominant role in statistics due to the central limit theorem, which states that the distribution of the sum of independent stochastic variables, having individual distributions with finite variance, converges rapidly to a Gaussian distribution [7].

The above expressions involving moments assume that the PDF decays faster than algebraically at large values of its argument or, equivalently, that the sums converge in principle as $N \to \infty$. However, 'long-tailed' PDFs may not satisfy this requirement, see chapter 3. A typical example of such a long-tailed PDF would be the Cauchy distribution, which has no finite standard deviation. While convergence never constitutes an issue for a given finite length data sample, it should still be a matter of concern, as the resulting values of the moments are meaningless if their value depends systematically on N.

Note that the analysis can be trivially extended to allow following the evolution of these quantities in time, if sufficient data are available. To do so, one subdivides the temporal arrays $\{x_i, i = 1 ..., N\}$ into $\lfloor N/m \rfloor$ subblocks of length $m \ll N$. For example, the evolving mean of x is then

$$\langle x \rangle_j^{(m)} = \frac{1}{m} \sum_{i=1}^{m} x_{(j-1)m+i}, \tag{2.6}$$

also called a 'boxcar average'. This procedure of analyzing by subblocks can be extended to higher order moments and, e.g., spectra and many of the other quantities discussed below. To improve the resulting graphs, one may consider overlapping the subblocks, although one should be aware that the amount of information obtained is not increased by overlapping! An extreme example of overlapping would be the $(2k + 1)$-point 'moving average' of a signal:

$$\bar{x}_i = \frac{1}{2k + 1} \sum_{j=i-k}^{i+k} x_j. \tag{2.7}$$

This equation is evidently only valid in the range $k + 1 \leqslant i \leqslant N - k$. The procedure can be generalized to the other statistical moments. By way of example, the $(2k + 1)$-point 'running RMS' is

$$\overline{\text{RMS}}(x_i)^2 = \frac{1}{2k + 1} \sum_{j=i-k}^{i+k} (x_j - \bar{x}_j)^2. \tag{2.8}$$

The signal can be decomposed as a sum of the moving average and a remainder:

$$x = \bar{x} + \tilde{x}. \tag{2.9}$$

This remainder, $\tilde{x}$, is called the fluctuating part of the signal and will be used often in the following, as it quantifies the fluctuations without influence from the equilibrium or mean of the signal. In particular, it does not contain slow signal drifts. To achieve an effective separation of fluctuations on the short timescales associated with turbulence and the slowly evolving component, the parameter k of equation (2.7) should be chosen such that $k \cdot \Delta t$ (where Δt is the sampling time interval) is well above the turbulence correlation time (by, e.g., two orders of magnitude), so that the turbulent fluctuations are essentially unaffected by subtracting $\bar{x}$ from x.

The slowly varying component $\bar{x}$ can also be estimated by applying a more sophisticated low-pass frequency filter[3] instead of a moving average; in view of the fact that the turbulence correlation time is typically of the order of 10 μs, a low-pass frequency filter with a cutoff frequency well below the inverse correlation time (at, say, 1 kHz) will effectively capture slowly varying components, almost without contribution from the turbulent timescales. However, the implied independence and

[3] A plethora of digital filters is available in literature and software packages, each with specific advantages and disadvantages.

separability of slow and fast timescales is somewhat artificial and may not always be correct, as will be discussed below.

2.1.1 The PDF of particle transport

Langmuir probes allow the measurement of a series of quantities that are relevant for turbulence studies in the plasma edge and SOL regions, depending on the configuration of the probe system: among these are the floating potential $V_f \simeq V_p - 3T_e$ (where V_p is the plasma potential and T_e the electron temperature), the ion saturation current $I_{sat} \propto n\sqrt{T_e}$ (where n is the density). Often, it is assumed that electron temperature fluctuations play a minor role, so that ion saturation current fluctuations can be considered proportional to density fluctuations, $\tilde{I}_{sat} \propto \tilde{n}$, although this approximation should be considered with some caution [8].

With three equally spaced probe pins aligned poloidally and measuring V_f, I_{sat}, and V_f, respectively, it is possible to estimate the fluctuating particle flux $\Gamma = \tilde{n}\tilde{v}_r = \tilde{n}\tilde{E}_\theta/B_\phi$, namely: $\Gamma \propto \tilde{I}_{sat}\tilde{E}_\theta$, where $E_\theta = (V_f^2 - V_f^1)/d$, d being the distance between the two V_f probe pins. As $\tilde{I}_{sat}$ and V_f are affected by T_e, so is this quantity. Nevertheless, it has been found that the relative phase of fluctuations is often such that Γ calculated on the basis of $\tilde{I}_{sat}$ and V_f instead of $\tilde{n}$ and V_p provides a good estimate of the particle flux [9, 10].

Among all the quantities that Langmuir probes can measure—many more than the ones cited above—, the particle flux Γ is especially interesting, as it provides information on radial transport. Therefore, it is not only sensitive to local fluctuations, but also to density 'blobs' or events arriving from positions further inward, so the flux truly reflects the complexity of turbulence. After all, turbulence is not only characterized by temporal fluctuations, but also by spatial structure.

Relative density fluctuations, $|\tilde{n}|/n$, tend to be small in the plasma core (a few percent) and to grow significantly towards the plasma edge, where the fluctuation amplitude can be as high as 50% [8]. Since the density is a positive definite quantity, its PDF is *a priori* not expected to be Gaussian under these conditions, but Poisson-like. Figure 2.2 shows an example of $\tilde{I}_{sat}$ measurements by a Langmuir probe in the edge region of the TJ-II stellarator [11]. The magnetic configuration of this stellarator is highly flexible and the magnetic well w can be varied over a significant range, $-1\% < w < 6\%$. A large, positive magnetic well provides stability to the dominant pressure-driven instabilities. By lowering the magnetic well, instabilities become enhanced, which is immediately evident from the data shown in the figure. Particularly in the low well (most unstable) case, the signal is clearly 'bursty'[4] and up-down asymmetric, so the skewness S will be non-zero. Deviations from zero skewness are observed systematically in the edge of a wide range of devices, both for the floating potential V_f and the ion saturation current I_{sat}. S tends to deviate more from its Gaussian value ($S = 0$) as one moves from the edge plasma into the SOL [12, 13].

So the PDFs of V_f and I_{sat} are often clearly asymmetric and show some evolution with plasma conditions and radial location. What about the PDF of the flux?

[4] The property of 'burstiness' will be addressed in detail in chapter 5.

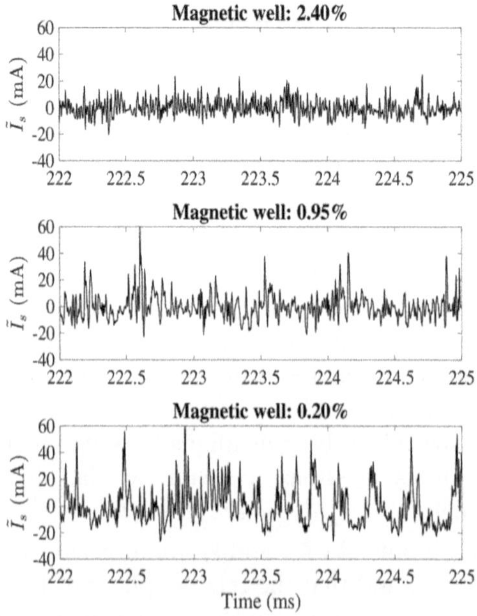

Figure 2.2. Ion saturation current fluctuations versus magnetic well in the plasma edge region of the TJ-II stellarator (shots 3571, 3573 and 3584). The level of fluctuations increases as the magnetic well is reduced. Measurements were taken at the normalized radius. $\rho \simeq 0.85$ (Reproduced with permission from [11].)

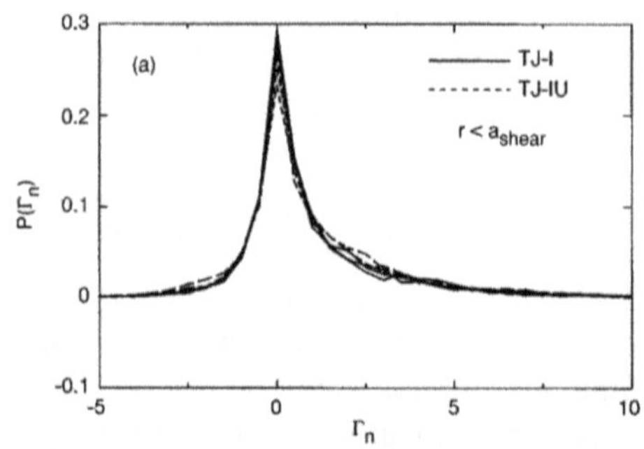

Figure 2.3. Probability distributions of the turbulent particle flux, Γ, in the edge regions of the TJ-I tokamak and the TJ-IU stellarator. (Reproduced from [12], with the permission of AIP Publishing.)

Figure 2.3 shows a comparison between two flux PDFs obtained in the plasma edge of two different small devices: the TJ-I tokamak and the TJ-IU stellarator. The most remarkable aspect of the figure is that the two experimental PDFs are nearly indistinguishable, in spite of the very different magnetic configurations. Is there a reason that the flux PDFs are similar in different devices?

Returning to the well scan experiments in TJ-II (figure 2.2), figure 2.4(a) shows the PDF of the fluctuating particle flux Γ on a double logarithmic scale. Only positive flux events are included in this plot. It is evident that the PDF has a similar

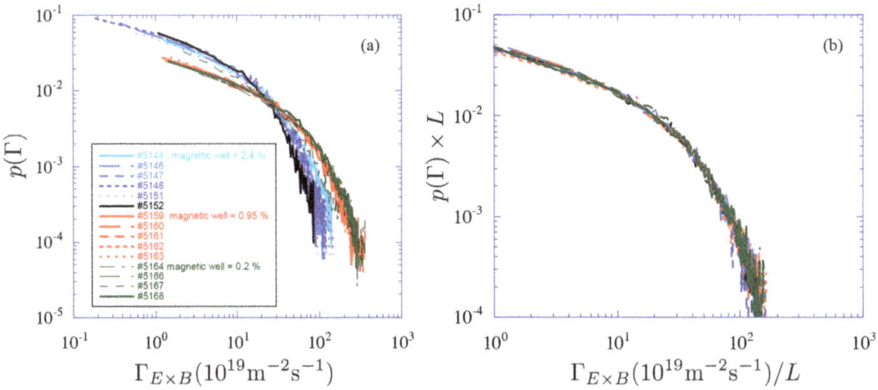

Figure 2.4. Probability distributions for (a) edge turbulent transport in discharges with a different value of the magnetic well in the TJ-II stellarator; (b) rescaled PDFs. (Reproduced with permission from [11]. © IOP Publishing. All rights reserved.)

shape for all cases. By means of a simple transformation, the curves can be made to collapse on top of each other: namely, if one assumes all curves have a similar shape $g(x)$, they can be matched with a single scale parameter L:

$$p(\Gamma) = L^{-1}g(\Gamma/L). \tag{2.10}$$

This expression is such that $\int p(x)dx = 1$ if $\int g(x)dx = 1$ (conservation of total probability). Figure 2.4(b) shows the curves after applying this rescaling operation: by choosing appropriate values of the rescaling parameter L for each case, the curves can be made to fall neatly on top of each other, in spite of the very different instability levels. Thus, although the size of the flux events may differ considerably in different situations, the shape of the PDF is similar. This empirical fact has a profound significance, as we will discuss in detail in section 3.2.4, where we address the self-similarity of stationary processes.

When attempting to understand the characteristic shape of the flux PDF, it is important to recall that the particle flux is a quadratic quantity, $\Gamma = \tilde{n}\tilde{v}_\mathrm{r} = \tilde{n}\tilde{E}_\theta/B_\phi$. Therefore, the PDF of the fluctuating particle flux Γ is not expected to be Gaussian. Let us assume, for simplicity, that the PDFs of $\tilde{n}$ and $\tilde{v}_\mathrm{r}$ are both Gaussian and that these two fields are correlated with correlation coefficient γ. Their joint PDF is then

$$p_{n,v}(\tilde{n}, \tilde{v}_\mathrm{r}) = \frac{1}{2\pi}\frac{\sqrt{1-\gamma^2}}{W_n W_v}\exp\left(-\frac{\tilde{n}^2}{2W_n^2} - \frac{\tilde{v}_\mathrm{r}^2}{2W_v^2} - \gamma\frac{\tilde{n}\tilde{v}_\mathrm{r}}{W_n W_v}\right), \tag{2.11}$$

where W_v is a parameter specifying the width of the PDF of $\tilde{v}_\mathrm{r}$, W_n likewise for $\tilde{n}$. This assumption is a simplification of the true situation, but turns out to be helpful to understand how the fact that the flux is quadratic affects the shape of the flux PDF. This function satisfies $\int\int p_{n,v}(\tilde{n}, \tilde{v}_\mathrm{r})d\tilde{n}d\tilde{v}_\mathrm{r} = 1$. The mean flux is

$$\langle\tilde{n}\tilde{v}_\mathrm{r}\rangle = \int\int p_{n,v}(\tilde{n}, \tilde{v}_\mathrm{r})\tilde{n}\tilde{v}_\mathrm{r}d\tilde{n}d\tilde{v}_\mathrm{r} = -\frac{\gamma}{1-\gamma^2}W_n W_v, \tag{2.12}$$

showing that the net mean flux depends on the turbulence amplitude (the W factors) and the correlation γ.

The PDF can be reworked into a PDF for the flux Γ by performing a change of variables using $\Gamma = \tilde{n}\tilde{v}_r$:

$$p_{\Gamma,v}(\Gamma, \tilde{v}_r)d\Gamma = p_{n,v}(\tilde{n}, \tilde{v}_r)d\tilde{n}. \tag{2.13}$$

Using $d\Gamma/d\tilde{n} = \tilde{v}_r$:

$$p_{\Gamma,v}(\Gamma, \tilde{v}_r) = p_{n,v}(\tilde{n}, \tilde{v}_r)/|\tilde{v}_r|. \tag{2.14}$$

Here, one should be careful to note that probabilities are positive definite, hence the need for the absolute value, $|\tilde{v}_r|$. The variable $\tilde{v}_r$ can be integrated out,

$$p_\Gamma(\Gamma) = \int_{-\infty}^{\infty} p_{\Gamma,v}(\Gamma, \tilde{v}_r)d\tilde{v}_r, \tag{2.15}$$

yielding [12]

$$p_\Gamma(\Gamma) = \frac{1}{\pi}\frac{\sqrt{1-\gamma^2}}{W_n W_v}K_0\left(\frac{|\Gamma|}{W_n W_v}\right)\exp\left(-\gamma\frac{\Gamma}{W_n W_v}\right), \tag{2.16}$$

where K_0 is the modified Bessel function. This expression explains the peaked PDF usually observed for Γ, as well as its asymmetry. Namely, it can be shown that γ is related to the cross phase between the two fluctuating fields $\tilde{v}_r$ and $\tilde{n}$, so that the asymmetry of the PDF is a measure of the cross phase α [12]:

$$\cos\alpha = \frac{\langle\tilde{v}_r\tilde{n}\rangle}{\langle\tilde{v}_r^2\rangle^{1/2}\langle\tilde{n}^2\rangle^{1/2}} = -\gamma. \tag{2.17}$$

Comparing equation (2.16) with equation (2.10), one observes that the heuristic scaling factor L actually corresponds to $W_n W_v$, a measure of the turbulence amplitude. Figure 2.5 shows a comparison between flux PDFs obtained in two devices with different magnetic configurations, different drives (heating and fuelling) and different machine sizes: JET and TJ-II. The figure shows that even such very different machines yield flux PDFs in the plasma edge with a very similar shape, and that these shapes conform rather well to equation (2.16), in spite of the underlying assumptions of the latter.

2.1.2 Self-similarity of the flux PDF

In the previous subsection, we have shown that equation (2.16) often provides a reasonable description of the flux PDF. On the other hand, it is also known that transport is subject to several timescales. The shortest relevant scale is the turbulence autocorrelation time (typically, about 10 μs) and the longest timescale the confinement time (several ms to many tens of ms, depending on the device [14]). It is natural to wonder whether flux events behave differently depending on whether they are related to fast (turbulent) timescales or slow timescales (associated with global

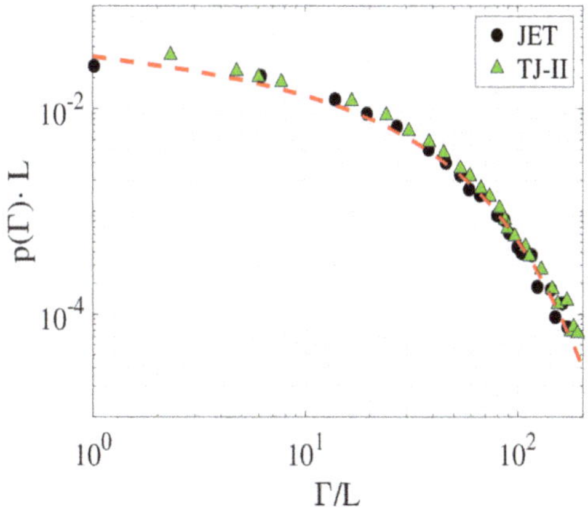

Figure 2.5. Comparison between the rescaled PDFs of edge turbulent transport measured in the TJ-II stellarator and in the JET tokamak. The dashed red line is a fit based on equation (2.16). (Adapted from [11].

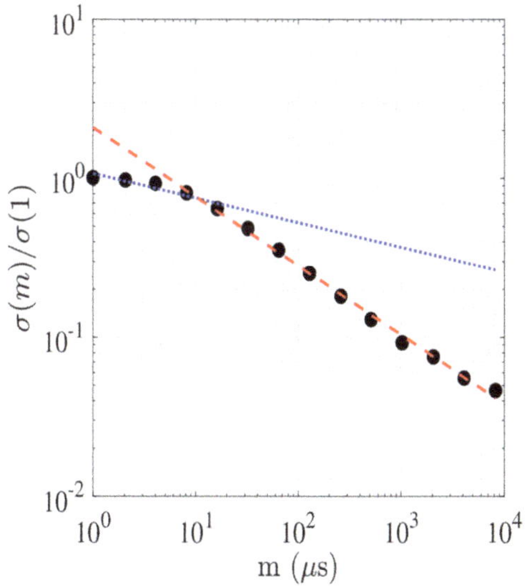

Figure 2.6. Standard deviation for each of the series $\Gamma^{(m)}$ derived from the particle flux as a function of m. The data are from discharge number 35427 of the W7-AS device. (Reproduced with permission from [15]. Copyright (1999) by the American Physical Society.)

transport and profile evolution)—in fact, it would be rather surprising if such very different timescales would produce similar flux events.

To look into this issue, we use equation (2.6) to compute the m-point time-averaged flux $\Gamma^{(m)}(j)$ from the original time series $\Gamma(i)$. Figure 2.6 shows the standard deviation $\sigma(m)$ for each of the $\Gamma^{(m)}$ series derived from particle flux data obtained in the plasma edge of the W7-AS stellarator (see appendix A.3). To relate m to a physical timescale, note that the sampling rate was 1 MHz in this case, so m points corresponds to m μs. As expected, $\sigma(m)$ decreases with m. Interestingly, the double logarithmic plot shows there may be two 'scaling' regions in which $\sigma(m)$ behaves as a power law of m (indicated by the two straight lines): $m < 30$ and $m > 60$. Although the power law scaling is perhaps not very clear for the small m region, it is rather clear for the large m region, extending over almost three orders of magnitude. If we fit functions $\sigma(m) \propto m^{H-1}$ to the data, we find $H = 0.84 \pm 0.11$ for the small m range, and $H = 0.57 \pm 0.02$ for the high m range.

As before, we can now use the rescaling according to equation (2.10) to make the PDFs of each m range overlap. The required scaling parameter L, which we have learned is a measure of the turbulence amplitude, can be taken proportional to $\sigma(m)$, i.e. we may set $L = m^{H-1}$. This procedure yields the two compound PDFs shown in figure 2.7, for $1 \leqslant m \leqslant 32$ (left) and for $64 \leqslant m \leqslant 1024$ (right).

This leaves us with the following, rather interesting picture: in the small m range, which we can call the 'fluctuation range' as these timescales are of the order of the turbulence correlation time, we have a characteristic PDF with the shape shown in figure 2.7 (left)—a shape described rather well by equation (2.16). Flux events are both positive (outward) and negative (inward). The central peak and asymmetric shape are consistent with the model of equation (2.16).

In the large m range, which we call the 'mesoscale range' because it lies between the 'fluctuation range' and the longest timescales corresponding to the confinement time, the PDF has a very different shape. It is rounded in the centre, looking a bit more like a Gaussian (an inverted parabola on this linear–logarithmic scale). There is still a positive 'tail' similar to the tail in the fluctuation range, although the negative tail (inward transport) has nearly disappeared.

One can think of the m-point average as a filtering procedure. Thus, the 'fluctuation range' results include both short-lived (fluctuation timescale) events and long-lived (mesoscale) flux events. The nearly symmetric shape of the corresponding PDF captures flux events produced by turbulent eddies, sometimes leading

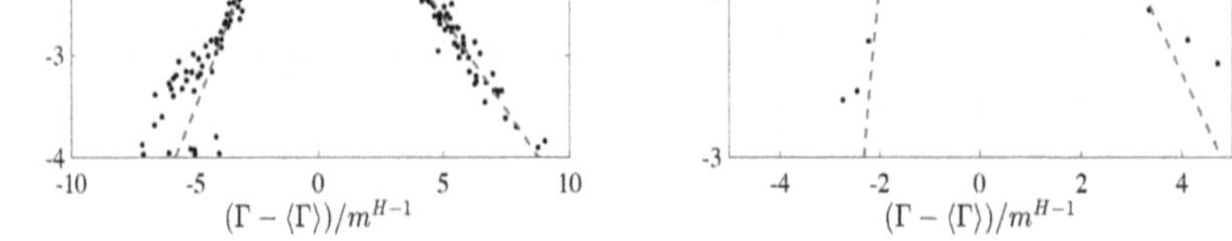

Figure 2.7. Left: probability distribution function of the turbulent fluxes for timescales in the range $1 \leqslant m \leqslant 32$. The red dashed curve is a fit based on equation (2.16). Right: probability distribution function of the turbulent fluxes for timescales in the range $64 \leqslant m \leqslant 1024$. The red dashed curve is a fit based on equation (2.18). (Reproduced with permission after [15]. Copyright (1999) by the American Physical Society.)

to inward and sometimes to outward transport events. However, in the mesoscale range, we have averaged (filtered) out the contributions from individual eddies and only long-lived events remain. These big events are seen to cause mostly outward transport. Due to the slowly decaying nature of the tails of the PDFs, the biggest events have an important impact on transport. In this respect, one should also note that the mesoscale time range corresponds to frequencies where the amplitude of the fluctuation power spectrum is large, as will be discussed in section 2.4.

The observation that the shape of the PDF of the flux remains approximately invariant over an extended range of m-values is considered evidence of self-similarity; an issue we will come back to later. In any case, this observation is significant in view of the fact that certain types of self-similarity are characteristic of so-called self-organized critical (SOC) systems [16]. SOC systems exhibit very characteristic transport properties, such that a significant fraction of transport is caused by the largest events the system can sustain (usually referred to as 'avalanches'). This is quite different from diffusive systems, in which transport events are microscopically small and system characteristics are essentially independent from the system size. This is not of mere academic importance, since a plasma governed by SOC dynamics would have quite different scaling behaviour for energy and particle confinement, usually worse than what would be expected for diffusive systems. We will return to these ideas in more detail in chapter 3, where we will highlight many other features of complex systems. In particular, the significance of the PDF rescaling property of equation (2.10) will be discussed in section 3.3, along with a discussion of the corresponding rescaling factor $L = m^{H-1}$ and the deeper meaning of the coefficient H.

In order to determine whether these results are reproducible, we have proceeded to check in several machines and discharges. Figure 2.8 shows rescaled $p(\Gamma)$

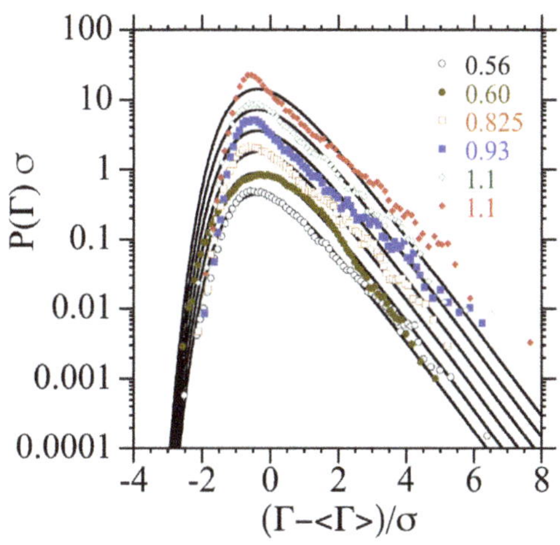

Figure 2.8. Rescaled turbulent flux PDFs for C-Mod data for increasing line integrated densities after smoothing with a 120 point sliding squared window. The curves correspond to different values of the line integrated density (as indicated by the labels, in units of 10^{20} m^{-3}). Except for the curve at $\bar{n}_e = 0.56 \times 10^{20}$ m^{-3}, all curves are displaced vertically by factors that are a multiple of 2. The continuous lines are the corresponding BHP curves. (Reproduced from [18], with the permission of AIP Publishing.)

distributions obtained in the SOL of the C-Mod tokamak (plasma current $I_p = 0.8$ MA; toroidal magnetic field $B_T = 5.3$ T) for various values of the line integrated density $\bar{n}_e$, applying a running average with $m = 120$. The distance of the Langmuir probe to the last closed flux surface (LCFS) was between 4 and 8 mm [17]. Similarly, figure 2.9 shows the rescaled flux PDFs with $m = 300$ for the three W7-AS discharges in the far SOL region (2–5 cm outside the LCFS).

All m-point averaged flux PDFs are compared to a fixed reference shape, the so-called Bramwell, Holdsworth, and Pinton (BHP) PDF [19, 20]:

$$p(y) = A(\exp(x - e^x))^a, \tag{2.18}$$

where $x = b(y - s)$, $A = \kappa/\sigma$, $\kappa = 2.16$, $a = 1.58$, $b = 0.934$, and $s = 0.373$, σ being the standard deviation. This PDF without free parameters has a shape that appears to match the general shape of the PDF of the m-point smoothed flux well. Note that the positive 'tail' of the BHP PDF decays exponentially (corresponding to a straight line on this linear–logarithmic scale), so that the non-exponential decay of the experimental PDFs is emphasized by comparison. This deviation from the exponential decay only becomes visible for the section of the tail that is farther than about 3σ from the peak (which we may denominate the 'far tail'). This implies that long data series are required to see this effect.

There has been some discussion about the nature of the flux PDF. It is obviously not Gaussian due to its strong asymmetry. The papers introducing the BHP distribution suggested the existence of an analogy between the statistical behaviour of a *confined turbulent flow* and that of a model of the critical behaviour of a ferromagnet, leading to similarly shaped PDFs for both systems [19]. As fusion plasmas are confined systems with turbulence and are suspected to be self-organized critical systems, this analogy seemed attractive, so that a description of the flux PDF based on the BHP distribution appeared plausible. In fact, it was even suggested that

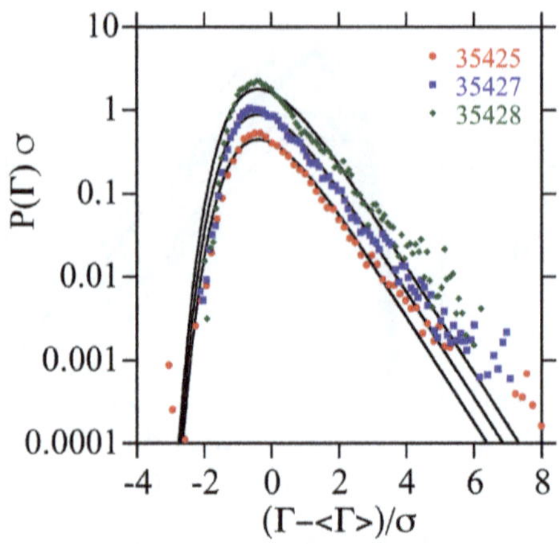

Figure 2.9. Rescaled flux fluctuation PDFs measured in the far SOL of W7-AS after smoothing with a 300 point sliding squared window; except for the curve for shot 35 425, all curves are displaced vertically by factors that are a multiple of 2. The continuous lines are the corresponding BHP curves. (Reproduced from [18], with the permission of AIP Publishing.)

the (rescaled) PDF of the fluctuations of *any global measure of activity* of a critical system should be distributed according to this PDF. Nevertheless, one should remember that the PDF shape and its statistical quantifiers, such as the skewness and kurtosis, vary radially, so that there is no reason to expect that a single function such as this (with fixed S and K) can describe all experimental PDFs precisely. Still, it could be that the BHP distribution acts as a sort of 'attractor' for the experimental PDFs of the m-point smoothed fluxes (noting that by smoothing, the local contributions to the fluctuations of the flux are removed—at least partially).

Other distributions have been suggested in literature, such as extreme value distributions[5] (in particular, the Gumbel distribution [21], which is similar to the BHP distribution) or the gamma or lognormal distributions [22]. In specific cases, the latter two distributions may provide a better fit to the data, because they have some parameters that allow modelling the gradual shape variations observed in experiment. Nevertheless, as there does not appear to be a convincing theoretical argument in favour of any of these distributions and it seems impossible to decide the underlying nature of the flux (or I_{sat}) PDF merely on the basis of experimental data, the question regarding the proper analytical shape of the PDF is unresolved.

In any case, based on our observations and some of the cited references we conclude that this heavy-tailed, asymmetric shape of the (smoothed) flux PDF occurs systematically in the SOL of both tokamaks and stellarators. Close to the LCFS, the PDF tends to become more Gaussian [18], but in the plasma edge region the PDF again acquires its characteristic asymmetric shape.

This variation of the PDF shape according to position has a physical basis. To clarify this statement, figure 2.10 shows two m-point smoothed particle flux PDFs,

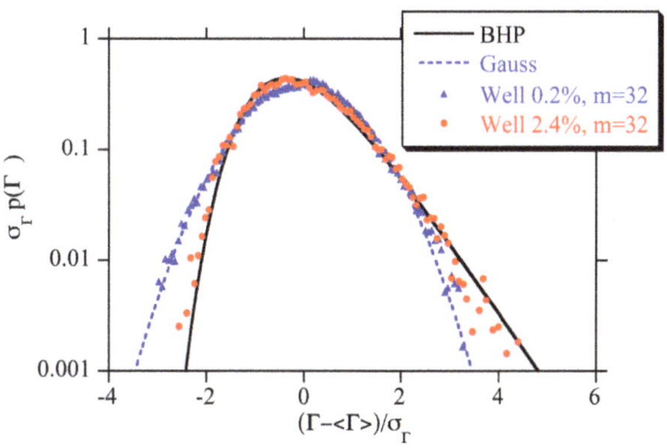

Figure 2.10. Probability distributions of the turbulence induced fluxes in the TJ-II stellarator for two different values of the magnetic well—a higher well implying more stability to pressure-driven instabilities. The non-Gaussian shape (similar to the BHP distribution) is lost when the instability level is too high. (Reproduced after [23]. © Transworld Research Network)

[5] An extreme value distribution is the distribution of the extremes of an underlying random variable. It is not clear why the particle flux should obey such a distribution, nor what the underlying random variable should be.

measured using Langmuir probes in the edge of the TJ-II stellarator, from the magnetic well scan discussed earlier. Recall that a high value of the magnetic well provides stability against pressure-driven instabilities. The shape of the smoothed PDF is observed to depend on the magnetic well: at low magnetic well (high instability levels), the PDF is nearly Gaussian, while at high magnetic well the PDF shape is close to the mentioned BHP shape. A possible explanation for this variation is that at low magnetic well and high instability levels, the system is 'overdriven' and not able to self-regulate (it is in a 'supercritical' state). In contrast, at high magnetic well, fluctuations are sufficiently damped that the system can self-regulate towards a critical state and the BHP-like shape becomes apparent.

Given the above evidence for the self-similarity of the flux PDF and the universality of its shape in the plasma edge and SOL regions, it is reasonable to ask how the PDF behaves well inside the plasma. Very few such observations exist due to the difficulty of measuring the particle flux inside the plasma. However, some measurements are available from the heavy ion beam probe at the TJ-II stellarator. This active diagnostic injects fast Cs^+ ions into the plasma and detects secondary Cs^{++} ions leaving the plasma, allowing measurement of the local plasma potential and electron density at the point of interaction. Using various diagnostic slits to simultaneously measure ϕ, $\tilde{n}_e$, and ϕ at three poloidally separated points inside the plasma, the particle flux Γ can be determined, in full analogy to the Langmuir triple probe system. Figure 2.11 shows $p(\Gamma)$ during L and H plasma phases, measured at normalized radius[6] $\rho = -0.5$ [24]. The PDF has the characteristic shape given by equation (2.16), showing that the flux PDF well inside the plasma behaves similarly to the flux PDF in the plasma edge. Note that the flux PDF is more asymmetric and broader in the L phase, implying a reduction of net outward transport in the H phase (as reflected by the fit parameters γ and $W_n W_v$ indicated in the figure caption).

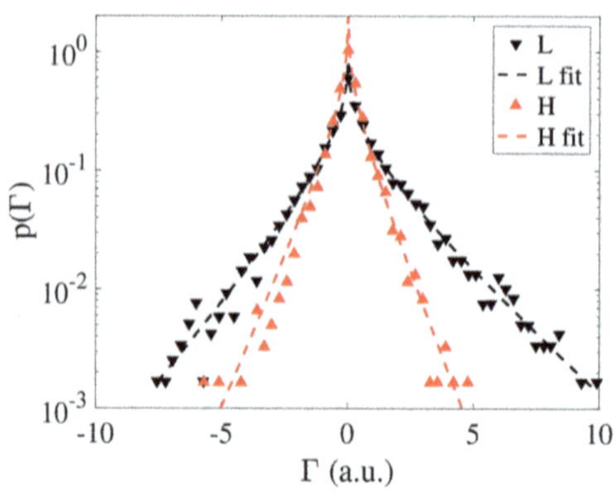

Figure 2.11. The PDF of the fluctuating particle flux during the L and H phases in the TJ-II stellarator, measured at $\rho = -0.5$ by the heavy ion beam probe. The curves fitted according to equation (2.16) yield the following parameter values: for L mode, $W_n W_v \simeq 2.0$ and $\gamma \simeq -0.13$; for H mode, $W_n W_v \simeq 0.9$ and $\gamma \simeq 0.05$. (Adapted with permission from [24]. © IOP Publishing. All rights reserved.)

[6] By convention, a negative value of $\rho = r/a$ corresponds to the high field side.

2.2 Correlation

The standard linear procedure for detecting a relation between two variables $x(t)$ and $y(t)$ is to calculate the correlation coefficient

$$C_{X,Y} = \frac{\langle (x - \langle x \rangle)(y - \langle y \rangle) \rangle}{\sigma(x)\sigma(y)}, \tag{2.19}$$

where $\langle x \rangle$ is the mean of x and $\sigma(x)$ is the standard deviation of x. Due to the normalization, $-1 \leqslant C_{X,Y} \leqslant 1$. The correlation between two variables only really detects a similarity of behaviour or waveform. Indeed, it attains its maximum value, $C_{X,Y} = 1$, in the case $y = a + b \cdot x$ (where $b > 0$ to obtain a positive value of $C_{X,Y}$, always assuming x and y are not trivial).

This coefficient can be calculated for zero time delay, but by introducing a time shift Δt of y, $C_{X,Y}$ can also be evaluated for a range of time delays: $C_{X,Y}(\Delta t) = C_{X(t),Y(t+\Delta t)}$. Usually, correlation values are only considered significant when $|C_{X,Y}| > 0.4$. Thus, the correlation coefficient may allow identifying a relation between variables by showing its value is high, often peaking at a specific non-zero time delay, which would suggest that variations in one variable precede those in the other. Often, this delay is the result of advection (propagation) from one measurement location to the other. Figure 2.12 shows an example of a typical correlation function encountered in plasma turbulence, calculated from two floating potential measurements obtained with a small poloidal separation (6 mm). The correlation peaks at a very small time delay $\Delta t \simeq 1\,\mu$s, which would suggest a propagation speed in the poloidal direction of about 6 km s^{-1}—noting that the error in this value is considerable due to lack of time resolution[7]; an alternative way of determining the propagation speed is the two-point $S(k, f)$ technique discussed in section 2.4. The decay of $C_{X,Y}(\Delta t)$ from its peak value allows establishing a 'correlation time' τ_c: namely, the difference of Δt values over which $C_{X,Y}(\Delta t)$ decays from its peak value

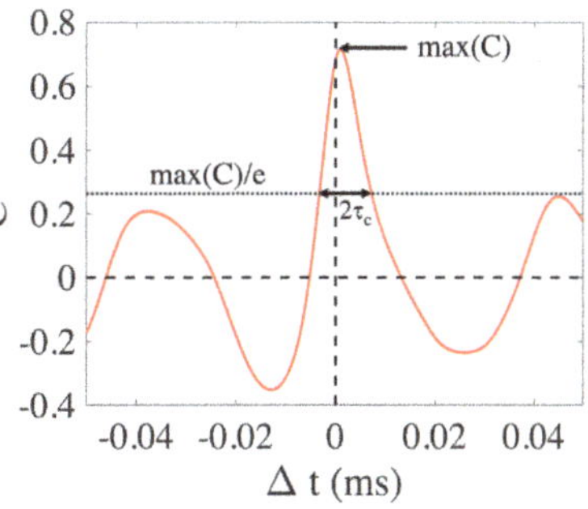

Figure 2.12. Illustration of a typical correlation function between two Langmuir probe pins measuring floating potential.

[7] The precision of the determination of the time delay of the correlation peak can sometimes be improved using a spline interpolation of the discretely sampled curve $C(\Delta t)$.

to $1/e$ times this peak value (this level is indicated by the dotted line in the figure). In this example, we obtain $\tau_c \simeq 5.2$ μs.

It is interesting to study the long range temporal behaviour of the correlation, in order to see whether the plasma oscillations contain 'memory effects' significantly exceeding the turbulence decorrelation time. Due to the oscillating nature of the correlation function, this is no easy task. However, it has been shown that the autocorrelation function $C_{X,X}(\Delta t)$ is related to the variance (σ^2) of the m-point averaged series $\langle x \rangle^{(m)}$ (see equation (2.6) for the definition of $\langle x \rangle^{(m)}$) [25]. Thus, if one is interested in the decay of $C_{X,X}(\Delta t)$ at large $\Delta t = m \cdot dt$, where dt is the sampling time interval, this is equivalent to studying $\sigma(\langle x \rangle^{(m)})$. The latter quantity is more robust than C, in a numerical sense, which is important for finite length time series. An example of $\sigma(\langle x \rangle^{(m)})$ as a function of m is shown in figure 2.6. This equivalence between $C_{X,X}(m \cdot dt)$ and $\sigma(\langle x \rangle^{(m)})$ can be generalized to the cross correlation [25].

2.2.1 Long range correlations

To investigate the existence of (spatial) long range correlations, the TJ-II stellarator disposes of two similar Langmuir probe systems mounted on fast reciprocating drives. The latter allow inserting and withdrawing the probes, so measurements can be made without burning the probe tips due to excessive contact with the hot plasma. Figure 2.13 shows the set-up. One of the probes (probe D) is located in a top window entering vertically through one of the corners of its beam-shaped plasma and at toroidal position $\phi = 35°$. Probe B is installed in a bottom window at $\phi = 195°$ and enters into the plasma through a region with a high density of flux surfaces (i.e. lower flux expansion) than probe D. It is important to note that the field line passing through one of the probes intersects the toroidal plane of the other probe at a poloidal distance of approximately 150° to that other probe, i.e. the probes are not connected directly by any field line, although field lines may connect the two probes after various toroidal turns (see below). The minimum toroidal distance between the two probes exceeds 5 m.

A biasing probe with a graphite electrode (12 mm high, 25 mm diameter) can be inserted to apply a biasing voltage to the plasma, and it has proved to be a valuable tool for controlling the edge plasma electric field and consequently for placing the plasma in an enhanced confinement regime. The electrode is typically inserted to 2 cm inside the last closed flux surface (LCFS) (at normalized radius $\rho \simeq 0.9$) and biased with respect to one of the poloidal limiters installed in TJ-II [26].

Figure 2.14 shows the time evolution of the line average plasma density and the cross correlation between ion saturation current and floating potential signals (as measured by the distant probes D and B, both located inside the plasma edge at $\rho = 0.93$ and 0.95, respectively). The cross correlation $C(\Delta t)$ was calculated over short time intervals with a length of 1 ms in order to track the time evolution of the correlation. Prior to biasing, the plasma density is below the critical value for the ion to electron root transition ($\bar{n}_e \simeq 0.6 \times 10^{19}$ m^{-3}), and during biasing the density is above that value. The long range correlation of the floating potential increases clearly during the biasing phase. The time delay Δt associated with the maximum

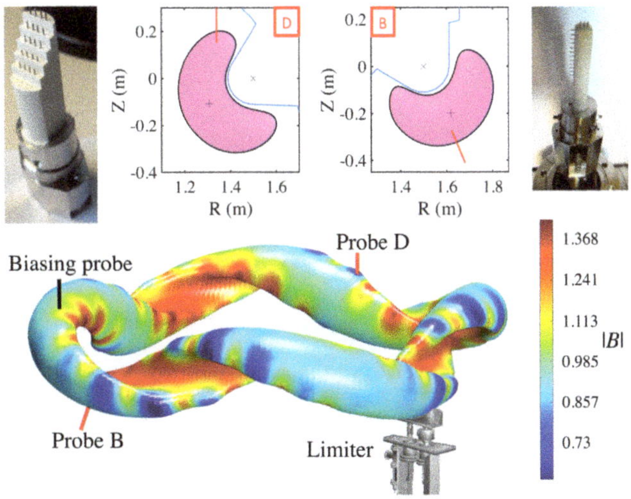

Figure 2.13. Schematic of reciprocating Langmuir probe locations in TJ-II. The top diagrams show poloidal cross sections at the toroidal angles of the two probes, and includes the Langmuir probe trajectories (red), the last closed flux surface (black outline), the magnetic axis (+), the location of the machine core ($R_0 = 1.5$ m, $Z = 0$, indicated by a ×) and part of the vacuum vessel (blue line). The bottom 3D figure shows the last closed flux surface (where colours represent the magnetic field strength), the limiter, the biasing probe and the Langmuir probe trajectories (red lines). (Reproduced with permission from [27]. © IOP Publishing. All rights reserved.)

cross correlation lies in the range of 5–10 μs. On the other hand, the degree of long range correlation of density (I_{sat}) fluctuations is negligible. Once biasing is turned off, the density decreases on the timescale of the particle confinement time ($\simeq$10 ms), whereas both the electric field (not shown) and the degree of long range correlation decrease on a much faster timescale [28]. The significant value of the cross correlation for probes that are located this far apart clearly indicates the existence of large scale structures. The fact that these correlations are enhanced when biasing is activated and particle confinement improves (as evident from the increase of the line average density) suggests that these long range structures are related to improved confinement.

Confinement transitions can be achieved in various ways at TJ-II. One method is raising the line average density using neutral beam injection (NBI). Figure 2.15 shows an example in which NBI heating was applied to an ECRH target plasma. When crossing the density value of $\bar{n}_e \simeq 2 \times 10^{19}$ m^{-3}, one observes an increase of the time derivative of the density ($d\bar{n}_e/dt$) and the stored plasma energy (not shown), a reduction in H_α emission (due to a decrease of the outward particle flux), and a reduction of the level of broadband fluctuations by a factor of 2–3 on a short timescale, a few tens of microseconds [29]. Jointly, these phenomena are considered characteristic of a transition to an improved confinement regime triggered by an edge bifurcation [30].

Figure 2.16 shows the toroidal cross correlation of the floating potential and the density fluctuation signals measured at different radial positions. In these

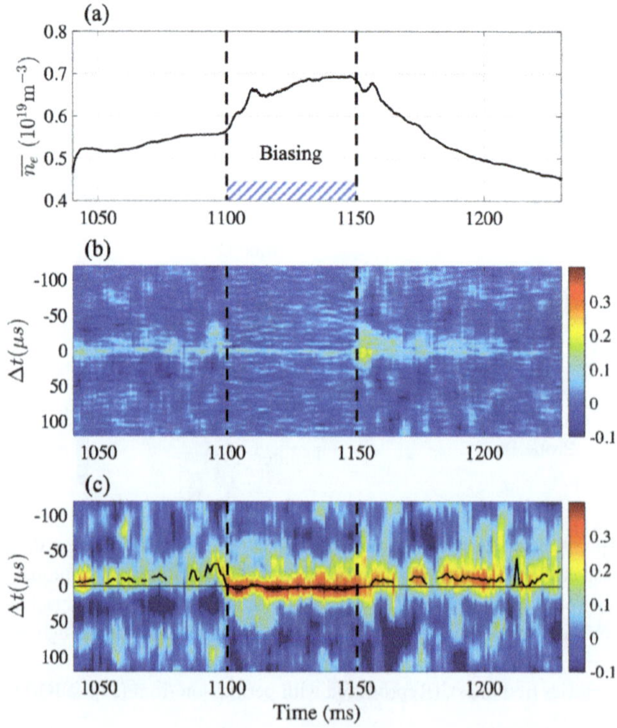

Figure 2.14. (a) Time evolution of the line average electron density. The density increases during the indicated biasing interval. The next two graphs show the cross correlation between (b) ion saturation current and (c) floating potential measurements taken at the two distant probes as a function of time. The overplotted solid line in (c) is the evolution of the time delay where the correlation is maximum. (Reproduced with permission after [28], Copyright (2008) by the American Physical Society).

experiments, probe D is moved radially while probe B is kept fixed at $\rho \simeq 0.9$. The toroidal correlation of density (I_{sat}) fluctuations is very low. On the other hand, the correlation between floating potential signals is significant (C up to 0.8); the maximum of the correlation of the floating potential is observed when the probes are approximately at the same radial location in the plasma edge. The radial decay length of long range correlations towards the SOL region is of the order of 1 cm. These experiments show that the long range correlations between the distant floating potential measurements occur at the time of the L–H (low to high) confinement transition and only in a specific radial range inside the plasma edge; peaking at $\rho \simeq 0.9$. The simultaneous reduction of fluctuation amplitude suggests that these long range structures are associated with turbulence suppression.

2.2.2 Filaments and zonal flows

The preceding results show clearly that confinement transitions at TJ-II are associated with the occurrence of long range correlations (LRCs) between probes that are separated by a long poloidal and toroidal distance, although both probes

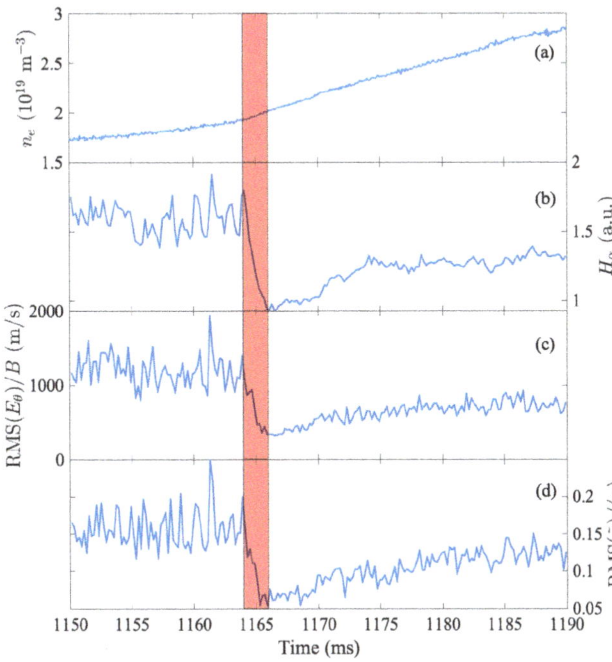

Figure 2.15. Time evolution of (a) plasma density, (b) H_α emission, and (c), (d) edge fluctuation levels during spontaneous bifurcations in pure NBI heated plasmas in the TJ-II stellarator. Edge fluctuations are measured at $\rho \simeq 0.9$ (discharge 18998). (Reproduced with permission after [29]. © IOP Publishing. All rights reserved.)

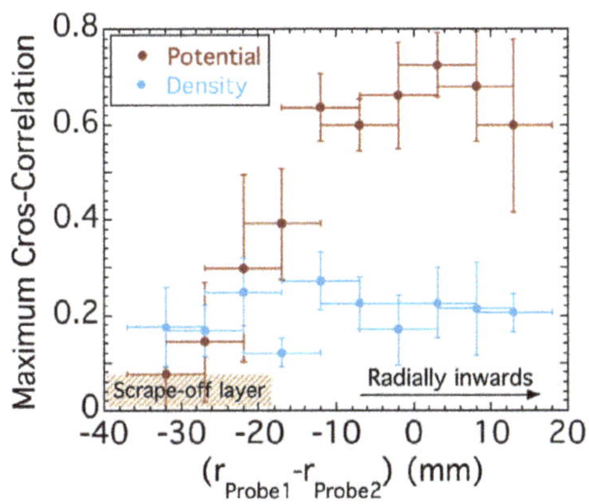

Figure 2.16. Radial profile of cross correlation of potential and density fluctuations in NBI plasmas at $\bar{n}_e \simeq 1.7 - 2 \times 10^{19}$ m^{-3}. (Reproduced with permission from [29]. © IOP Publishing. All rights reserved.)

should be positioned at roughly the same flux surface to detect this. Theory suggests that these LRCs are due to the existence of a zonal flow (ZF), which has exactly this form: namely, a structure with very small poloidal and toroidal wavenumbers ($k_\theta \simeq 0$ and $k_\phi \simeq 0$), but a relatively large radial wavenumber (k_r) [31]. Such a structure (of plasma potential and the poloidal velocity) generates a local gradient in the radial electric field (E_r) that would lead to turbulence suppression and the creation of an edge transport barrier, explaining the improved confinement.

However, filamentary structures are also quite common in the edge plasma of many fusion devices (e.g. MAST [32] or W7-X [33]). These are structures that are extremely elongated along the magnetic field lines (or $k_{\parallel} \simeq 0$), but have a relatively large wavenumber in the direction perpendicular to the field lines ($k_{\perp}$). Figure 2.17 shows an example of such filaments, as captured by a fast camera in the JET tokamak. To improve the visualization of the filamentary structures that are embedded in the rotating plasma, it is convenient to subtract the background image (i.e. the mean over various movie frames) and normalize to that same background. This clearly reveals the filamentary structures, stretching along the magnetic field lines in the plasma edge. These structures arise due to the virtually unimpeded movement of electrons along field lines, while their perpendicular motion is strongly restrained.

The question is, therefore, whether it is possible to distinguish between LRCs caused by zonal flows and LRCs due to filamentary structures. For this purpose, we performed a set of experiments, using the capacity of the TJ-II stellarator to modify the rotational transform within a single discharge [34]. As the rotational transform is modified, the connection length measured along the field lines between the two distant Langmuir probes is changed, so that the contribution to LRCs from filaments is varied—whereas the contribution due to zonal flows should not experience such a variation.

The two distant probes (B and D) were placed inside the plasma edge, at approximately the same normalized radius $\rho = r/a$. The rotational transform profile, $\iota\!\!\!-(\rho)$, was varied linearly in time during a discharge, and the correlation between floating potential measurements at the distant probes was monitored. Several scans were made, in different $\iota\!\!\!-$ ranges, and scanning both upward and downward. The scans

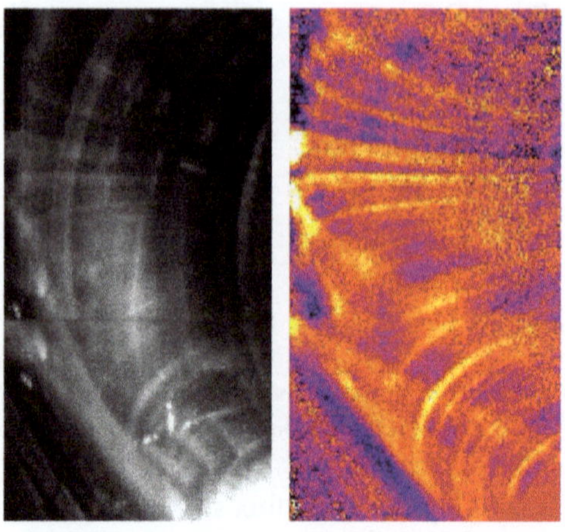

Figure 2.17. Image from a fast camera installed in the JET tokamak, discharge 96358. Left: raw image frame. Right: moving filamentary structures are enhanced by removing the temporal mean over several frames (the background) and dividing the result by this temporal mean image. (Image courtesy of E de la Cal.)

were repeated several times to verify reproducibility. We use the edge rotational transform, $\iota(1)$, as a control parameter for the rotational transform profile. As the vacuum magnetic field geometry is known for each value of $\iota(1)$, and the actual geometry in the edge plasma is known to be very close to the vacuum geometry, it is possible to make an analytical estimate of the correlation between the distant probes on the basis of a simple geometric model.

For this purpose, we define two geometric distances $d_\perp$ and $L_\parallel$ as follows. Starting from the nominal position of the reference probe (probe D), the magnetic field line is followed both in the positive and negative toroidal direction, using a numerical model of the magnetic field. Each toroidal turn of the field line is assigned a label, i. In a given toroidal turn of the magnetic field line, the point of closest approximation to probe B is found. The distance of closest approximation is $d_\perp(i)$. The corresponding distance to the reference probe D, measured along the field line, is called $L_\parallel(i)$. Thus, $d_\perp(i)$ is the distance between the field line passing through probe D and probe B in the perpendicular direction in turn i, and $L_\parallel(i)$ is the distance between probe D and the point of closest approach to probe B in turn i.

As noted, the distances $d_\perp(i)$ and $L_\parallel(i)$ depend on the configuration (i.e. on $\iota(1)$). Thus, for each value of $\iota(1)$ in the scans discussed below, $d_\perp(i)$ and $L_\parallel(i)$ must be recomputed.

The simplest possible model for the total correlation between probes B and D is to assume that the correlation decays exponentially in the directions perpendicular and parallel to the field line, and considering the contributions from all close approximations of the field line:

$$C_{\mathrm{DB}} = A\sum_i \exp\left(-\frac{d_\perp(i)^2}{\lambda_\perp^2} - \frac{L_\parallel(i)^2}{\lambda_\parallel^2}\right), \tag{2.20}$$

where A is a proportionality constant and i enumerates the close approximations. Thus, in this simple model, the correlation between probes D and B is a function of only three parameters (A, $\lambda_\perp$, and $\lambda_\parallel$), assumed fixed during each ι scan for simplicity.

First, we performed a set of experiments at low line average density ($\bar{n}_e \simeq 0.4 \times 10^{19}\ \mathrm{m}^{-3}$), in order to ensure that zonal flows are absent or insignificant and to test the correlation model. Figure 2.18 shows the mean experimental correlations for upward and downward scans in the range $1.630 \leqslant \iota(1) \leqslant 1.672$ and a joint least-squares fit with the function C_{DB}, equation (2.20), yielding $\lambda_\perp = 0.069 \pm 0.007$ m, $\lambda_\parallel = 35 \pm 5$ m. The fit function is computed considering 80 toroidal turns in each direction from the reference probe. Figure 2.19 shows the mean experimental correlations for upward and downward scans in the range $1.593 \leqslant \iota(1) \leqslant 1.630$ and a joint fit with the function C_{DB}, equation (2.20), yielding $\lambda_\perp = 0.079 \pm 0.002$ m, $\lambda_\parallel = 59 \pm 8$ m. These values should be compared to the approximate poloidal circumference, $2\pi\langle a\rangle \simeq 1.3$ m and to the approximate length of one toroidal turn, $2\pi R \simeq 9.4$ m, respectively. The shot to shot variation of the measured correlation is indicated in the figures, showing that the measured correlation is quite reproducible. Although the model curves do not lie fully within the range of shot to shot variation indicated in the figures, it should be noted that this

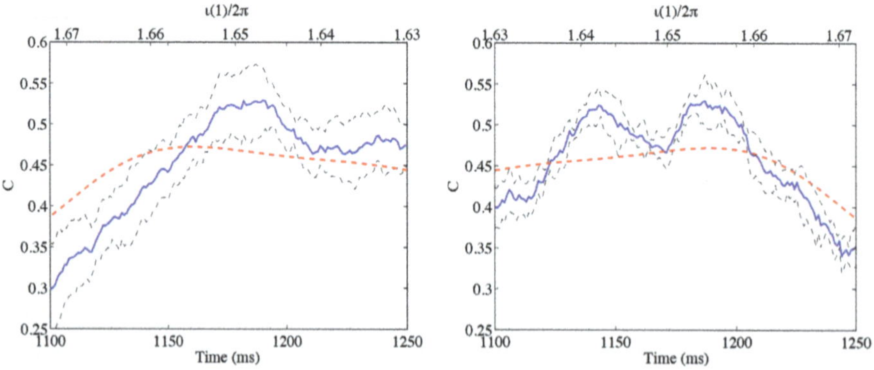

Figure 2.18. Left: downward iota scan. Right: upward iota scan over the same iota range ($1.630 \leqslant \iota(1) \leqslant 1.672$). Thick blue lines: mean of experimental data. Thin black dashed lines indicate the range of shot to shot variation. Red dashed lines: model curves, equation (2.20), with $\lambda_{\perp} = 0.069 \pm 0.007$ m, $\lambda_{\parallel} = 35 \pm 5$ m. (Reproduced with permission from [34]. © IOP Publishing. All rights reserved.)

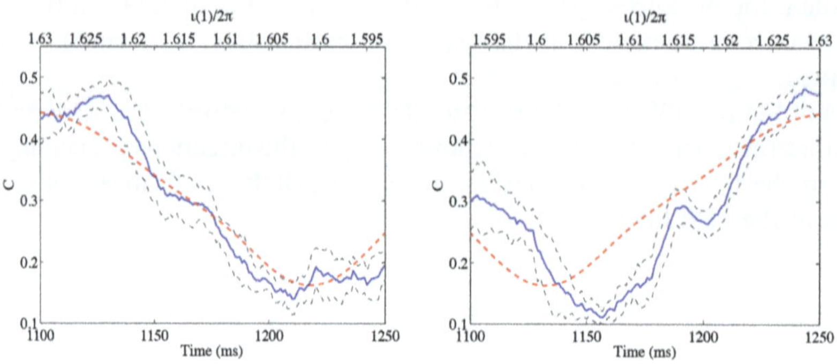

Figure 2.19. Left: downward iota scan. Right: upward iota scan over the same iota range ($1.593 \leqslant \iota(1) \leqslant 1.630$). Thick blue lines: mean of experimental data. Thin black dashed lines indicate the range of shot to shot variation. Red dashed lines: model curves, equation (2.20), with $\lambda_{\perp} = 0.079 \pm 0.002$ m, $\lambda_{\parallel} = 59 \pm 8$ m. (Reproduced with permission from [34]. © IOP Publishing. All rights reserved.)

variation is not the same as the measurement error. The actual measurement error may be larger than the indicated variation, due to, e.g., systematic errors in the external coil currents, systematic errors in the probe position, contamination from minor mode activity, etc. Furthermore, it may be that the model parameters $\lambda_{\perp}$ and $\lambda_{\parallel}$ are not fixed, as assumed here for simplicity, but vary slightly as ι is scanned (e.g. they may vary according to whether the local value of ι is a low-order rational or not).

Next, a set of 22 additional scans were performed (upward in the iota range $1.593 \leqslant \iota(1) \leqslant 1.630$), in which the density was gradually raised on a shot to shot basis, while attempting to keep the density constant within each shot. The motivation for performing this experiment is that it is known that an electron to ion root confinement transition occurs at $\bar{n}_e \simeq 0.6 \times 10^{19}$ m^{-3}, which is accompanied by zonal flows in the edge region. Thus, we can study how zonal flows affect the long range correlation between distant probes. The resulting evolution of the correlation

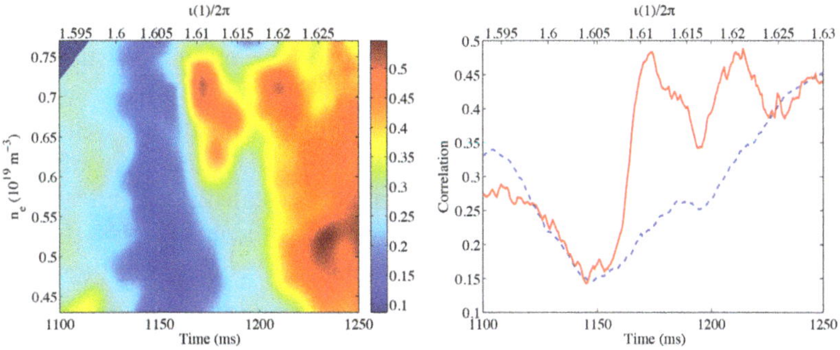

Figure 2.20. Upward iota scans in which the density was raised on a shot to shot basis. Left: correlation versus time and line integrated density, $\bar{n}_e$. Colour scale: probe correlation. Right: mean correlation. Dashed blue line: mean correlation at low density, $\bar{n}_e < 0.6 \times 10^{19}$ m^{-3}. Red continuous line: mean correlation at the critical density, $0.65 \leqslant \bar{n}_e \leqslant 0.73 \times 10^{19}$ m^{-3}. (Reproduced with permission from [34].)

with time and density is displayed in figure 2.20. At low densities, the temporal evolution of the correlation is nearly identical with the evolution shown in figure 2.19 (right). At higher density values, an additional correlation is observed, mainly in the time window $1170 < t < 1220$ ms and density values in the range $0.65 \leqslant \bar{n}_e \leqslant 0.73 \times 10^{19}$ m^{-3}. This additional correlation, on top of the 'background' correlation due to filaments, is ascribed to zonal flows. It is interesting to note that this zonal flow correlation depends on the edge rotational transform: it appears the value of ι has an impact on the zonal flow.

The above study provides clear evidence that the observed total correlation between remotely placed measurement systems (here: Langmuir probes) may be due to several mechanisms, operating simultaneously. The basic drift-wave model correlation mechanism (filaments) [35] appears to provide a correlation 'baseline' on top of which the correlations due to other mechanisms, such as zonal flows, are mounted. Consequently, in general one should be careful not to interpret a detected long range correlation as being due to a single mechanism, unless further work is performed to clarify the precise cause of the correlation.

2.3 Conditional averages

A conditional average refers to the expectation or mean value of one measurement, $y(t)$, given that another quantity, $x(t)$, satisfies some condition: symbolically, $E(Y|X)$. The condition placed on X could be, for example, that $x(t)$ is larger than a threshold, $(x(t_i) > \theta)$, or that x crosses a threshold $(x(t_{i-1}) < \theta \wedge x(t_i) \geqslant \theta)$. These conditions are met for a set of index values $\{i(k), k = 1, \ldots, K\}$, corresponding to 'trigger' times $t_{i(k)}$. Then, the mean of Y is taken over time intervals $-M \leqslant m \leqslant M$ around these trigger times:

$$\bar{y}(m) = \frac{1}{K}\sum_{k=1}^{K} y(t_{i(k)+m}). \tag{2.21}$$

The resulting average $\bar{y}$ reflects the mean behaviour of $y(t)$ whenever $x(t)$ satisfies the given condition. This technique is very useful to determine the structures of large or extreme events, precursors to major events, etc.

For this to work properly, it is essential that the chosen condition is well defined, does not occur too often (which would invalidate the approach), nor too infrequently (in which case the mean loses its usefulness due to lack of statistics). Even when it is well chosen, false triggers due to noise in x may make it difficult to apply this technique. False triggers can be detected and removed, simply by comparing the shape of the individual waveforms $y(t_{i(k)+m})$ with the result $\bar{y}(m)$ [36]. Figure 2.21 shows a section of data from one of the pins of the TJ-II rake probe that measures the floating potential at a series of radially spaced locations in the plasma edge (an image of the rake probe can be seen in figure 2.13, top right). As condition, we imposed that the trigger points were local maxima ($x(t_i) > x(t_{i-1}) \wedge x(t_i) > x(t_{i+1})$) such that the peak amplitude $x(t_i)$ exceeded the value 0.2. These points are indicated by red circles in the figure.

Figure 2.22 shows the conditionally averaged time traces $\bar{y}$ for five floating potential measurements, aligned radially along the rake probe. The conditional average reaches the highest value when the averaged trace equals the reference trace (labelled 'LOP109'). For other traces, the amplitude decays with distance, indicating that the detected events have a clear radial size. Figure 2.23 shows the data from figure 2.22 as a contour plot, emphasizing the radial and temporal structure. The radial decay from the peak indicates the typical structure size in the radial direction,

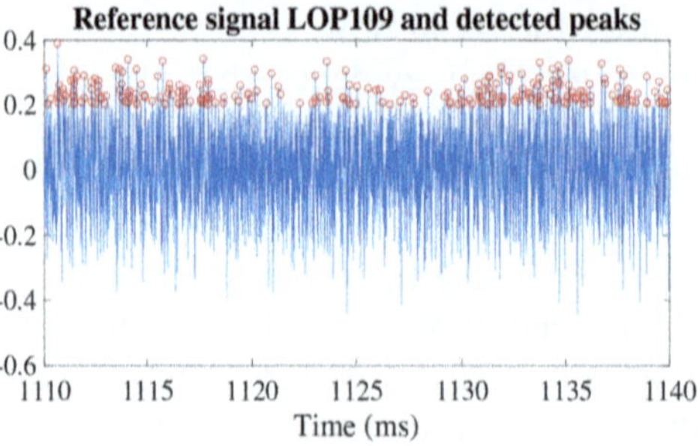

Figure 2.21. Reference signal for conditional average. Red circles are the detected events.

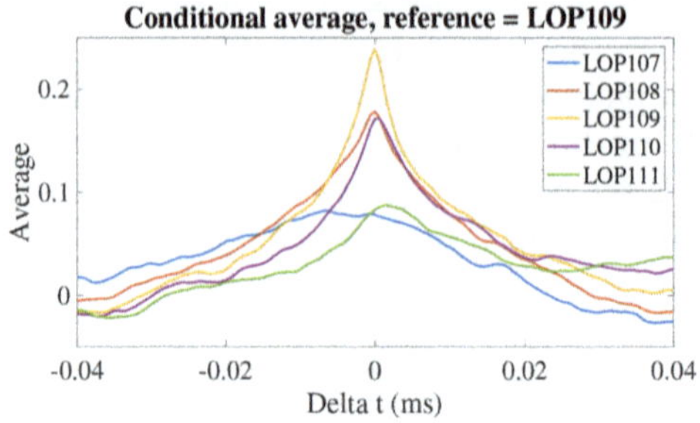

Figure 2.22. Conditionally averaged floating potential measurements.

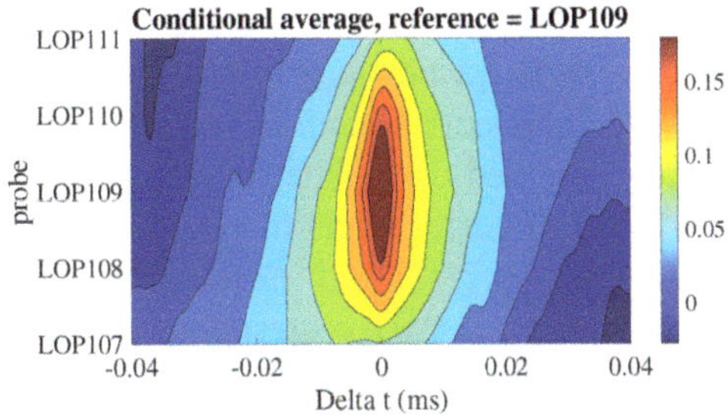

Figure 2.23. Same as figure 2.22, represented as a contour plot. The vertical axis indicates the rake probe pin, corresponding to a radial position (bottom: inside; top: outside).

whereas the temporal decay possibly indicates the structure lifetime—consistent with the turbulence decorrelation time. The tilt of the elliptic contours could be due to outward propagation of the structures ('LOP107' being the innermost pin and 'LOP111' the outermost; the radial distance between pins is 3 mm). It appears that the tilt angle of the contours with respect to the vertical is larger for bigger structures, which would imply that larger events propagate more slowly.

The interpretation of these results in terms of lifetime, propagation, and tilt is however strongly affected by the poloidal rotation of the plasma. The apparent 'lifetime' could in part be due to the poloidal structure size Δ_θ divided by the poloidal velocity v_θ. Likewise, the apparent 'outward propagation velocity' v_r could in part be due to an intrinsic tilt of the structures in the (r, θ) plane in combination with the poloidal velocity v_θ. This interpretational ambiguity arises due to the radial layout of the rake probe. To resolve these issues, a 2D probe pin distribution in the (r, θ) plane is required, as in [37].

2.4 Spectra

Fourier analysis is a standard tool by means of which a time series is decomposed in orthonormal modes, more specifically monochromatic oscillations (sines and cosines). We briefly summarize the main concepts for convenience. The discrete complex Fourier transform of a time series X: $\{x_i = x(t_i), i = 0, \ldots, N-1\}$ (N even) can be written

$$\hat{x}_j = \sum_{k=0}^{N-1} x_k \exp(-2\pi i(j/N)k) \tag{2.22}$$

and its inverse

$$x_k = \frac{1}{N}\sum_{j=0}^{N-1} \hat{x}_j \exp(2\pi i(j/N)k). \tag{2.23}$$

The inverse transform of the Fourier transform of X again yields X, so this transformation preserves information. Different authors may make different choices for the $1/N$ factor, putting it in the forward transform instead of the inverse transform, or even assigning $1/\sqrt{N}$ to both. Assuming a fixed sampling rate, the sampling time interval is $\Delta t = t_i - t_{i-1}$, and the sampling frequency is $f_s = 1/\Delta t$. The Nyquist frequency is half the sampling rate, $f_N = f_s/2$, and corresponds to the highest frequency that can be resolved using the discretely sampled data. The frequencies assigned to the index j in equation (2.22) are such that $f(0)$ corresponds to zero frequency, positive frequencies $0 < f < f_N$ correspond to $1 \leqslant j \leqslant N/2$ and negative frequencies $-f_N < f < 0$ to $N/2 + 1 \leqslant j \leqslant N - 1$. The value $j = N/2$ corresponds to both $f = f_N$ and $f = -f_N$. The Fourier transform is evaluated using standard fast algorithms.

Parseval's theorem states

$$\sum_{i=0}^{N-1} |x_i|^2 = \frac{1}{N} \sum_{j=0}^{N-1} |\hat{x}_j|^2. \tag{2.24}$$

Thus, the mean spectral power (the mean square absolute value of the Fourier coefficients) is directly proportional to the square of the standard deviation of RMS value of the time series (i.e. the fluctuation power).

It is not possible to assign an error to the individual Fourier coefficients $\hat{x}_j$. A practical upper bound estimate of the error in the spectral power can be obtained by subdividing a stationary time series of length N in $\lfloor N/M \rfloor$ sections of length M, calculating the spectrum for each section, and averaging the results. The variation of the spectral power at each frequency then provides the mentioned error estimate. Of course, the frequency resolution (Δf) of the resulting average spectra is lower by a factor $\lfloor N/M \rfloor$, as compared to the spectrum of the original time series. The error estimate is an upper bound because it is the result from both statistical errors and physical variations that one might not consider errors.

The spectral cross power between two time series X and Y is $S^2_{X,\,Y} = |\hat{x}||\hat{y}|$ and the spectral cross phase is defined by $\exp(i\Delta\phi) = \hat{x}\hat{y}^*/|\hat{x}||\hat{y}|$. The coherence is

$$\gamma_{X,Y}(f) = \frac{|\langle \hat{x}(f)\hat{y}^*(f)\rangle|}{\sqrt{\langle |\hat{x}|^2(f)\rangle\langle |\hat{y}|^2(f)\rangle}}. \tag{2.25}$$

The angular brackets imply an average over equivalent realizations; typically, these would correspond to successive time intervals of the stationary signals X and Y (i.e. the subsections of length M of the preceding paragraph). The coherence $0 \leqslant \gamma_{X,Y}(f) \leqslant 1$ can reach 1 for a given frequency f when the cross phase $\Delta\phi$ is constant over the realizations implied in the average. If the cross phase is uniformly distributed over $(-\pi, \pi]$, $\gamma_{X,Y}(f)$ will be approximately zero. The mean (frequency-averaged) coherence, weighted with cross power, is

$$\bar{\gamma}_{X,Y} = \frac{\int S^2_{X,\,Y}(f)\gamma_{X,Y}(f)df}{\int S^2_{X,\,Y}(f)df}. \tag{2.26}$$

Of great interest is also the joint wavenumber–frequency spectrum $S(k, f)$, i.e. the square of the k, f harmonic obtained by performing a double space-time Fourier transform on a one-dimensional field, $A(x, t)$. In principle, one would need measurements of A with both spatial and temporal resolution to determine it. However, a good approximation can be obtained using the 'two-point technique', based on the assumption that the fluctuations obey a dispersion relationship $k = k(f)$ [38]. In this case, one makes use of two measurements of the same quantity (say, a floating potential), X and Y, spaced by a distance d. Then the phase shift of the cross spectrum, $\Delta\phi(f)$, is related to the wavenumber by $k = \Delta\phi(f)/d$. It is implicitly assumed that all relevant wavenumbers are smaller than the resolution imposed by the distance between the two points, i.e. $|k| < \pi/d$. Higher physical wavenumbers, if present in the data, will be aliased down to lower wavenumbers. The joint spectrum $S^2(k, f)$ can then be constructed by dividing the k-range, $-\pi/d \leqslant k \leqslant \pi/d$, in n_k bins. For each of the $\lfloor N/M \rfloor$ sections of length M (see above), one determines the cross spectrum and hence $\Delta\phi(f)$. Then, the quantity $S^2_{X,Y}(f)$ is added to the bin corresponding to the wavenumber $k = \Delta\phi(f)/d$ and the frequency f—noting there are $M/2$ frequency bins. The resulting joint spectrum is normalized by $1/\lfloor N/M \rfloor$. Since the spectral amplitude decays rapidly for high f (see section 2.4.1), it is sometimes convenient to add '1' instead of $S^2_{X,Y}(f)$ to the bins, which (after normalization) would result in a joint wavenumber–frequency probability distribution, $p(k, f)$. Figure 2.24 shows examples of the joint wavenumber–frequency probability distribution, $p(k, f)$, computed from floating potential measurements from the edge of the TJ-II stellarator in the improved confinement regime. The $p(k, f)$ spectrum from closely spaced poloidally separated pins, figure 2.24(a), is dominated by frequencies below 300 kHz and reveal a local phase velocity ($v = \omega/k = 2\pi f/k$) of about $(1 - 5) \times 10^3$ m s^{-1}, in agreement with the expected $E \times B$ drift velocity [39]. It should be noted that the short-range wavenumber is close to zero at the lowest frequencies ($f < 40$ kHz), where long range correlations exist. The long range $p(k, f)$ distribution, computed from toroidally separated potential measurements during the improved confinement phase, figure 2.24(b), is dominated by frequencies below 40 kHz; this distribution

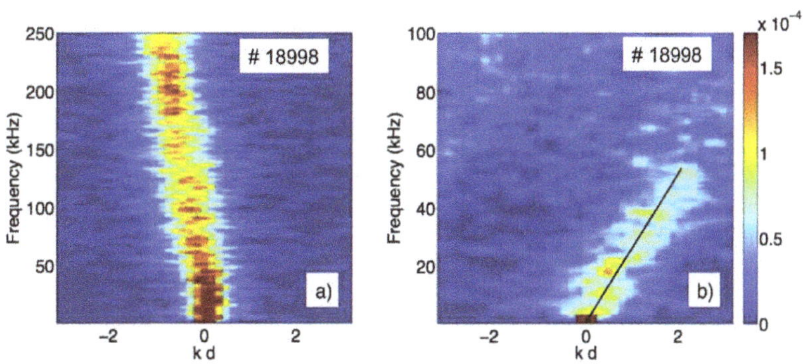

Figure 2.24. Joint wavenumber–frequency probability distribution, $p(k, f)$, obtained in an NBI-induced improved confinement phase at TJ-II (shot 18998), computed from floating potential measurements taken by Langmuir probes in the plasma edge. (a) $p(k, f)$ from two poloidally spaced pins of a single probe (poloidal distance $d \simeq 0.003$ m); (b) $p(k, f)$ from two pins in distant probes. (Reproduced with permission from [29]. © IOP Publishing. All rights reserved.)

indicates propagation, understood to be due to the radial propagation of structures with a significant toroidal extent [29].

2.4.1 Turbulent frequency spectra

In fluid turbulence, spectra often exhibit the Kolmogorov power law decay, $S^2(k) \propto k^{-5/3}$, where k is the wavenumber [40]. Self-organized critical systems often exhibit a $S^2(f) \propto 1/f^{\alpha}$ decay, with $0 < \alpha < 1$, at least over part of the frequency range (see chapter 3). So in plasma turbulence, what shape do (spatial and temporal) turbulent spectra have?

Figure 2.25 shows the spectrum of ion saturation current fluctuations from W7-AS data (at a radial location where the flow velocity is approximately zero) on a double logarithmic scale. The spectrum is calculated using a record of length 20 ms, and it is averaged over three reproducible discharges. Its shape is typical for plasma edge turbulence: rather flat at low frequencies and dropping off steeply for very high frequencies (above ~100 kHz). Occasionally, more or less narrow peaks are seen on top of such a background spectrum, corresponding to MHD modes or Alfvén modes, although this is not the case in this example. In this particular spectrum, the narrow spikes seen near 700 kHz are instrumental. Very similar spectra were also obtained, e.g., in DIII-D [42] and several other devices (TJ-I, JET, TJ-IU, and W7-AS)—i.e. in both tokamaks and stellarators. The similarity can be highlighted using a rescaling operation similar to the one we used for PDFs, equation (2.10). Of course, spectra are not normalized as PDFs are (their integral is not 1), hence the frequency and amplitude scale of the spectra are not linked. Assuming the spectra follow some universal shape function $g(f)$, the rescaled spectra would be [43]

$$S^2(f) = P_0 g(\lambda f), \tag{2.27}$$

with two adjustable parameters instead of one. Figure 2.26 shows how the turbulent spectra for very different devices and plasmas can be made to fall on top of each

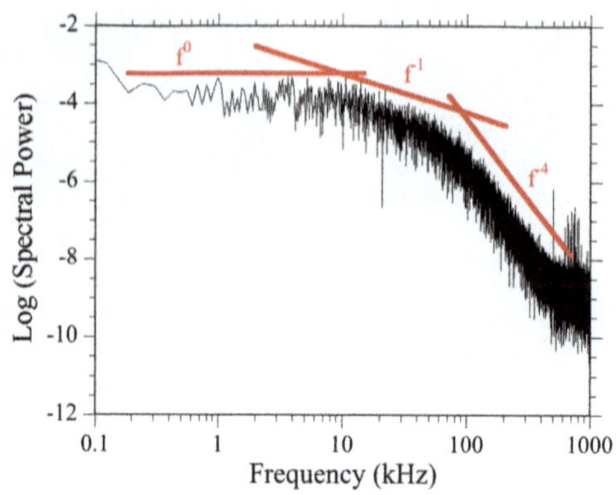

Figure 2.25. The ion saturation current spectrum for W7-AS, at the radial position where the flow velocity is zero. We used a record of length 20 ms, and the spectrum is averaged over three reproducible discharges. (Reproduce after [41], with the permission of AIP Publishing)

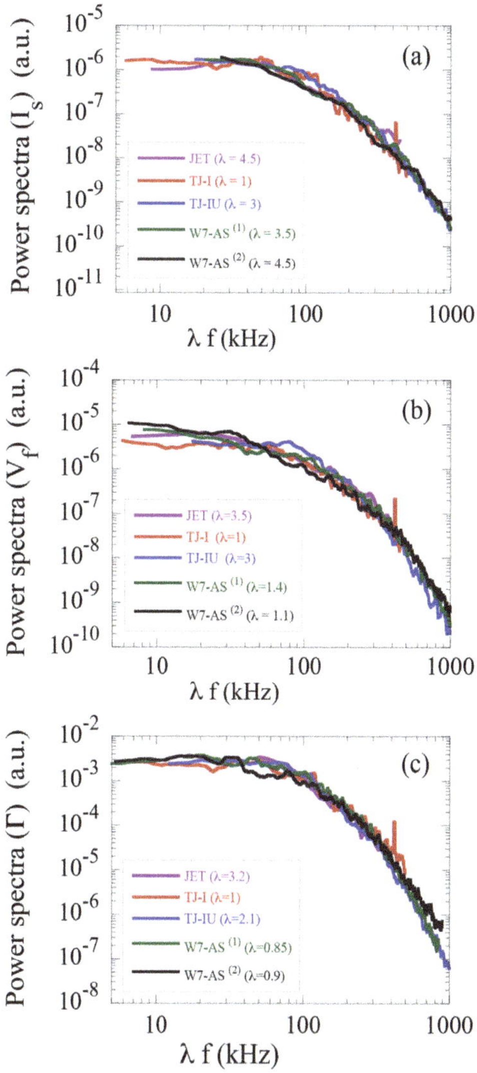

Figure 2.26. Rescaled frequency spectra of fluctuations for the (a) ion saturation current, (b) floating potential, and (c) turbulent transport. (Reproduced with permission after [43], Copyright (1999) by the American Physical Society.)

other [43]. Care was taken to obtain the spectra in regions with very low poloidal plasma rotation velocity, to minimize the deformation of spectra through Doppler shift effects.

Coming back to the spectrum shown in figure 2.25, one may distinguish three regions. Assuming the spectrum decays like $S^2(f) \propto f^{-\alpha}$ in each region, one observes that the three regions are characterized by approximate decay indices α that are consistent with 0, 1, and 4, respectively, as indicated by the red lines. We will delay the discussion about the meaning of these exponents until section 3.5, where we will be able to give an interpretation in the context of complex dynamics. To improve the identification of this central region, we may reduce the noise on the spectrum by averaging it using frequency bins that grow in proportion to f (or using equal size bins for the logarithm of the frequency, i.e. 'logarithmic binning'). Once an estimate of the decay index α in the central frequency region is available, we multiply the spectrum $S^2(f)$ by the inverse of the decay in the intermediate region, i.e. we calculate $S^2(f)f^{\alpha}$. Figure 2.27 shows the result. This straight-forward procedure has two advantages: it makes it much easier to see that the spectrum is indeed rather straight in the intermediate region (on a double logarithmic scale), and it allows determining the boundaries of the intermediate region with better precision.

The intermediate region is 'special' for a reason. Recall that the fluctuating flux is $\Gamma = \tilde{n}\tilde{E}_\theta/B_\phi$. Figure 2.28 shows the cross power spectrum $S^2_{X,\,Y}$ for $X = n_e \propto I_{sat}$ and $Y = E_\theta \simeq (V_f^2 - V_f^1)/d_\theta$ for a few cases. Interestingly, the cross power peaks precisely in the intermediate spectral region identified before (figure 2.27). In other words, in this intermediate region the spectrum not only has a decay index of $\alpha \simeq 1$, it is also where most of the power of the fluctuating flux is concentrated.

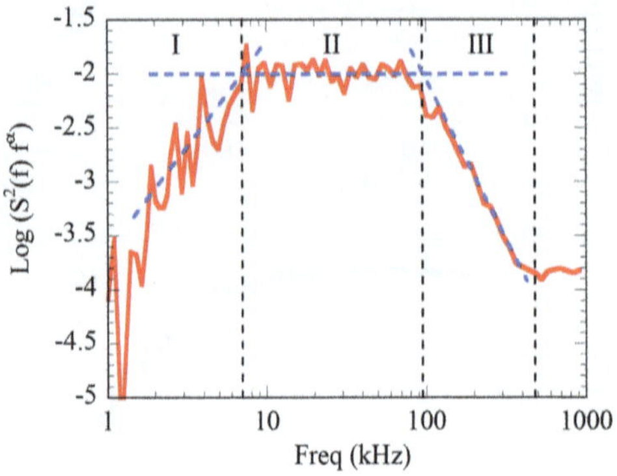

Figure 2.27. Spectrum of the ion saturation current fluctuation from a measurement in W7-AS near $V_\theta = 0$. A 20 ms data record has been used. The spectrum $S^2(f)$ is multiplied by f^{α} to obtain a flat intermediate region and to identify the three spectral regions. (Reproduced after [44], with the permission of AIP Publishing.)

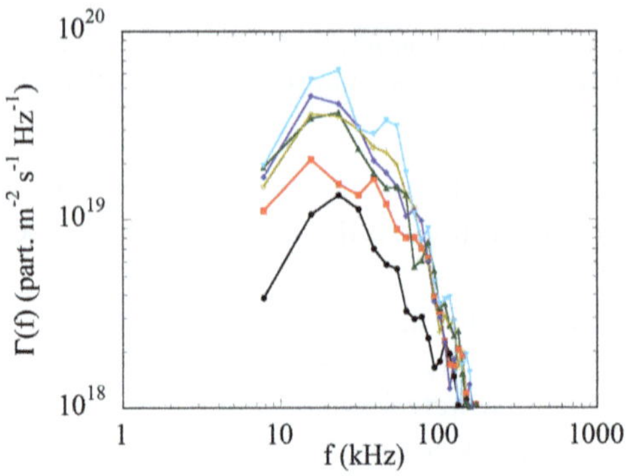

Figure 2.28. Cross power spectral function for 20 ms data sections near the flow shear layer at W7-AS. (Reproduced after [44], with the permission of AIP Publishing.)

Next, we come back to the impact of the Doppler shift on the frequency spectra. This is a serious difficulty for the determination of spectra in fusion plasmas, as plasmas tend to rotate at rather high speeds, whereas the measurements are performed in the laboratory frame of reference. Theoretical predictions of spectra tend to be valid for the rest frame of the plasma only.

Assuming that (i) the plasma is rotating at a fixed poloidal velocity v_θ and that (ii) the mean dispersion relation $\bar{k}_\theta(f)$ is known, the Doppler shift will modify the frequency as follows [44]:

$$\hat{f} = f + \bar{k}_\theta(f) v_\theta, \tag{2.28}$$

where f is the frequency in the plasma rest frame and $\hat{f}$ the detected frequency in the laboratory reference frame. As the frequency shift depends on the frequency (through $\bar{k}_\theta(f)$), the spectral decay indices are modified as well. Figure 2.29 shows an example of a Doppler shift correction on the basis of this idea. We note that the dispersion relation $\bar{k}_\theta(f)$ can be estimated using the two-point technique for determining the joint wavenumber–frequency spectrum $S(k, f)$ described above. The figure illustrates that a Doppler shift can significantly modify the measured spectra and decay indices. Therefore, considerable care should be taken when comparing spectra from different devices, different locations in a given device, or numerical simulations and experiments.

2.4.2 Turbulent wavenumber spectra

In the preceding section, it was noted that turbulence in fusion plasmas is usually advected by the (poloidal) plasma velocity, at great speed, leading to an almost linear dispersion relation $k(f)$. Thus, to the first approximation, the measured frequency spectra $S^2(f)$ are mostly just a reflection of the wavenumber spectra $S^2(k)$. The wavenumber spectrum can be measured directly using techniques such as reflectometry, in which the reflection of a microwave beam off turbulence at a specific reflection layer depends on the local wavenumber.

At TJ-II, the Doppler reflectometer has access to about one decade of wavenumbers, $2 < k_\perp < 15$ cm^{-1} in the density gradient region, typically $0.6 < \rho < 0.8$,

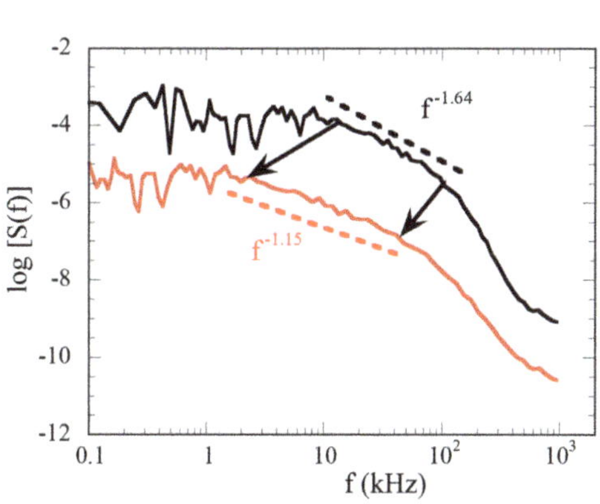

Figure 2.29. Correction of frequency spectrum for Doppler shift using equation (2.28); a poloidal velocity of $v_\theta = 800$ m s^{-1} was assumed. (Reproduced after [44], with the permission of AIP Publishing.)

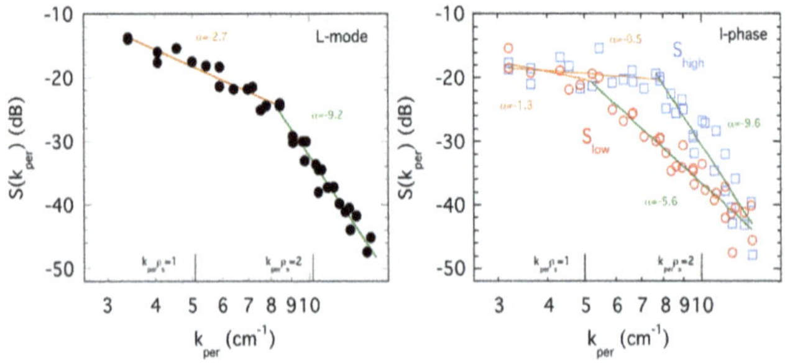

Figure 2.30. Density fluctuation wavenumber spectra measured during the L mode at line averaged central density $\bar{n}_e \simeq 1.8 \times 10^{19}$ m^{-3} (left) and during the intermediate oscillatory phase $\bar{n}_e \simeq 2.3 \times 10^{19}$ m^{-3} (right). The extreme values of the turbulence level measured during the I phase are represented: maxima (in blue) and minima (in red). The spectral indexes are shown for each spectrum and the normalized wavenumbers $k_\perp \rho_s = 1$ and 2 are marked in the figures. (Reproduced with permission from [46].

through variation of the microwave beam angle and frequency [45]. Figure 2.30 shows an example of the measured wavenumber spectra in the L phase and the intermediate oscillatory phase or I phase, in which the turbulence amplitude oscillates cyclically as a consequence of predator–prey dynamics [46]. These predator–prey oscillations are slow with respect to turbulent timescales, having an oscillation period that varies but is typically of the order of 0.5–1 ms.

To compare these results with the typical frequency spectrum observed in fusion devices, one must take into account that the y-scale of figure 2.30 is expressed in dB, so that the decay indices must be divided by 10 to compare with the indices given on the double logarithmic scale of figure 2.25. Thus, the logarithmic decay indices for the low-$k_\perp$ interval ($k_\perp \lesssim 7$ cm^{-1}) are well below 1. On the other hand, the logarithmic decay indices for the high-$k_\perp$ interval ($k_\perp \gtrsim 7$ cm^{-1}) are between 0.5 and 1. These values seem to match the indices in the low and intermediate frequency ranges of figure 2.25. In spite of this rough agreement, the interpretation of Doppler reflectometry wavenumber spectra remains complicated [47].

2.5 Structure detection: the biorthogonal decomposition

The preceding sections have shown that the turbulent plasma contains structures with significant spatial and temporal scales. Is it possible to extract these structures from the data and perhaps visualize them? For this specific purpose, so-called 'data-driven' techniques such as the singular value decomposition (SVD) are ideally suited, as they do not make any prior assumptions regarding the specific data structure. In particular, these linear techniques can be very useful for the extraction of normal modes[8] from multipoint measurements. In the following, we will briefly describe some of the features of the biorthogonal decomposition or SVD, which are well-known, so the reader is referred to the references for more detailed information [48, 49].

[8] Natural oscillation modes, eigenmodes, or resonances.

Let us assume one has multipoint measurements that constitute a data matrix $Y(i, j)$, where the index $i = 1, \ldots, N$ labels the time and $j = 1, \ldots, M$ the detector. Typically, the time corresponding to the time index, $t(i)$, is equally spaced, since measurements are usually taken at a fixed sampling rate, although this is not strictly necessary. The physical location of the detectors, $\vec{x}(j)$, however, is often dictated by practical convenience or space limitations and will not necessarily correspond to a regular grid (although they could, e.g. in the case of the pixels of a two-dimensional camera frame). To facilitate the physical interpretation of the results, the measurements Y performed at the various detectors j should ideally be of the same physical quantity and be given in the same units.

The BOD method decomposes the data matrix $Y(i, j)$ as follows:

$$Y(i, j) = \sum_k \lambda_k \psi_k(i) \phi_k(j), \tag{2.29}$$

where ψ_k is a 'chrono' (a temporal function) and ϕ_k a 'topo' (a spatial or detector-dependent function), such that the chronos and topos satisfy the following orthogonality relation [48]:

$$\sum_i \psi_k(i) \psi_l(i) = \sum_j \phi_k(j) \phi_l(j) = \delta_{kl}. \tag{2.30}$$

An alternative normalization (indicated by the superscript r) is

$$\frac{1}{N} \sum_i \psi_k^r(i) \psi_l^r(i) = \frac{1}{M} \sum_j \phi_k^r(j) \phi_l^r(j) = \delta_{kl}, \tag{2.31}$$

such that the root-mean-square (RMS) value of the topos and chronos equals 1. In this case,

$$Y(i, j) = \sum_k \lambda_k^r \psi_k^r(i) \phi_k^r(j) \tag{2.32}$$

with $\lambda_k^r = \lambda_k / \sqrt{NM}$. With this normalization, the number λ_k^r represents the contribution of mode k to the total RMS 'fluctuation amplitude'. Thus, the fractional contribution of a given mode k to the total 'fluctuation energy' (proportional to the square of the RMS) can be computed from

$$e_k^2 = \frac{\lambda_k^2}{\sum_{k'} \lambda_{k'}^2}. \tag{2.33}$$

The superscript 'r' is irrelevant in this expression.

The combination chrono/topo at a given k, $\psi_k(i)\phi_k(j)$, is called a spatiotemporal 'mode' of the fluctuating system, and is constructed from the data matrix without any prejudice regarding the mode shape.

The decomposition is performed by computing the singular value decomposition (SVD) of the data matrix $Y(i, j)$. Note that this is always possible for any real-valued rectangular matrix Y, and standard software packages are available for this purpose, e.g. [50]. The $\lambda_k \geqslant 0$ are the eigenvalues (sorted in decreasing order), where $k = 1, \ldots, \min(N, M)$.

A threshold may be set for cutting off the expansion in BOD modes (based on, e.g., a noise level) and so keeping only the dominant contribution of modes to the data matrix for further analysis and ignoring noisy or minor contributions. When the BOD expansion is cutoff at a certain mode $k_{\max}$, the reconstructed data, $Y_{\rm rec}$, are defined by equation (2.29) while restricting the sum to $k \leqslant k_{\max}$. In this case, the reconstructed data $Y_{\rm rec}$ minimize the least-squares error with respect to the actual data Y among all possible sets of $k_{\max}$ spatiotemporal modes.

If the physical system under study contains normal mode oscillations and is sufficiently well-sampled (spatially and temporally), the probability that the BOD modes correspond closely to the said normal modes is high; the BOD is particularly sensitive to resonant linear normal modes [49]. On the other hand, the BOD technique is perhaps less suited for the analysis of non-linear systems in which no normal modes exist, as the BOD modes are unlikely to correspond to any meaningful physical modes in this case. An advantage of the BOD analysis is that no *a priori* assumption is made regarding the mode shape or spectral properties, unlike standard analysis techniques such as Fourier decomposition. This has the mentioned advantage that the BOD modes concentrate a maximum of fluctuation power in the lowest modes, but the disadvantage that the modes are not guaranteed to coincide exactly with the normal modes of the system (if known).

Furthermore, it should be noted that the SVD decomposition is not unique. First, the sign of the chrono and topo of a given mode k can trivially be inverted without affecting their product, i.e. without affecting the reconstruction according to equation (2.29). Second, if two SVD modes k_1 and k_2 have the same eigenvalues, a rotation in the two-dimensional vector space spanned up by the chronos k_1 and k_2 and a compensating rotation in the space spanned up by the corresponding topos can be defined so that the reconstructed data, according to equation (2.29), are unchanged, giving rise to a (rotational) indeterminacy. As will be clarified below, this may occur in the case of propagating modes. As such, it does not imply a great disadvantage, since such modes will be considered pairwise anyway.

2.5.1 Covariance and the biorthogonal decomposition

The covariance between a signal $Y(i, j_1)$ and a signal $Y(i, j_2)$ (i being the temporal and j the spatial index) is defined as

$$\mathrm{Cov}(j_1, j_2) = \frac{1}{N}\sum_i Y(i, j_1)Y(i, j_2). \tag{2.34}$$

Since the signals are expanded according to equation (2.29), this expression can be simplified, using equation (2.30), to

$$\begin{aligned}\mathrm{Cov}(j_1, j_2) &= \frac{1}{N}\sum_k \lambda_k^2 \phi_k(j_1)\phi_k(j_2) \\ &= \sum_k (\lambda_k^r)^2 \phi_k^r(j_1)\phi_k^r(j_2),\end{aligned} \tag{2.35}$$

i.e. the covariance between two signals is simply the sum of the products of the corresponding topos, weighed by the square of the eigenvalues. In other words, the topos reflect the covariance between the measurement signals; and the contribution of each mode k to the covariance $\mathrm{Cov}(j_1, j_2)$ is simply $(\lambda_k^r)^2\phi_k^r(j_1)\phi_k^r(j_2)$.

From equation (2.35) one immediately obtains, again using equation (2.30),

$$\sum_{j_2} \mathrm{Cov}(j_1, j_2)\phi_k(j_2) = \frac{\lambda_k^2}{N}\phi_k(j_1). \tag{2.36}$$

Thus, the topos ϕ_k are the eigenvectors of the covariance matrix Cov, and λ_k^2/N the corresponding eigenvalues. This suggests that the topos ϕ_k and eigenvalues λ_k can be found by computing the eigenvectors and eigenvalues of the covariance matrix Cov, after which the corresponding chronos ψ_k can be obtained by simple matrix multiplication [49]:

$$\lambda_k\psi_k(i) = \sum_j Y(i, j)\phi_k(j). \tag{2.37}$$

This view of the BOD mode decomposition may be helpful when interpreting the meaning of the BOD modes.

The fact that the topos are the eigenvectors of the covariance matrix immediately suggests an alternative approach: instead of analyzing the raw data $Y(i, j)$, one could first normalize the data to their RMS value

$$y(i, j) = \frac{Y(i, j)}{\sqrt{\mathrm{Cov}(j, j)}}. \tag{2.38}$$

When computing the SVD of $y(i, j)$ instead of $Y(i, j)$, the resulting topos ϕ_k will then be the eigenvectors of the *correlation* matrix C.

2.5.2 Correlation and the biorthogonal decomposition

By definition, the correlation between measurements j_1 and j_2 is obtained by normalizing the covariance, equation (2.34), to the RMS of each signal:

$$C(j_1, j_2) = \frac{\mathrm{Cov}(j_1, j_2)}{\sqrt{\mathrm{Cov}(j_1, j_1)\mathrm{Cov}(j_2, j_2)}}. \tag{2.39}$$

From equation (2.35), it is clear that this quantity only depends on the eigenvalues and the topos. The contribution of a given mode k to the correlation is (see [51])

$$C_k(j_1, j_2) = \frac{(\lambda_k^r)^2\phi_k^r(j_1)\phi_k^r(j_2)}{\sqrt{\mathrm{Cov}(j_1, j_1)\mathrm{Cov}(j_2, j_2)}}. \tag{2.40}$$

Thus, it is possible to quantify the contribution of a specific BOD mode k to the so-called *long range correlation* (LRC). To achieve this, assume we may subdivide the set of measurements $j = 1, \ldots, M$ into two complementary sets: $\{j \in S_1\}$, with M_1 elements, and $\{j \in S_2\}$, with M_2 elements, such that measurements in set S_1 are taken at a remote location from measurements in set S_2. The LRC is then defined as the correlation between any pair of measurements taken at remote locations. We define the contribution to the *mean* LRC of a given mode k by averaging the correlation due to this mode among any two remote measurement points:

$$C_k^{\mathrm{LR}} = \frac{1}{M_1 M_2} \sum_{\{j_1 \in S_1\}} \sum_{\{j_2 \in S_2\}} C_k(j_1, j_2). \tag{2.41}$$

This definition may lead to low values in the case of partial cancellations due to the existence of a mixture of correlations and anti-correlations between individual measurements. To measure the overall correlation 'intensity' regardless of sign, we also define the mean absolute LRC:

$$C_k^{\mathrm{LR,abs}} = \frac{1}{M_1 M_2} \sum_{\{j_1 \in S_1\}} \sum_{\{j_2 \in S_2\}} |C_k(j_1, j_2)|. \tag{2.42}$$

These quantities are particularly interesting in the framework of the detection of zonal flows [52].

2.5.3 Mode identification with the biorthogonal decomposition

One important aspect of the BOD method is that it assumes that all 'modes' are separable in space and time. Standing waves are of this type. However, one often has to contend with propagating structures, such as running waves. It is important to be able to recognize and distinguish these different structures.

A typical example of a running wave is $Y(t, x) = \cos(kx - \omega t)$. This can trivially be decomposed into a sum of two standing wave patterns:

$$\cos(kx - \omega t) = \cos(kx)\cos(\omega t) - \sin(kx)\sin(\omega t). \tag{2.43}$$

By Fourier's theorem, this result can be generalized to any linearly propagating spatiotemporal structure of argument $kx - \omega t$. Thus, linearly propagating or 'travelling' waves or structures generate a pair of BOD modes, with similar eigenvalues and a mutual phase difference (in space and time) of around 90° [48].

Consequently, a method is needed to identify such mode pairs. If the propagating mode has a clearly defined frequency ω, the Fourier power spectra $|\hat{\psi}_k|^2$ of the concerned chronos will have a peak at this frequency (here, the circumflex indicates the Fourier transform), and the cross phase between the chronos k_1, k_2 is easily determined as the phase of the complex cross spectrum $\hat{\psi}_{k_1}\hat{\psi}^*_{k_2}$ at the frequency ω, where the star signifies complex conjugate.

As the low-frequency modes of interest do not always have a clearly defined spectral peak, we will base quadrature detection on the Hilbert transform [53]. In this framework, we must assume that the BOD topos and chronos fluctuate symmetrically about zero, which can be achieved by subtracting their mean or applying a suitable high-pass frequency filter prior to analysis. This is necessary as it is a well-known fact that the Hilbert transform does not produce a reasonable estimate of the quadrature if the mean of the signal is not zero or is drifting [53].

With this additional assumption, it is feasible to compute the quadrature of a given topo or chrono by means of the Hilbert transform.

Thus, we can use the full set of topos and chronos and compute the following spatial and temporal quadrature matrices:

$$\begin{aligned} Q_x(k_1, k_2) &= \sum_j \tilde{\phi}_{k_1}(j) H(\tilde{\phi}_{k_2}(j)) \\ Q_t(k_1, k_2) &= \sum_i \tilde{\psi}_{k_1}(i) H(\tilde{\psi}_{k_2}(i)), \end{aligned} \tag{2.44}$$

where H is the Hilbert transform and the tilde refers to the removal of the mean (or lowest frequencies) mentioned above. The elements of these quadrature matrices are restricted to the range $[-1, 1]$; the absolute value of a given element $Q_{x,t}(k_1, k_2)$ will differ significantly from 0 when the corresponding modes are in approximate quadrature. By means of this technique, it becomes possible to identify linearly propagating structures by finding pairs of modes with similar eigenvalues, $\lambda_{k_1} \simeq \lambda_{k_2}$, such that $Q_x(k_1, k_2)$ and $Q_t(k_1, k_2)$ are significantly different from zero. By default, modes that do not occur in pairs then correspond to standing wave structures.

2.5.4 Zonal flow detection in the TJ-II stellarator

In this section, we will apply the techniques outlined above to see whether we can identify a zonal flow with some degree of confidence, using appropriately located potential measurements. TJ-II disposes of a double reciprocating probe system, allowing the simultaneous measurement of the floating potential on various radially separated probe pins on two probes with a large toroidal separation [54]. Fluctuating structures with a high level of correlation between the toroidally separated probes (LRC) can possibly be identified with ZF structures [51]. The radially spaced pins on each of the two probes provide information about the radial wavenumber of the detected structures.

However, ZFs are not the only phenomena that can give rise to LRCs. In particular, it has been shown that other modes (e.g. drift waves or MHD oscillations) may also produce LRCs [34]. The question, therefore, is whether one can distinguish between the different origins of observed LRCs: on the one hand, ZF-like behaviour and on the other, MHD or other (rotating/propagating) oscillations. ZFs are electrostatic potential fluctuations having toroidal mode number $n = 0$, poloidal wavenumber $m = 0$, and a finite radial wavenumber [31]. Due to this symmetry, the ZF potential structure may fluctuate—relatively slowly—but *does not rotate*. The latter property allows one to distinguish a ZF from, e.g., a low-frequency MHD mode, which (with very rare exceptions) usually does rotate.

The techniques discussed above, i.e. the quantification of the contribution of each mode to LRCs and the quadrature detection technique, should enable one to make this distinction based on the experimental data: ZFs are long range correlated 'standing wave structures', while MHD, rotating or propagating modes are 'travelling wave structures'.

TJ-II is a heliac type stellarator with four field periods, having major radius $R = 1.5$ m, minor radius $a < 0.2$ m, and toroidal magnetic field $B_0 < 1$ T. The plasmas considered here are heated by the electron cyclotron resonant heating (ECRH) system, consisting of two beam lines with an injected power of up to 400 kW each. Electron density is controlled by means of a gas puffing system and the line average electron density typically reaches values of about $\bar{n}_e \simeq 0.6 \times 10^{19}$ m^{-3}. The central electron temperature is typically $T_e(0) \simeq 300 - 800$ eV, and the central ion temperature is around $T_i(0) \simeq$ 120–150 eV.

Langmuir probe D is located in a top port at toroidal position $\phi \simeq 35°$, while probe B is located in a bottom port at $\phi \simeq 195°$. Thus, the two probe systems are remote, separated toroidally by a distance of ~5 m (about half a toroidal turn) [55]. In the experiments analyzed here, each probe is fitted with a 'rake' probe head measuring floating potential at ~10 radially spaced pins. Probe pins are separated radially by 3 mm in the D probe, and by 1.7 mm in the B probe; probe D covers a radial range of about 3.5 cm, while probe B covers about 1.5 cm.

We analyze shot 36012 in the time interval $1090 \leqslant t \leqslant 1250$ ms. In this discharge, the rotational transform, $\iota/2\pi$, has a value of around 1.455 at the magnetic axis and 1.55 at the edge, so that the 3/2 rational surface is located at about $\rho \simeq 0.73$. In the mentioned time interval, both the line average electron density and the plasma energy content are fairly constant; $\bar{n}_e \simeq 0.63 \times 10^{19}$ m^{-3}, near the critical density of the electron to ion root confinement transition at TJ-II [56, 57]. Figure 2.31 shows the mean spectrum of all probe pins. Several spectral peaks are visible, which we will identify in the following.

Figure 2.32 shows the BOD eigenvalues and the LRC. To calculate the LRC, equations (2.41) and (2.42), the probe pins have been subdivided into two sets, namely pins corresponding to probe D or probe B. Figure 2.33 shows the spectra of the first three modes. The first BOD mode suggests a strong long range *anti*-correlation. The second and third BOD modes form a propagating pair (confirmed by quadrature analysis, not shown); the spectrum shows that this propagating mode has a clear frequency peak at 5 – 6 kHz. Regarding the negative value of C_1^{LR}, it

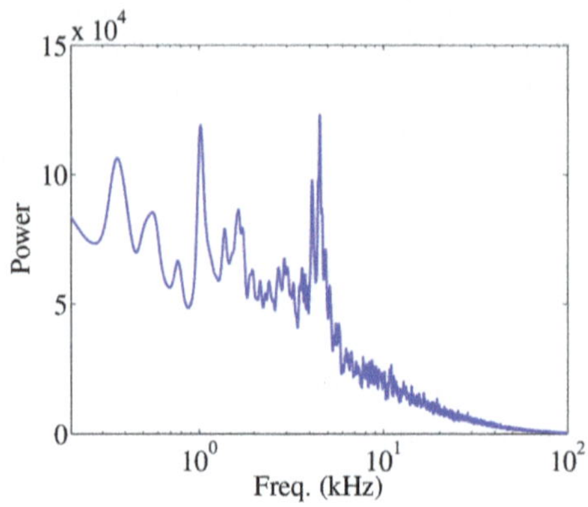

Figure 2.31. Mean spectrum of the probe pins in discharge 36012. (Reproduced with permission from [52].

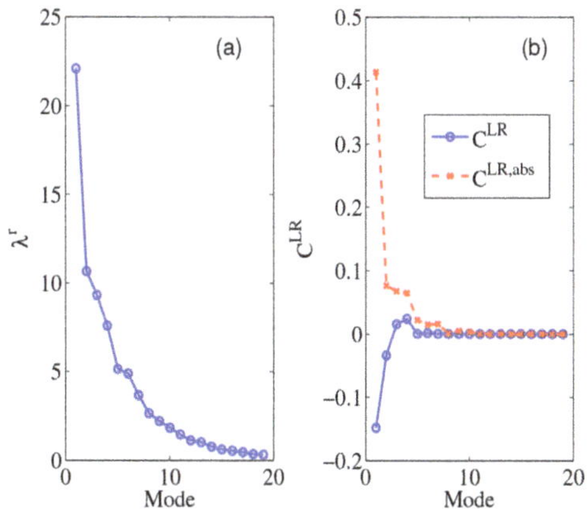

Figure 2.32. (a) BOD eigenvalues λ_k^r and (b) LRC contributions C_k^{LR} and $C_k^{LR,abs}$ (shot 36012). (Reproduced with permission from [52]. © IOP Publishing. All rights reserved.)

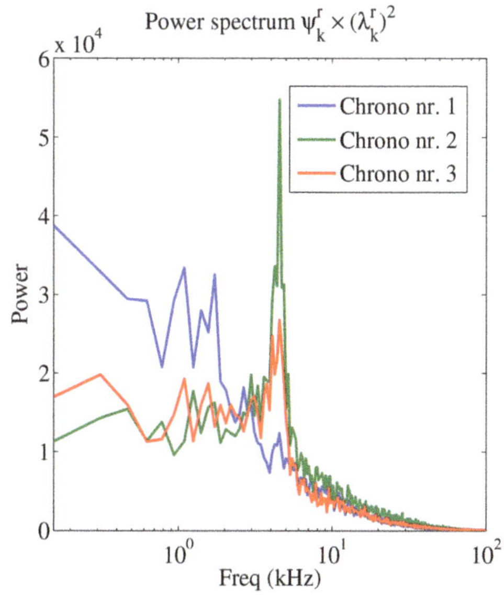

Figure 2.33. Spectra of the chronos ψ_k^r of the first 3 BOD modes (shot 36012). (Reproduced with permission from [52]. © IOP Publishing. All rights reserved.)

should be noted that topo 1 exhibits a complex radial structure, see figure 2.34. Thus, the global character of the definition of C^{LR} may not capture the details of this structure. The two peaks appearing in topo 1 at $\rho \simeq 0.86$ are in fact *correlated* and correspond to a zonal flow-like structure, indicated schematically by a grey area in figure 2.34. This structure is considered zonal flow-like for the following reasons: (i) it is LRC (recall equation (2.35)); (ii) the structure has the same value of floating potential in the two remote probes, suggesting an $m = 0$, $n = 0$ global structure; (iii) the spectral characteristics of the structure show no clear peak and spectral

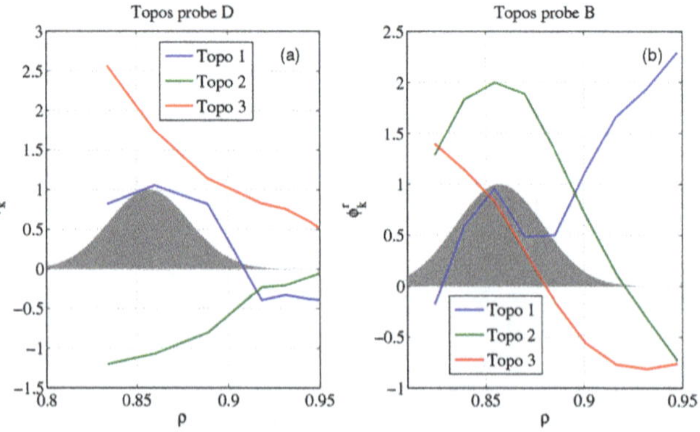

Figure 2.34. Topos ϕ_k^r of the first 3 BOD modes for (a) probe D and (b) probe B (shot 36012). (Reproduced with permission from [52]. © IOP Publishing. All rights reserved.)

power is concentrated at a very low frequency, see figure 2.33; and (iv) the radial position and radial width are very similar for the two remote probe systems (as indicated by the grey area in figure 2.34). The radial width of the ZF-like structure is rather small, namely about one centimetre. The four points mentioned above constitute the most unambiguous identification of a ZF-like structure yet reported in the literature.

Further outward ($\rho \gtrsim 0.9$), the floating potentials in the two probes are predominantly *anti*-correlated such that topo 1 has an opposite sign for the two probes. The temporal variation of this anti-correlated spatial structure is linked to the ZF fluctuations, as is evident from its inclusion in the same chrono. We speculate that the fluctuations of the ZF intensity (at $\rho \simeq 0.86$) cause a variation of outward transport (in the range $\rho \gtrsim 0.9$) with some toroidal/poloidal asymmetry—probably associated with the different local curvature and flux expansion at the two probe locations.

At this point, we would like to emphasize that the radial structure revealed by this technique is very hard to extract using traditional methods based on, e.g., two-point correlation functions, due to the fact that the signals contain contributions from both the ZF-like structure and the propagating modes. Traditional techniques cannot separate these contributions clearly, unless a hypothesis is made about the spectral distribution and frequency filters are applied. Here, no hypotheses are needed and the results follow directly from the multipoint analysis of the raw data. It appears that the BOD is a powerful technique to separate ZF-like, global oscillations from other (propagating) oscillations and to extract their mode structure.

2.6 Wave interactions: the bicoherence

Fluctuations in a turbulent system can always be decomposed into component waves, based on the Fourier theorem. To account for the temporal evolution of turbulence and the selective growth of specific modes, the corresponding wave evolution equations include non-linear terms. One could therefore imagine that there is a hierarchy of interactions, such that the time rate of change of a given wave depends on a linear term (growth), plus a quadratic term (wave interaction), plus higher order terms. If one assumes that higher-order terms are less important than lower-order terms, one can hope to learn more about the energy exchange between waves (see the Kolmogorov energy cascade) by determining second-order (wave–wave) interactions.

Unfortunately, this program only appears to work when wave interactions are weak. In the case of strong turbulence (frequently encountered in fusion plasmas), the lifetime and spatial extension of the assumed waves are too small to be able to fruitfully detect their existence and interaction, while it is also not obvious that the hierarchy of interactions is as assumed. Therefore, calculating wave–wave interaction quantifiers such as the bicoherence does not appear to be a fruitful way to approach strongly developed turbulence in general.

Still, the bicoherence is an interesting tool for the following reason. In the sea of turbulence, sometimes dominant waves appear, in particular related to the existence of large scale structures, such as ZFs or MHD modes. If they do, the bicoherence can tell us more about the non-linear interactions giving rise to or associated with these structures.

2.6.1 Fourier bicoherence

Given two variables $X(t)$ and $Y(t)$, and their complex Fourier transforms $\hat{X}(\omega)$ and $\hat{Y}(\omega)$, the cross bispectrum is defined as [58]

$$B_{X,Y}(\omega_1, \omega_2) = \langle \hat{X}(\omega_1)\hat{X}(\omega_2)\hat{Y}^*(\omega)\rangle, \tag{2.45}$$

such that $\omega = \omega_1 + \omega_2$. This 'sum rule' expresses a wave resonance condition. The angular brackets indicate an average over equivalent realizations; in practice, this is an average over successive time intervals of the signals, which are assumed stationary. If the three spectral components at the specified frequencies are statistically independent, then the bispectrum will converge to zero as the number of realizations is increased. On the other hand, if these three waves maintain their relative phases, the converged bispectrum will have a finite value, indicating phase locking or coupling between the three waves. This could happen if the wave at ω decays into two 'daughter' waves ω_1, ω_2, or when two waves at ω_1, ω_2 resonate to produce a third wave at ω. In actual fact, for waves to resonate, not only their frequencies should satisfy the sum rule, but their wave vectors as well ($k = k_1 + k_2$). This can be taken into consideration when spatially resolved data are available [59].

The bicoherence is a normalization of the bispectrum. We note that the literature mentions various alternative normalizations. We prefer:

$$b_{X,\,Y}^2(\omega_1, \omega_2) = \frac{|\langle \hat{X}(\omega_1)\hat{X}(\omega_2)\hat{Y}^*(\omega)\rangle|^2}{\langle|\hat{X}(\omega_1)\hat{X}(\omega_2)|^2\rangle\langle|\hat{Y}(\omega)|^2\rangle}. \tag{2.46}$$

It can be shown that $0 \leqslant b^2 \leqslant 1$. The auto-bicoherence is $b_X^2 = b_{X,\,X}^2$.

As noted, the signals X and Y are assumed stationary. If they are not, it may be necessary to remove slow drifts to avoid spurious results. In any case, to avoid spurious high frequencies in the Fourier transforms, it is convenient to apply a

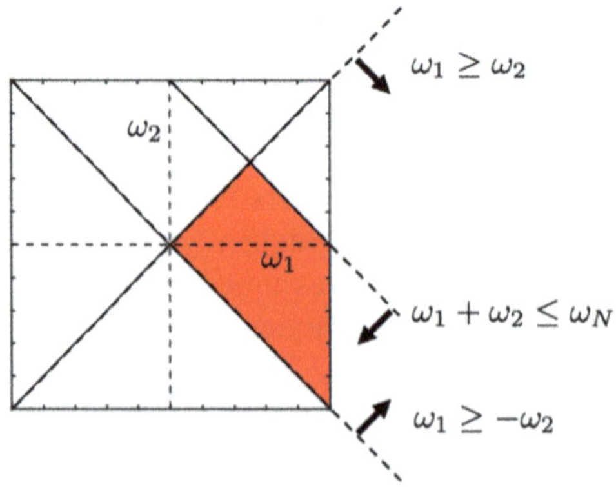

Figure 2.35. Sketch of the area in the (ω_1, ω_2) plane with non-redundant information for the auto- and cross-bicoherence. The outermost square is set by the Nyquist frequency, $|\omega_1| \leqslant \omega_N$ and $|\omega_2| \leqslant \omega_N$. The two diagonal lines through (0, 0) delimit $|\omega_1| \geqslant |\omega_2|$.

window to the data of each realization (such as a Tukey window) before applying the transform. If these precautions are not taken, characteristic horizontal and diagonal lines may appear at low frequencies in the two-dimensional bicoherence diagram that may lead to erroneous interpretations.

It is not necessary to represent the whole (ω_1, ω_2) plane. Due to the interchange symmetry of ω_1 and ω_2 in the calculation of the auto- and cross-bicoherence, the region of the (ω_1, ω_2) plane with independent information is delimited by the inequality $|\omega_1| \geqslant |\omega_2|$. Furthermore, the sampling theorem limits all frequencies to values below the Nyquist frequency: $|\omega_1|$, $|\omega_2|$, $|\omega_1 + \omega_2| \leqslant \omega_N$. These inequalities delimit the characteristic polygonal area in which the bicoherence is usually represented, shown as the red area in figure 2.35. Some authors only display the triangular area $\omega_1 \geqslant \omega_2$ and $\omega_1 + \omega_2 \leqslant \omega_N$ and $\omega_2 \geqslant 0$, but this does not represent all information available.

It is convenient to introduce the 'summed bicoherence',

$$b_{\mathrm{sum}}^2(\omega) = \langle b^2(\omega_1, \omega_2) \rangle_{\omega=\omega_1+\omega_2}, \tag{2.47}$$

which is a mean of the bicoherence such that the sum rule is satisfied, and the 'total bicoherence'

$$b_{\mathrm{tot}}^2 = \langle b^2(\omega_1, \omega_2) \rangle, \tag{2.48}$$

which is simply the mean over all bicoherence values. Naturally, the numerical values of these quantities depend on the chosen calculation grid and are therefore not fundamental. But they serve to compare cases computed under the same numerical conditions since they summarize the information conveniently, as will be seen later.

An important issue is the error in the bicoherence estimate. If the average appearing in equation (2.46) is over N realizations, an upper estimate of the error in b is $1/\sqrt{N}$ [58]. This estimate can be easily verified by computing the bicoherence numerically for independent random signals.

2.6.2 Bicoherence and confinement transitions at TJ-II

In the flexible heliac TJ-II, confinement transitions are observed, which can either be induced (by applying biasing) or spontaneous. The improved confinement state is characterized mainly by a higher (energy and particle) confinement time, which is reflected in, e.g., a higher value of the electron density, and correspondingly higher values of the density gradient near the edge. The confinement transition is associated with the formation of a sheared flow layer in the edge plasma [60]. As pointed out in [61], not much is known about the behaviour of the bicoherence during confinement transitions. Such transitions are characterized by a critical point (a bifurcation). The experimental difficulty resides in the fact that the critical point is unstable and typically, plasmas are either far below or far above the critical condition, and if the transition is made, it occurs very fast. The stellarator TJ-II is in a unique position to investigate this issue due to the specific magnetic configuration of this device: the transitions at TJ-II are 'soft' and the plasma can remain close to the transition for some time and cross the critical point 'slowly', thus allowing a study of the evolution of parameters during the transition.

The plasmas studied here have a toroidal magnetic field of $B_T = 1$ T, a major radius of $R = 1.5$ m, a mean minor radius of $\langle a \rangle = 0.22$ m, and a rotational transform of $\iota(a) \sim 1.6$–1.8. Heating is performed using ECRH, with power $P_{\mathrm{ECRH}} =$ 200–400 kW. The forced confinement transitions are induced using external biasing [60]. The spontaneous confinement transitions are related to slow changes in the mean plasma density [62].

The data are taken with a reciprocating probe system sampling at 2 MHz, described in more detail in [54]. The probe head contains three probe sets, each set consisting of three pins. In each set, two pins measure the floating potential, and one the ion saturation current.

The signals analyzed here are the floating potential V_{f}, the ion saturation current I_{sat}, the poloidal electric field $E_\theta = (V_{\mathrm{f}}^2(p_2) - V_{\mathrm{f}}^1(p_2))/\Delta_\theta$ (where V_{f}^1 and V_{f}^2 are measured at two probe tips of probe set p_2, separated by a distance of approximately $\Delta_\theta = 3$ mm in the poloidal direction), and the fluctuating particle flux $\tilde{\Gamma} \propto \tilde{I}_{\mathrm{sat}}\tilde{E}_\theta$ (where the tilde means the fluctuating part of the signal). In addition, we compute the radial electric field E_{r} by $E_{\mathrm{r}} = (V_{\mathrm{f}}^2(p_1) - V_{\mathrm{f}}^2(p_2))/\Delta_{\mathrm{r}}$, where Δ_{r} is the radial separation of probe set 1 and probe set 2. We note that the alignment of these two probe sets is only approximately radial, and furthermore p_1 is partially shaded.

To perform the biasing experiments, a biasing electrode was inserted about 2 cm into the plasma and biased with respect to a poloidal limiter tangent to the last closed flux surface (LCFS). Reference [60] provides more information about this type of experiments. Measurements were taken with the triple Langmuir probe described above. On a shot by shot basis, the normalized radial probe position $\rho = r/a$ was varied (r being an effective radius, $r = a\sqrt{\psi}$, where a is the plasma minor radius and ψ the normalized poloidal flux). Biasing was applied in the time window $1100 < t < 1150$ ms (for the shots having $\rho < 0.99$) or $1100 < t < 1170$ ms (for the shots having $\rho \geqslant 0.99$). A selection of raw data is shown in figure 2.36.

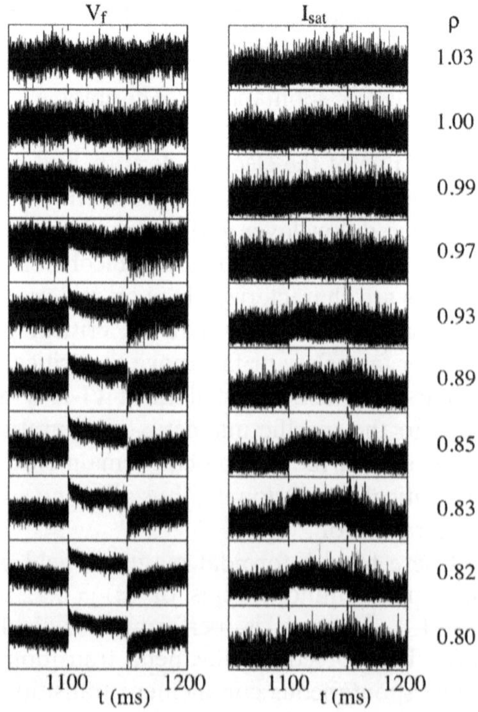

Figure 2.36. Raw signals versus time of the biasing experiments: V_f (left) and I_{sat} (right). Top: outermost position (SOL), bottom: innermost position (shot numbers: 16006 through 16015; radial position, ρ, indicated on the right). (Reproduced with permission from [63]. © IOP Publishing. All rights reserved.)

The auto-bicoherence was computed for V_f^1, V_f^2, I_{sat}, E_θ, and Γ, with different temporal and frequency resolutions. Similar computations were made for various signal combinations (yielding cross-bicoherence estimates), involving, e.g., estimates of the radial and poloidal electric fields. Here, we present a selection of the most significant results. We note that the time trace of the auto-bicoherence of I_{sat} shows no significant time variation at all, i.e. the auto-bicoherence of I_{sat} is not sensitive to biasing. This contrast with the results reported below reinforces the confidence that the bicoherence is not due to some instrumental effect that would affect all raw probe data equally. The auto-bicoherence traces of the floating potentials $V_f^{1,2}$ did show a significant time variation, but it was observed that the signals were contaminated by a weak low-frequency mode (at 1.07 kHz), related to the regulation system of the power supply of the magnetic field coils, so that these signals were discarded for further analysis. It is noted that E_θ, in particular, was largely free from this contamination due to the cancellation implied by $E_\theta \propto V_f^1 - V_f^2$.

To put the following results into context, In figure 2.37 we show the RMS variation of E_θ versus time and position, normalized to the mean RMS of the same data over the whole time window shown. It is clearly seen how $\mathrm{RMS}(E_\theta)/\langle\mathrm{RMS}(E_\theta)\rangle$ is reduced during the biasing across all radii. The fluctuation level is high after the biasing is switched off, because the density has risen during biasing (and therefore

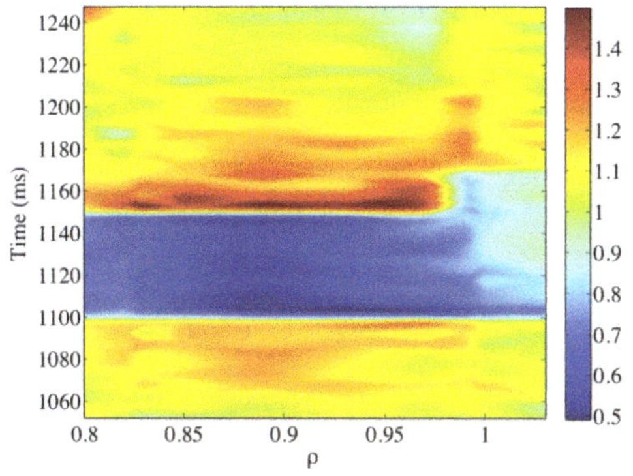

Figure 2.37. $\mathrm{RMS}(E_\theta)/\langle\mathrm{RMS}(E_\theta)\rangle$ versus time and position, clearly showing the suppression of fluctuations during biasing, applied between 1100 and 1150 ms for $\rho < 0.99$ and between 1100 and 1170 ms for $.\rho \geqslant 0.99$ (Reproduced with permission from [63]. © IOP Publishing. All rights reserved.)

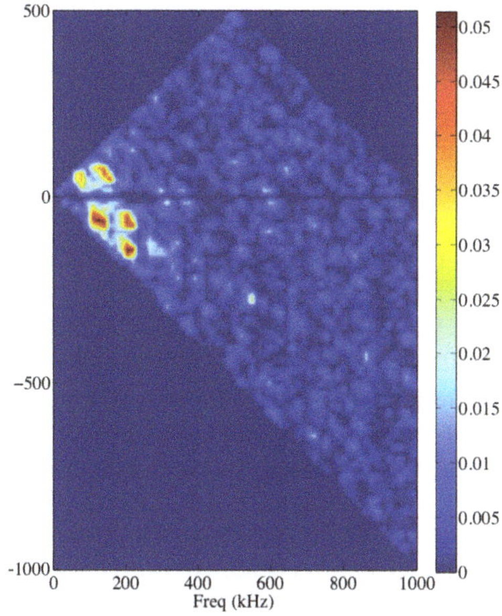

Figure 2.38. Auto-bicoherence plot in a time interval and radial position with high bicoherence values (shot 16009, signal E_θ, $1125 < t < 1150$ ms, $\rho = 0.97$). (Reproduced with permission from [63]. © IOP Publishing. All rights reserved.)

the associated fluctuation level). Afterwards, the fluctuation level decays as the density decays.

A typical result for the bicoherence of E_θ at a radial position and in a time window where the bicoherence is high is shown in figure 2.38. We note that high values of the bicoherence are mainly concentrated at relatively low frequencies. Figure 2.39 shows the summed auto-bicoherence (see equation (2.47)) for the same data selection, along with the summed auto-bicoherence computed in a later time window (after biasing), and the corresponding statistical error level. This plot shows

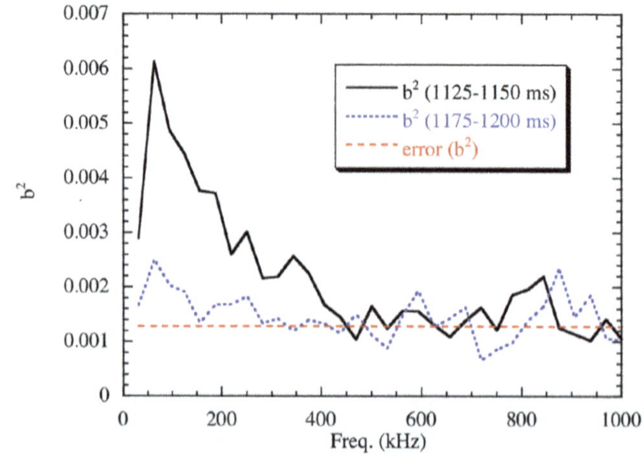

Figure 2.39. Summed bicoherence plot corresponding to the 2D plot in figure 2.38 (shot 16009, signal E_θ, $1125 < t < 1150$ ms, $\rho = 0.97$), i.e. a time window during biasing with high bicoherence, and a later time window ($1175 < t < 1200$ ms), i.e., a time window after biasing with low bicoherence, and the statistical error level (horizontal dashed line). (Reproduced with permission from [63]. © IOP Publishing. All rights reserved.)

that the auto-bicoherence is statistically significant at this radial position during biasing, at least for relatively low frequencies, while it is not so after biasing is switched off.

The high values of the two-dimensional bicoherence in a period of relatively high total bicoherence are always concentrated in the low-frequency range with $|f_1|, |f_2|, |f_1 + f_2| < 250$ kHz. This justifies the calculation of the total bicoherence using the expression of equation (2.48) while summing only over the frequencies below this limit. Performing this analysis for the shots shown in figure 2.36, we evaluate the total low-frequency bicoherence versus time and space. The result is shown in figure 2.40. The response to the biasing, applied between 1100 and 1150 ms for $\rho < 0.99$ (or between 1100 and 1170 ms for $\rho \geqslant 0.99$), is clearly visible in the radial range $0.9 < \rho < 1$. The fact that the bicoherence is high across several radial positions and independent analysis time windows heightens our confidence in the significance of these results. Note that this is also the radial range where the biasing potential produces a gradient in the floating potential, as can be seen directly from the floating potential measurements (figure 2.36). Recall that the biasing electrode was inserted about 2 cm into the plasma, i.e. the biasing voltage is applied at $\rho \simeq 0.9$. High values of the bicoherence occurring in the SOL region ($\rho > 1$) are presumably due to the very different nature of transport here, as compared to the confinement region ($\rho < 1$); see [13]. This plot clearly suggests that the applied external biasing induces three-wave coupling, detected by the bicoherence, in a narrow radial range (presumably, the flow shear layer). Comparing this plot with figure 2.37, it is evident that the detected bicoherence is not just caused by a reduction of turbulence, for the turbulence is reduced over a broad radial range and the bicoherence is only increased in a narrow range. Conversely, it is evident that the turbulence is suppressed in a far wider radial range than the range where the bicoherence is high.

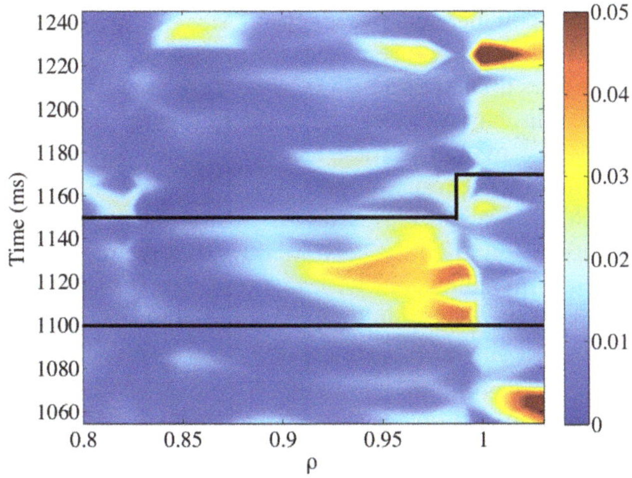

Figure 2.40. Low-frequency ($f < 250$ kHz) total auto-bicoherence plot versus time and radial position computed for the shot series of figure 2.36 (signal E_θ). The response to the biasing, applied between 1100 and 1150 ms (or 1100 and 1170 ms for $\rho \geqslant 0.99$), indicated by the black lines, is clearly visible in the radial range $0.9 < \rho < 1$. (Reproduced with permission from [63].)

Spontaneous confinement transitions are routinely observed at TJ-II [60, 64]. As the line average density increases, a velocity shear layer is generated at the edge of the plasma, leading to turbulence suppression and confinement improvement. The transition to improved confinement occurs near a specific value of the line average density (gradient). The transition point itself is characterized by, among others, relatively large values of local fluctuation levels [60, 65]. The transition point is also a point where (the radial derivative of) the radial electric field, E_r, and the poloidal flow velocity, v_θ, are near zero. For line average density values above this transition point, turbulence levels are lower, and (energy and particle) confinement is improved with respect to values below the transition point. Actually, as we will see, the transition is not as sharp as the word 'point' suggests.

The critical value of the line average density, marking the transition, depends in a specific way on certain plasma parameters [65–67]. It may be assumed that the 'true' critical parameter underlying the transition is some local plasma parameter for which the line average density (easily accessible experimentally) acts as a proxy. This local parameter might be, e.g., the (normalized) density or pressure gradient at a specific radial position in geometric or magnetic co-ordinates (e.g. associated with a specific value of the rotational transform), or the (particle) flux [62].

We have analysed a set of discharges in which the plasma density evolves slowly, triggering a confinement transition as a critical value of the line integrated density is exceeded, and the corresponding back-transition as the line integrated density decays [60, 62]. These discharges are 12378 (TJ-II magnetic configuration 100_44_64, $\rho = 0.91$) and the series (configuration 100_40_63) 12890 ($\rho = 0.87$), 12894 ($\rho = 0.98$), 12895 ($\rho = 0.94$), 12896 ($\rho = 0.92$), and 12897 ($\rho = 1.03$). The differences in magnetic configuration are small and do not affect the results presented here.

During the enhanced confinement phase, the RMS variation of the experimental signals (such as I_{sat}) is reduced with respect to the normal confinement phase (see [60]).

Since the density goes up first and then decays, there are two passages through the critical point in these discharges. Here and in the following, the line average density is given in units of 10^{19} m^{-3}. The total bicoherence was again calculated for low frequencies ($f < 250$ kHz). From the previous section we have learned that high auto-bicoherence values are concentrated in a narrow radial range. Indeed, we find that the bicoherence responds only very weakly, except in the case of discharge 12890 ($\rho = 0.87$, virtually coincident with the approximate location of the spontaneous velocity shear layer reported in [60]). This radial position is different from the position of maximum bicoherence in the biasing cases (previous section), which may either be ascribed to the different discharge conditions (e.g. the density is much higher here), or to the fact that the location of the biasing electrode dictates the shear layer position in the biasing experiments. The top panel of figure 2.41 shows the

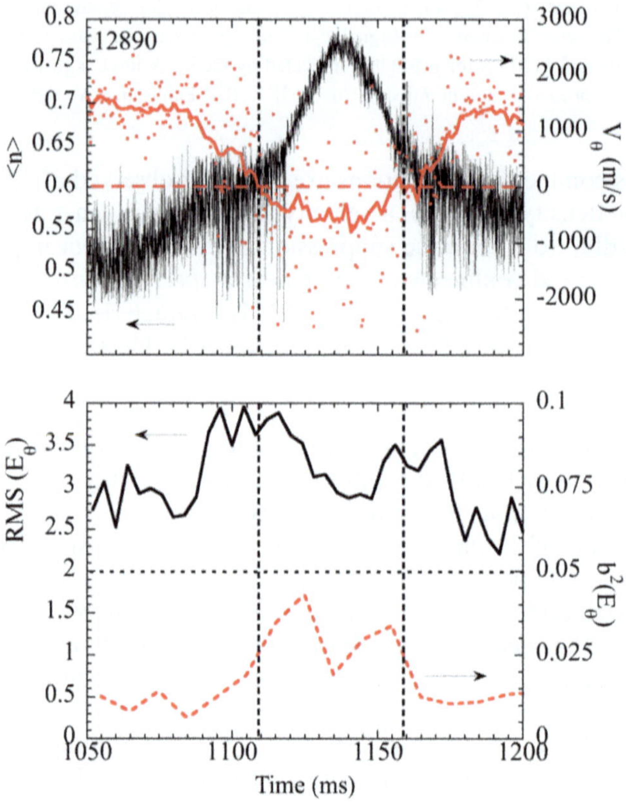

Figure 2.41. Top: line average density and poloidal phase velocity versus time for shot 12890 ($\rho = 0.87$). The phase velocity shows much scatter, so the smoothed phase velocity is shown as a continuous line to guide the eye. Bottom: the RMS variation of E_θ, showing how the fluctuation level increases during the confinement transition, but decreases in the high confinement state (between the vertical dashed lines), and total low-frequency auto-bicoherence versus time for V_f, I_{sat}, and E_θ. The vertical dashed lines indicate the approximate times of the zero crossings of v_θ. (Reproduced with permission from [63]. © IOP Publishing. All rights reserved.)

evolution of the line average density and the poloidal phase velocity (calculated from the floating potential measured at poloidally separated probe pins). The vertical dashed lines mark the zero crossings of the phase velocity. The time interval between these two vertical lines is the improved confinement region, as evidenced by a drop in fluctuation level (visible in the density time trace) and a sharp increase of the line density. The bottom panel shows the RMS fluctuation level of E_θ. The confinement transitions (marked by the vertical dashed lines) are characterized by an increased RMS level (in agreement with the results reported in [60]). In the improved state (between the dashed lines), the RMS fluctuation level drops. In this case, the fluctuation level does not drop to values below the low confinement state, because the plasma makes a back-transition before this can happen. In the time interval corresponding to the improved state, the bicoherence reaches a level that is three to four times above the level outside of this time interval. The rising and falling flanks of the bicoherence traces correspond to the maxima of RMS (E_θ). This is evidence that the bicoherence spikes near the confinement transitions: to be more specific, at the forward transition the spike in bicoherence occurs when the RMS is decaying (and reversely for the back-transition).

The spontaneous transition experiments described above were repeated at a later date (reflected by much higher discharge numbers). As in the preceding experiment, significant bicoherence was only observed for probe locations in a quite narrow radial range close to the edge. Even so, significant bicoherence was detected during the transition, confirming the reproducibility of the results reported above regarding spontaneous transitions [63]. To clarify the structure of the bicoherence graph at times of increased total bicoherence, we show the corresponding two-dimensional auto-bicoherence plots of E_θ in figure 2.42. As in the biasing case, the significant values of the bicoherence are mostly concentrated at relatively low frequencies. However, horizontal and diagonal lines are observed in the graph, indicating the coupling of a broad range of frequencies to a low frequency. As has been signalled

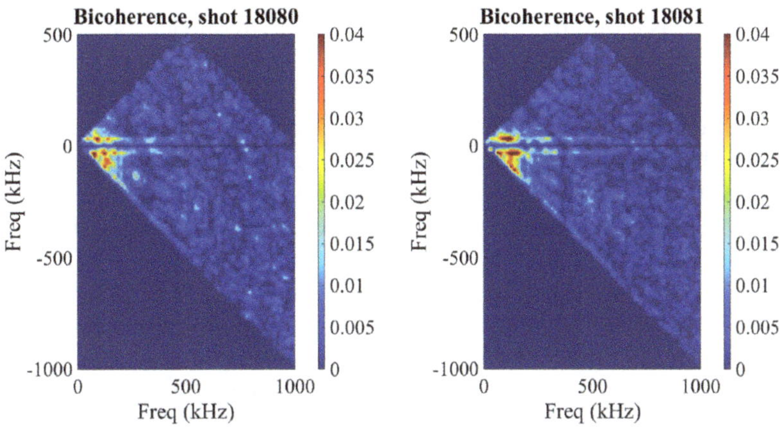

Figure 2.42. Two-dimensional auto-bicoherence of E_θ for shots 18080 ($\rho = 0.92$, time interval $1120 < t < 1150$ ms) and 18081 ($\rho = 0.89$, time interval $1110 < t < 1150$ ms). (Reproduced with permission from [63].

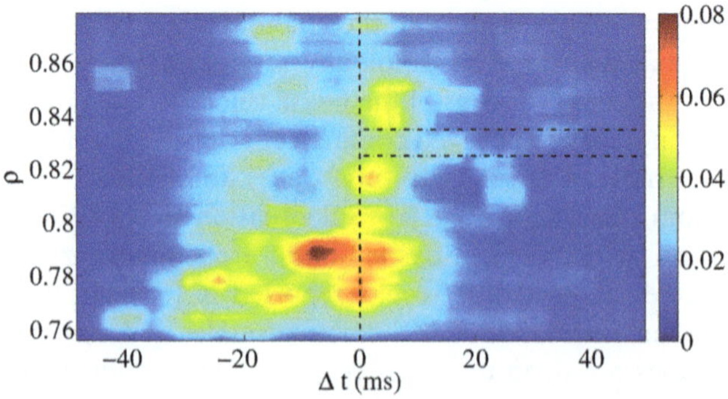

Figure 2.43. Mean auto-bicoherence versus time and normalized radius, calculated from Doppler reflectometry across an L–H transition. The vertical dashed line indicates the L–H transition; the horizontal dashed lines indicate the approximate position of the Er shear layer in the H phase. (Reproduced with permission from [70].)

by other authors, this is expected to occur when a zonal or shear flow is present and coupled to turbulence [61, 68, 69].

Finally, the bicoherence was also studied during L–H transitions at TJ-II, using Doppler reflectometry signals [70]. For details please refer to the cited article. Here, we simply note that the bicoherence again displays a short-lived burst at the time of the transition, at a specific radial location—very similar to the observations reported above (figure 2.43).

2.6.3 Wavelet bicoherence

In order to allow the study of phenomena that due to their transient nature may last only for a short time (as may be the case with ZFs associated with confinement transitions), it is important to dispose of appropriate analysis tools that can follow these variations. In this context, wavelet analysis offers an interesting alternative to Fourier analysis, as it optimizes the trade-off between frequency and time resolution. However, orthogonal wavelets are not suited for the calculation of the bicoherence, as the frequencies (inverse scales) of orthogonal wavelet coefficients form a set $\omega \in \{c \cdot 2^n\}$, such that it is not possible to fulfil the frequency sum rule $\omega_1 + \omega_2 = \omega$ in general. Thus, estimating the bicoherence using wavelets requires using the continuous wavelet transform, which allows obtaining coefficients for arbitrary values of ω, although unfortunately it is slower to evaluate than the orthogonal wavelet transform [71].

The (continuous) wavelet transform is based on a choice of analyzing wavelet. Many different wavelet types are available; however, to ease interpretation and remain close to the Fourier transform, we prefer the 'Morlet' wavelet which is simply a sine/cosine of frequency $1/a$, windowed by a Gaussian with width na:

$$\Psi_a(t) = \frac{1}{(na\sqrt{\pi})^{1/2}} \exp\left(2\pi i\frac{t}{a} - \frac{1}{2}\left(\frac{t}{na}\right)^2\right). \tag{2.49}$$

This wavelet is shown in figure 2.44, along with its Fourier spectrum. The factor $n = 1$ for the standard Morlet wavelet, but can be modified as required. When $n > 1$, the frequency resolution is improved at the expense of time resolution. The spectrum of this wavelet peaks at $f = 1/a$, while the width of the peak is $\Delta f \sim f/4n$. The clear single peak of the spectrum of the wavelet allows the interpretation of the wavelet scale a in terms of frequency, $f = 1/a$. We note that the frequency resolution $\Delta\omega = 2\pi/(4na)$ and the time resolution $\Delta t = 2na$ satisfy $\Delta\omega\Delta t = \pi$, independent of either n or a. One observes an analogy with the Heisenberg uncertainty principle—known as Benedick's theorem.

The wavelet transform of a variable X is defined by

$$W_X(a, \tau) = \int X(t)\Psi_a(t - \tau)dt. \tag{2.50}$$

For $n \gg 1$, this transform is similar to a Fourier transform, which yields maximum frequency resolution with poor time resolution. At the other extreme, for $n \downarrow 0$, the wavelet approaches a δ-function, so that the wavelet transform returns the original signal, providing optimal time resolution but no frequency resolution. In this sense, the wavelet transform provides a useful trade-off between frequency and time resolution. We will generally use $n = 1$, as it is a suitable midway point for most purposes.

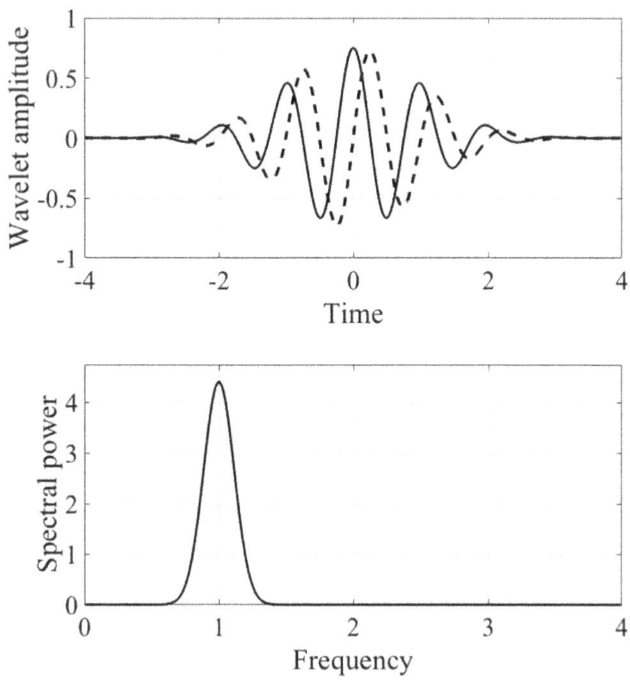

Figure 2.44. Wavelet of the 'Morlet' type, equation (2.49), with scale $a = 1$ (corresponding to frequency $f = 1$) and $n = 1$. Top: real and imaginary waveforms (continuous and dashed lines, respectively). Bottom: Fourier spectrum of the real waveform.

One should be aware that this wavelet family is not orthogonal. This implies a certain redundancy in the wavelet transform coefficients which must be taken into account upon interpreting the results. This disadvantage is compensated by the mentioned conceptual closeness to the Fourier transform. The redundancy is actually a necessity when calculating higher-order spectra, as will become clear later. Even though the wavelets are not orthogonal, the inverse wavelet transform can be calculated (for almost all t) by

$$X(t) = \frac{1}{c_\Psi} \int \int W_X(a, \tau) \Psi_a^*(\tau - t) \frac{da d\tau}{a^2}, \tag{2.51}$$

where c_Ψ is a normalization constant. This completes the analogy with the Fourier transform.

With Fourier analysis, a signal is decomposed in 'everlasting' monochromatic oscillations (sines and cosines), whereas wavelet analysis decomposes the signal in elementary oscillations with a finite lifetime, which becomes shorter as their frequency increases. This seems better suited to physical non-linear and/or turbulent systems with dissipation than Fourier analysis, simply because the elementary oscillations are more similar to the observed oscillations.

Seen from another perspective, one may consider the problem of decomposing the oscillations of a system into modes. When the system is linear and the modes used are eigenmodes of that same system, the decomposition is likely to reveal much information. However, when the modes are not eigenmodes, the information is likely to be poorly represented on the basis of these non-eigenmode functions and each mode coefficient of the decomposition corresponds to a number of eigenmodes—in other words, the information is scrambled, making interpretation difficult. In the non-linear systems with dissipation we are concerned with here, we know that in general no eigenmodes exist that allow a unique decomposition. Nevertheless, locally one can often perform a mode decomposition, since in a local environment of a given point the equations describing the system can be linearized, provided the non-linearity of the system is not too strong. This local decomposition is precisely what the wavelet analysis intends to do, and this explains the success of the wavelet analysis for turbulence analysis.

The wavelet transform coefficients can be treated analogously to the Fourier coefficients in order to obtain useful quantities. For example, the wavelet cross spectrum for two variables X and Y can be defined as

$$P_{X,Y}^{w}(a, t_0) = \int_T W_X^*(a, \tau) W_Y(a, \tau) d\tau, \tag{2.52}$$

where T indicates a time interval of length T centred on t_0: $t_0 - T/2 \leqslant \tau \leqslant t_0 + T/2$. The integration over the time interval T has two advantages: it reduces the statistical uncertainty of a single wavelet coefficient by averaging over several such coefficients, and it allows establishing an error level [72]. We will use that idea to estimate the error of the wavelet bicoherence below.

By setting $Y = X$, one obtains the wavelet auto-power spectrum:

$$P_X^w(a, t_0) = P_{X,X}^w(a, t_0). \tag{2.53}$$

Note that the wavelet power spectrum can also be written in terms of the Fourier power spectra of the wavelet and $f(t)$ when $T \to \infty$:

$$P_X^w(a) = \frac{1}{2\pi} \int P_{\Psi_a}(\omega) P_X(\omega) d\omega. \tag{2.54}$$

Thus, in this limit the wavelet power spectrum equals the Fourier power spectrum averaged by the power spectrum of the wavelet filter [73].

By analogy to the Fourier cross bispectrum, equation (2.45), one can define the wavelet cross bispectrum in a similar way as the wavelet cross spectrum, equation (2.54):

$$B_{X,Y}^w(a_1, a_2, t_0) = \int_T W_X^*(a, \tau) W_Y(a_1, \tau) W_Y(a_2, \tau) d\tau, \tag{2.55}$$

where a, a_1, a_2 again satisfy the sum rule, here expressed as

$$\frac{1}{a} = \frac{1}{a_1} + \frac{1}{a_2}. \tag{2.56}$$

This leads to the following definition of the wavelet cross-bicoherence:

$$(b_{X,Y}^w(a_1, a_2, t_0))^2 = \frac{|B_{X,Y}^w(a_1, a_2, t_0)|^2}{\int_T |W_Y(a_1, \tau) W_Y(a_2, \tau)|^2 d\tau \int_T |W_X(a, \tau)|^2 d\tau}. \tag{2.57}$$

It goes without saying that the symbolic integrals over T in the above equations are in fact summations over a range of digitally sampled time points, so that the integrals should be understood as summations over N_T points. The wavelet coefficients at a given frequency f (or scale $a = 1/f$) at these N_T points are not statistically independent, as the wavelet type we are using is not orthogonal. Each wavelet coefficient is calculated by evaluating equation (2.50), integrating (summing) over the whole available data range—not just the mentioned N_T points of the interval T. Because of the periodicity a of the wavelets, two statistically independent estimates of the wavelet coefficients are separated by a time $a/2$ or a number of points $M(a) = af_{\text{samp}}/2$, where f_{samp} is the sampling frequency.

This has consequences for the calculation of the 'statistical error level' mentioned in section 2.6.1—i.e. the level of bicoherence obtained in the absence of true bicoherence due to the fact that the number of independent estimates is finite. It is easily estimated numerically by calculating the bicoherence for uncorrelated random noise, but as before, an analytical estimate can be made on the basis of the law of large numbers, according to which the value of any bicoherence coefficient for uncorrelated noise must converge to 0 as $1/\sqrt{N}$, where N is the number of independent estimates.

With wavelets, the summations over T appearing in the calculation of the bicoherence are not carried out over N independent wavelet coefficient estimates (as in the Fourier case), but only over $N / \max_a(M(a))$, where the maximum is taken over the various values of a that come into play for the evaluation of a specific value of the squared bicoherence. This consideration leads to the following estimate of the upper bound of the statistical noise level in the bicoherence b^w, expressed in terms of frequency f instead of scale a:

$$\epsilon(b^w) \simeq \left[\frac{1}{N} \frac{f_{\text{samp}}/2}{\min(|f_1|, |f_2|, |f_1 + f_2|)} \right]^{1/2}. \quad (2.58)$$

It is an upper bound because the statement that two independent wavelet coefficient estimates at the same scale a are separated by a time $a/2$ is a maximum estimate. Observe that at low frequencies, this statistical noise level will dominate the bicoherence, so that significant interpretations based on the wavelet bicoherence must limit themselves to (relatively) high frequencies. At higher frequencies, though, the noise level drops rapidly, and in this sense the wavelet bicoherence is a significant enhancement over the Fourier bicoherence. In the following, we show a practical example of this feature.

In order to compare the wavelet and Fourier-based bicoherence, we analyzed measurements taken at the Advanced Toroidal Facility (ATF) [74] with Langmuir probes and with known strong Fourier bicoherence levels (see [75] for a more detailed description of these data). The data record consists of 16 384 data points sampled at 1 MHz. The first data point is taken to occur at 0 ms. Figure 2.45 shows the wavelet and the smoothed Fourier spectrum (calculated for a data section running from 1 to 9.192 ms) showing the base-10 logarithm of the spectral power. The spectrum is clearly turbulent with no dominant modes. The shape of the Fourier and wavelet spectra is similar, as expected from equation (2.54). Figure 2.46 shows both the Fourier and wavelet bicoherence graphs, calculated on a frequency grid with 64 grid points from 0 to 500 kHz from a data section running from 1 to 16 ms. Although the wavelet graph shows much less detail, as is expected from the lower frequency resolution, the overall features are similar, and the values of bicoherence

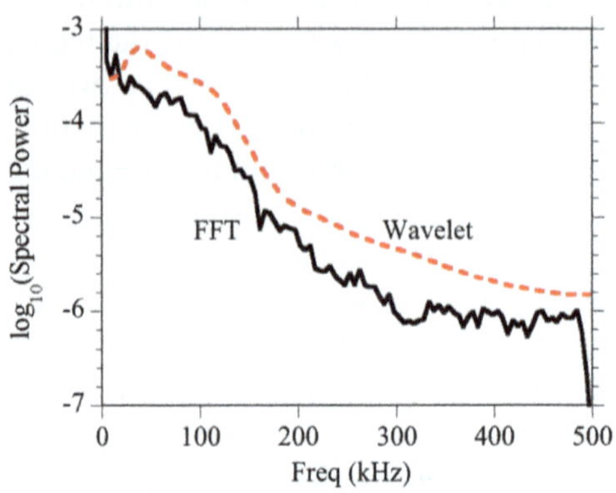

Figure 2.45. Spectral analysis of data (from 1 to 9.192 ms) taken with Langmuir probes at ATF; shown is the base-10 logarithm of the spectral power for both the smoothed Fourier spectrum and the wavelet spectrum (with arbitrary offset). (Reproduced after [76], with the permission of AIP Publishing.)

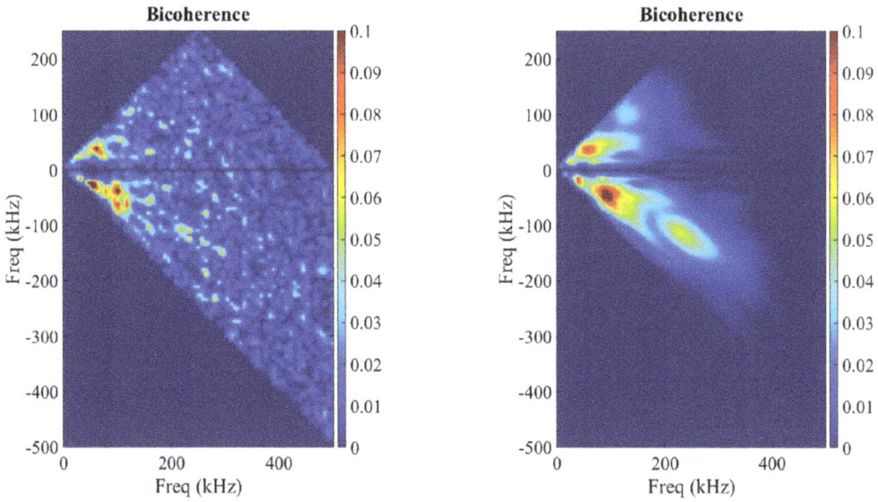

Figure 2.46. Fourier and wavelet bicoherence graphs, calculated on a frequency grid with 64 grid points from 0 to 500 kHz on a data section from 1 to 16 ms. (Reproduced after [76], with the permission of AIP Publishing.)

agree quite well. Figure 2.47 (top) shows the summed bicoherence for both the Fourier and wavelet bicoherence. Due to better statistics, the noise level in the wavelet graph is much lower than in the Fourier graph, see equation (2.59). Here, one observes a major advantage of the wavelet approach: results appear more robust and less noisy, at least at relatively high frequencies, when the above approach is followed (the mentioned integration over short time intervals T). The price to be paid for this benefit is less frequency resolution.

Comparing the two bicoherence graphs in the range $f < 150$ kHz, where the Fourier bicoherence is still above noise level, one sees that the values agree quite well with the wavelet bicoherence. This shows that the bicoherence result is roughly independent from the choice of analyzing function. As noted, the wavelet graph shows less detail due to its lower frequency resolution.

To show that the bicoherence in both cases is detecting something that is not evident from the Fourier spectrum, we performed a fast Fourier transform (FFT) on the raw data, scrambled the phase information (i.e. maintaining the FFT transform coefficient amplitudes, we gave each coefficient a random phase), and performed an inverse FFT to obtain a new data series. It is expected that this new series exhibits no coherence of any kind, even though its power spectrum is identical to the original series. Effectively, as is shown in figure 2.47 (bottom), both the Fourier and wavelet bicoherence drop to their corresponding noise levels.

2.6.4 Bicoherence and L–H transitions

We have analyzed data taken with a broadband X-mode heterodyne reflectometer [77] in the edge region of the Wendelstein 7-AS stellarator [78]. This is a low-shear modular stellarator with major radius $R_0 = 2$ m and minor radius $a \simeq 0.18$ m.

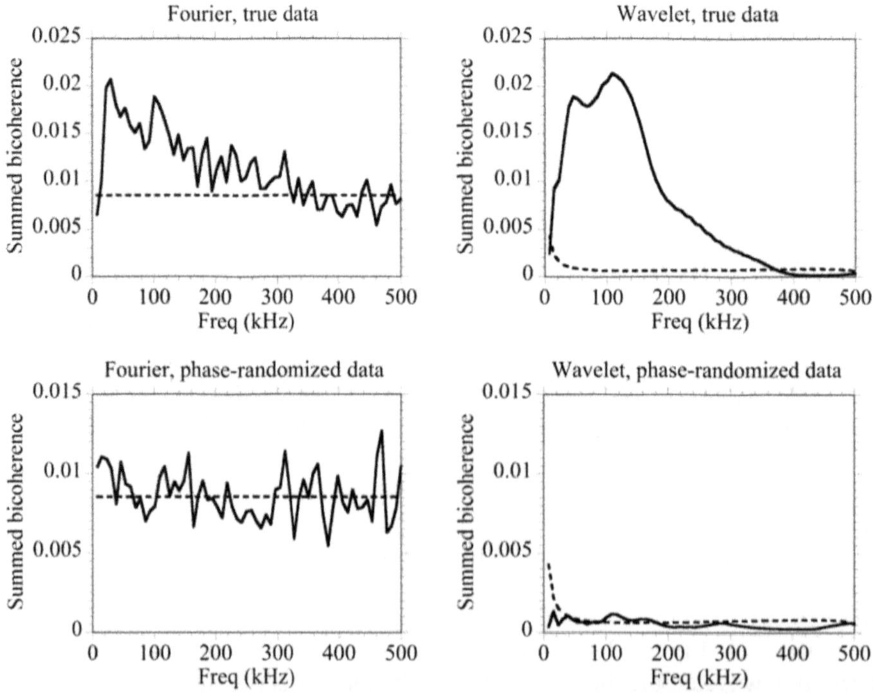

Figure 2.47. (Top) The summed bicoherence for both the Fourier and wavelet bicoherence. (Bottom) The summed bicoherence for both the Fourier and wavelet bicoherence after scrambling the phase information in the data (see text). As expected, the bicoherence drops to noise level. (Reproduced after [76], with the permission of AIP Publishing.)

The toroidal magnetic field is $B = 2.5$ T. In the discharge investigated, the plasma was heated with $P_{ECH} = 400$ kW of ECRH power at 140 GHz. The central electron density was $n_e = 8 \times 10^{19}$ m^{-3}, while the central electron temperature was $T_e = 0.8$ keV. The rotational transform at the plasma edge was $\iota\!\!\!- = 0.53$. We studied the L–H transition that is generally observed in this kind of discharge [79], i.e. a transition from a low (L) to high (H) energy confinement mode, which is always accompanied by a sudden reduction in the measured fluctuation levels of, e.g., the electron density. Here we present the data from one discharge only which we deem sufficient for purposes of illustration of the data analysis technique.

Data are taken at various radial positions, but in the following we focus on the radial position of about 15.5 cm, which is where the strongest reduction in the fluctuation level is observed during the L–H transition. This position is about 1 cm inside the separatrix. Figure 2.48 shows the time development of the total bicoherence and the RMS level of the reflectometry data, as well as the signal from the H_α-light emission detector. A sharp drop in the signal from the H_α detector is generally considered to be indicative of an L–H transition. The transition is seen to occur at 757.5 ms. The spikes in the H_α signal in the H phase are caused by edge localized modes (ELMs), whose nature is only partly understood. Observe that the L–H transition is clearly seen both on the total bicoherence and RMS signals.

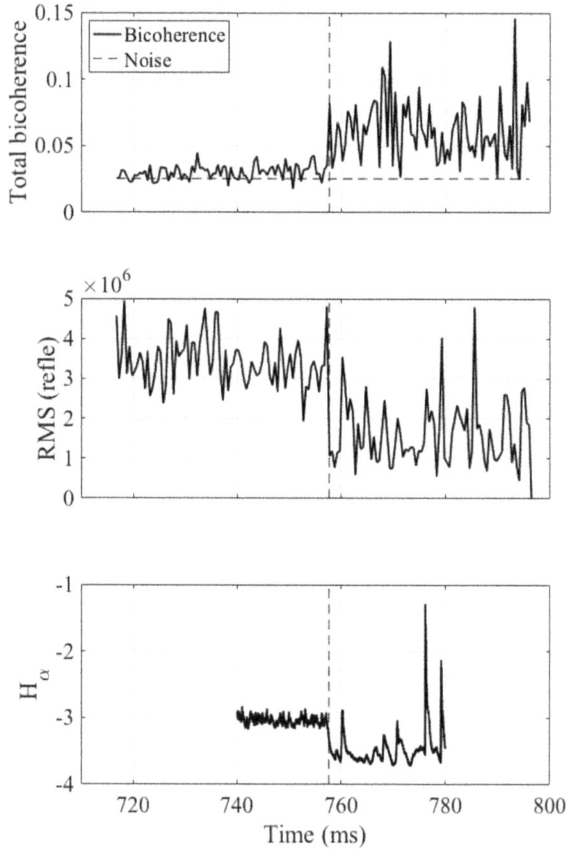

Figure 2.48. Plots of the total bicoherence and RMS of the raw reflectometry signal taken at Wendelstein 7-AS, and the fast H_{α} signal. The L–H transition is marked by a vertical dashed line and shows up clearly in all graphs. (Reproduced after [76], with the permission of AIP Publishing.)

Note that before the transition (in the L phase), the RMS value is high and the bicoherence is at noise level (dashed line). During the H phase the bicoherence is clearly augmented (by a factor of 2–3), whereas the RMS level has dropped by a similar factor. Figure 2.49 shows that the level of bicoherence increases at all radial positions in the plasma bulk side of the last closed flux surface (LCFS) position [80]. The high level of bicoherence is due to the enhanced non-linear interaction in a wide range of frequencies (10–500 kHz) [76].

The apparent lack of correlation in figure 2.48 between the RMS level or the bicoherence and the occurrence of ELMs is entirely due to the low time resolution of these calculations; in reality, a clear correlation between the RMS level and the occurrence of ELMs exists. In the case of the bicoherence, the problem is more complicated, since the data corresponding to one ELM do not allow a calculation of the bicoherence with sufficient statistics to obtain reliable results; nevertheless, we investigated this question in a little more detail, the results of which are presented below.

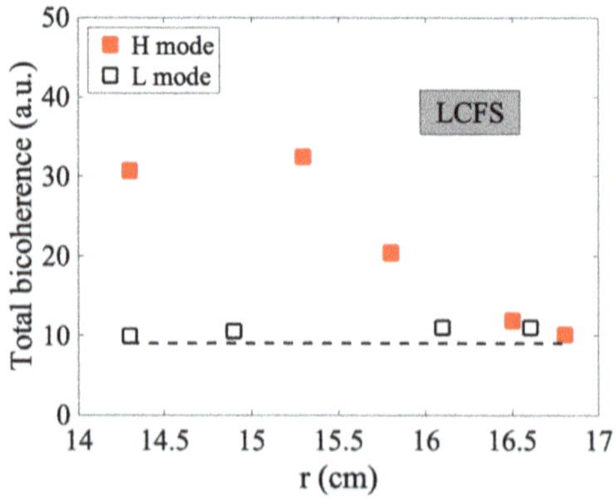

Figure 2.49. Comparison of total bicoherence in L and H modes in the W7-AS stellarator. (Reproduced after [80], with the permission of AIP Publishing.)

To gain some insight into the characteristics of the L and H phase and the ELMs, we calculated the bicoherence and the wavelet spectra over time sections with a length of 1 ms and summed the results. For the L phase, five such sections were selected arbitrarily but before the transition at 757.5 ms. For the ELMs, these sections were chosen centred on the spikes in the RMS level, corresponding to the five major spikes in the H_α, signal (figure 2.48); finally, for the H phase, five sections were selected in between the ELMs. The bicoherence and summed wavelet bicoherence resulting from these calculations is presented in figure 2.50. Mean wavelet spectra are presented in figure 2.51, showing that the wavelet power spectrum drops almost by an order of magnitude when going from L to H, in accordance with the drop in the RMS level seen earlier. The spectrum corresponding to the ELMs is in between, and a bit closer to the L phase than the H phase level. From the summed bicoherence graphs in figure 2.50 it is clear that the bicoherence is close to noise level in the L phase, as expected from the earlier observation presented in figure 2.48. The H mode exhibits a high bicoherence at nearly all frequencies, with a peak at 50 kHz. There is a weak indication that this frequency couples to higher frequencies, but the graph is by no means conclusive. Finally, the ELMs exhibit a high bicoherence over a broad range of frequencies as well, and several broad peaks can be discerned (at 80 and 220 kHz), but the two-dimensional graph shows that the bicoherence here is quite different from the H phase case, and it exhibits a complex structure.

Summarizing, although we cannot make any statement about the nature of the ELMs, we believe that we can maintain that an ELM is not simply a short return to L mode conditions (which has been a working supposition for some time), but a rather more complex phenomenon involving non-linear modes or phase coupling effects. This statement is based on the wavelet bicoherence analysis which shows that the phase coupling processes in the turbulence during ELMs are more similar to the situation in the H phase than the L phase, even though the RMS levels are more similar to the L phase than the H phase [76].

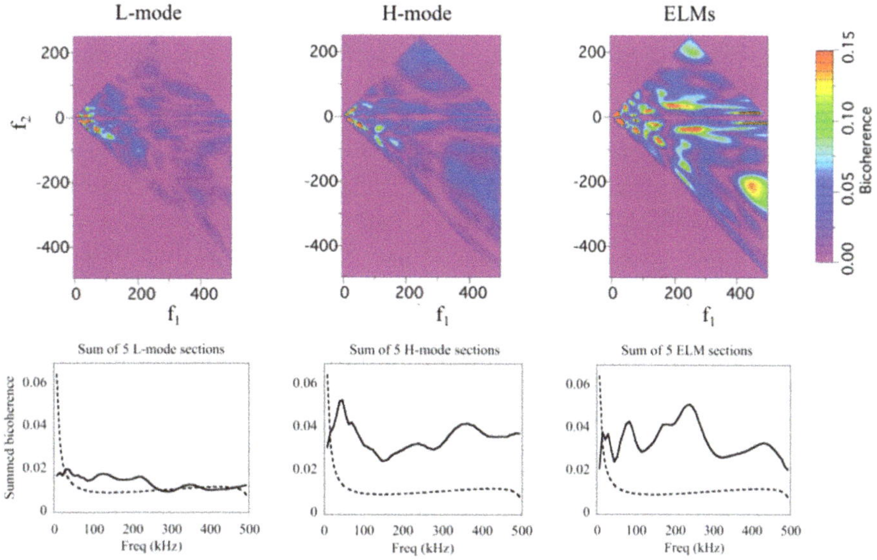

Figure 2.50. Bicoherence (top) and summed bicoherence (bottom) graphs obtained by averaging the bicoherence data from several time windows of 1 ms each: five L mode sections (742–743 ms; 744–745 ms; 746–747 ms; 748–749 ms; 750–751 ms); five H mode sections 058–759 ms; 761–762 ms; 762.2–763.2 ms; 763.4–764.4 ms; 779.6–780.6 ms), and five sections containing ELMs (760–761 ms; 767.7–768.7 ms; 770.2–771.2 ms; 775.9–776.9 ms; 778.6–779.6 ms). (Reproduced after [76], with the permission of AIP Publishing.)

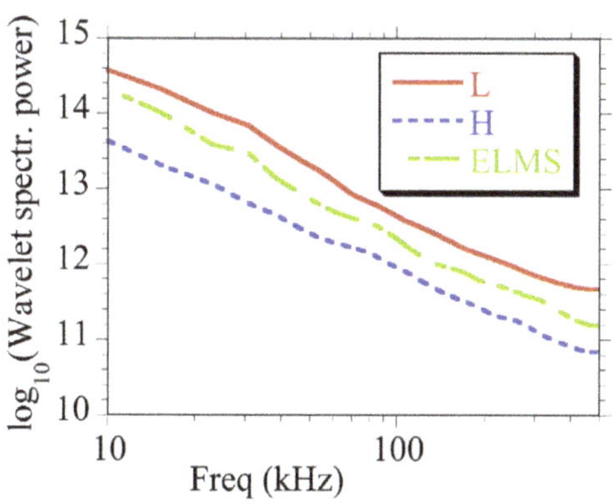

Figure 2.51. Wavelet spectra, averaged over the time windows as given in the caption of figure 2.50. (Reproduced after [76], with the permission of AIP Publishing.)

Next, we look at some data from the continuous current tokamak (CCT), with major radius $R = 1.5$ m, and minor radius $a = 0.35$ m. It was operated with toroidal magnetic field $B_T = 0.25$ T, plasma current $I_p \simeq 40$ kA, electron density $n_e = 2 \times 10^{18}$ m^{-3}, $V_{\text{loop}} = 1.2 - 1.4$ V, central electron temperature $T_e(0) > 150$ eV, and central ion temperature $T_i(0) > 100$ eV. H modes were induced by biasing a small electrode located about 0.1 m inside the limiter radius with respect to the vessel wall [81].

For studies of edge turbulence and transport, a poloidal Langmuir probe array was used [82]. It was configured such that one of the probe tips was recording the floating potential locally. The sampling rate was 2.5 MHz. Here we focus on data from a Langmuir probe located at the outside midplane. Prior to the initiation of H mode, the probe was located just outside the LCFS. During H modes, the increasing plasma pressure causes a slow movement of the plasma column out towards the low-field side. Thus, when the H mode electrode bias is suddenly turned on, the outside midplane probe records a slow increase in the negative dc floating potential. This enabled us to reconstruct the radial profile in the H mode of the quantities measured by the probe using an estimate of the instantaneous probe position relative to the LCFS. Knowing the value of the radial electric field E_r, the radial position r of the probe can be estimated as $r = \Phi_f / E_r$, where Φ_f is the floating potential measured by the probe. From Doppler shift measurements, we estimated the electric field to be $E_r \sim 100$ V/cm. Uncertainties in this estimate translate into an uncertainty of 30%–50% in the absolute reconstructed position, although the relative position is much more accurate since E_r does not vary significantly during the measurements. Figure 2.52 shows the RMS values and the bicoherence of the measured ion saturation current (I_{sat}) for the outside midplane probe. The H mode period (to the right of the vertical dashed line) shows a slight reduction of the RMS and a gradual increase of the bicoherence as the plasma moves outward. A broad range of frequencies is involved in the production of the high bicoherence around $t = 80$ ms, with predominance of frequencies around 250 and 500 kHz.

The L/H transition at W7-AS (discussed above) showed an abrupt increase of the bicoherence and decrease of the RMS at the transition. The difference with the present apparently smooth transition may be explained by the fact that here the probe location is initially at the LCFS and moves gradually inward, whereas in the measurements reported earlier the measurements were taken well inside the LCFS. This would imply that the gradual change observed is due rather to the existence of a bicoherence profile than to a slow temporal change.

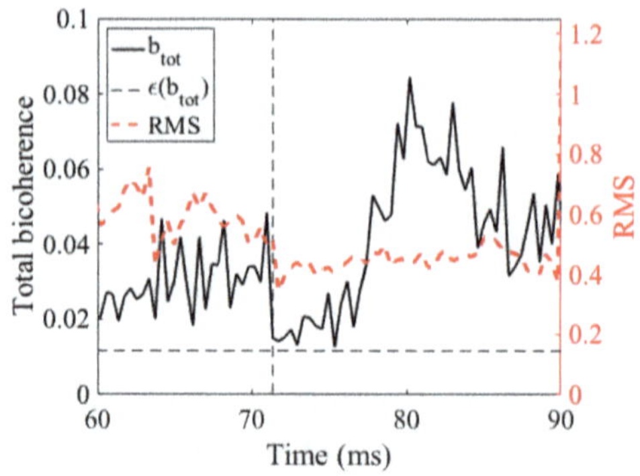

Figure 2.52. Analysis of the saturation current I_{sat} measured by a Langmuir probe positioned on the outside midplane at CCT. Red dashed line: RMS(I_{sat}). Continuous black line: total bicoherence of I_{sat}. Short dashes: Noise level of the bicoherence. The H mode period induced by probe biasing occurs to the right of the vertical dashed line. (Reproduced after [84], with the permission of AIP Publishing.)

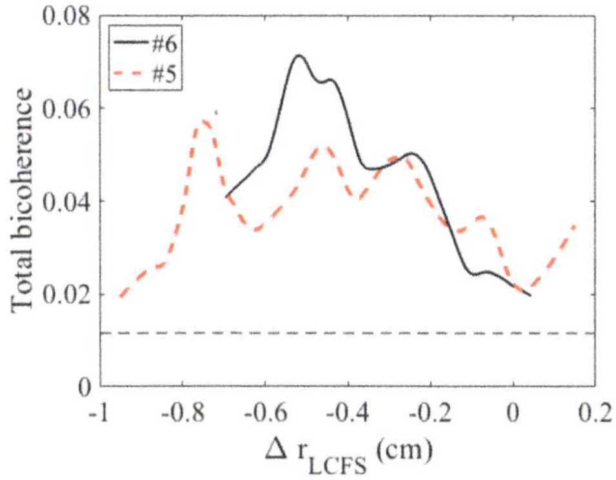

Figure 2.53. Bicoherence profile in the H mode deduced from the outside midplane probe signal shown in figure 2.52 and an estimate of the probe position relative to the LCFS (continuous black line). Also included is a similar profile for another, similar discharge (red dashed line). The statistical noise level of the bicoherence is indicated by the straight dashed line. (Reproduced after [84], with the permission of AIP Publishing.)

Using the above-mentioned estimate of the probe position, the profile of bicoherence was reconstructed during the H mode phase from the signal of the outside midplane probe. The result is shown in figure 2.53, along with a similar profile obtained from a different but similar discharge. We recall that a high value of the bicoherence may either indicate the presence of non-linear interactions or of (quasistatic) structure [83]. It is interesting that these features should occur a small distance inside and not at the LCFS, which is possibly related to a (decorrelating) velocity shear layer at the LCFS. The results also seem to indicate a maximum of bicoherence about half a centimetre inside the LCFS—likely due to the interaction between turbulence and sheared flows.

2.6.5 Non-linear energy transfer

The bicoherence can be taken a step further by assuming that the waves in the plasma obey a quadratic evolution equation [85, 86]:

$$\frac{\partial \phi(k, t)}{\partial t} = (\gamma_k + i\bar{\omega}_k)\phi(k, t) + \frac{1}{2}\sum_{k_1,k_2} \Lambda_k^Q(k_1, k_2)\phi(k_1, t)\phi(k_2, t), \tag{2.59}$$

where $k = k_1 + k_2$. In other words, the rate of change of the spectral component at k is considered to be due to a linear term plus a quadratic wave-interaction term. We note that this is a very strong assumption that will only hold in very specific situations: interactions between a dominant structure such as a zonal flow and broadband turbulence, or—possibly—weakly developed turbulence in which higher order terms can be ignored. Based on this assumption, one can proceed to estimate the coupling coefficient Λ_k^Q from the bispectrum, and the resulting energy transfer rate between coupled waves. The main practical issue complicating the application of this technique is that very long, stationary records are required to obtain reliable estimates of the various quantities needed, which is no easy task in fusion plasmas.

Consequently, only few applications of this technique to experimental data have been reported [87].

2.7 Summary

In this chapter, we have started characterizing plasma turbulence by using standard tools like probability distribution functions, correlations, and spectra. These standard techniques revealed that plasma turbulence has some interesting properties that are generic to many devices. Plasma turbulence contains large scale structures, of which the most obvious are filaments and zonal flows that extend poloidally and toroidally over distances as large as the machine size. The mean properties of these structures can be determined using standard techniques (cross correlation and conditional averages). Some other techniques were also found to be useful, among which the biorthogonal decomposition for structure detection based on multipoint measurements and the bicoherence for the detection of wave–wave interactions. Both techniques were applied to the detection of zonal flows. The detected zonal flows were found to peak in a narrow radial layer in the plasma edge, where they generate long range correlations that can be distinguished from the correlations caused by filaments, involving quadratic wave interactions that intensify during confinement transitions (whether spontaneous or induced by external biasing).

The characteristic peaked shape of the central part of the PDF of the particle flux could be understood from the fact that it is a quadratic quantity—a product of two fluctuating variables with given correlation. Interestingly, PDFs of the particle flux were seen to have 'long tails' (decaying less than exponentially fast, far from the centre of the PDF), which suggests that events propagating radially are 'bursty' (see chapters 3 and 5), i.e. occasionally these radial events can be very large and possibly have a significant radial extent. The existence of a 'long' (algebraic) tail also implies that large events contribute significantly to overall transport (see chapter 3).

The PDF of the particle flux was found to exhibit self-similar behaviour over a range of timescales (m), see section 2.1.2. Likewise, turbulent spectra were found to have similar shapes in many devices, while the spectrum could roughly be divided into three regions with different decay indices. It was shown that measured frequency spectra in the laboratory frame of reference can vary from their shape in the plasma frame of reference due to the Doppler effect, so that the comparison of turbulent frequency spectra should always be done carefully and ideally in radial regions with low plasma rotation velocity. Taking account of this Doppler effect, the decay index of the central spectral region in the plasma rest frame was found to be close to 1, which is a characteristic property of self-organized systems [88], while it was shown that this particular spectral region is dominant for radial particle transport.

The study of long range spatial correlations in the radial direction is complicated, because the radial extent of fusion plasmas is limited physically, particularly in small devices. In addition, the plasma edge and core regions are characterized by different physical phenomena, which reduces the size of the suitable region with homogenous behaviour further. Finally, suitable radially spaced measurements, providing simultaneous data with sufficient temporal and spatial resolution are rare due to the difficulties posed by the hostile plasma environment. Recently, however, significant progress has been made in this direction, as we will report in chapter 4.

In chapter 3, we will see that the 'long tails' and the self-similarity of the flux PDF, and the self-similarity of the power spectra are the likely manifestations of an underlying self-organized state of the plasma. The consequences of this situation for our understanding of turbulent transport could be profound: namely, the spatial and temporal connectedness of the plasma due to the existence of long range structures challenges the idea that turbulence is a strictly 'local' phenomenon, and temporal self-similarity implies that a clear separation of scales between rapid turbulent fluctuations and slow transport effects is not possible. Rather, these observations point to the idea that phenomena on all relevant temporal and spatial scales are connected and influence each other: turbulence and transport are not independent phenomena.

References

[1] Garbet X, Idomura Y, Villard L and Watanabe T H 2010 Gyrokinetic simulations of turbulent transport *Nucl. Fusion* **50** 043002

[2] Hutchinson I H 2002 *Principles of Plasma Diagnostics* (New York: Cambridge University Press)

[3] Carreras B A, Llerena Rodríguez I and García L 2013 A topological analysis of plasma flow structures *J. Phys. A: Math. Theor.* **46** 375501

[4] Mancho A M, Wiggins S and Curbelo J 2013 Lagrangian descriptors: a method for revealing phase space structures of general time dependent dynamical systems *Commun. Nonlin. Sci. Numer. Simul.* **18** 3530–57

[5] Béranger B, Duong T, Perkins-Kirkpatrick S E and Sisson S A 2019 Tail density estimation for exploratory data analysis using kernel methods *J. Nonparam. Stat.* **31** 144–74

[6] Sattin F, Agostini M, Cavazzana R, Serianni G, Scarin P and Vianello N 2009 About the parabolic relation existing between the skewness and the kurtosis in time series of experimental data *Phys. Scr.* **79** 045006

[7] van Kampen N G 1981 *Stochastic Processes in Physics and Chemistry* (New York: North-Holland)

[8] Balbín R, Hidalgo C, Pedrosa M A, García-Cortés I and Vega J 1992 Measurement of density and temperature fluctuations using a fast-swept Langmuir probe *Rev. Sci. Instrum.* **63** 4605

[9] Lin H, Bengtson R D and Ritz C P 1989 Temperature fluctuations and transport in a tokamak edge plasma *Phys. Fluids* B **1** 2027–30

[10] Hidalgo C, Balbín R, Pedrosa M A, García-Cortés I and Ochando M A 1992 Experimental evidence of significant temperature fluctuations in the plasma edge region of the TJ-I tokamak *Phys. Rev. Lett.* **69** 1205

[11] Hidalgo C *et al* 2002 Empirical similarity in the probability density function of turbulent transport in the edge plasma region in fusion plasmas *Plasma Phys. Control. Fusion* **44** 1557

[12] Carreras B A, Hidalgo C, Sánchez E, Pedrosa M A, Balbín R, García-Cortés I, van Milligen B, Newman D E and Lynch V E 1996 Fluctuation-induced flux at the plasma edge in toroidal devices *Phys. Plasmas* **3** 2664

[13] Hidalgo C *et al* 2000 Statistical characterization of fluctuation wave forms in the boundary region of fusion and nonfusion plasmas *Phys. Plasmas* **7** 1408
[14] ITER Physics Expert Groups on Confinement and Transport *et al* 1999 Ch 2: Plasma confinement and transport *Nucl. Fusion* **39** 2175
[15] Carreras B A, van Milligen B, Hidalgo C, Balbín R, Sánchez E, García-Cortés I, Pedrosa M A, Bleuel J and Endler M 1999 Self-similarity properties of the probability distribution function of turbulence-induced particle fluxes at the plasma edge *Phys. Rev. Lett.* **83** 3653
[16] Bak P, Tang C and Wiesenfeld K 1987 Self-organized criticality: an explanation of the 1/*f* noise *Phys. Rev. Lett.* **59** 381
[17] Carreras B A, Lynch V E and LaBombard B 2001 Structure and properties of the electrostatic fluctuations in the far scrape-off layer region of Alcator C-Mod *Phys. Plasmas* **8** 3702–7
[18] van Milligen B P, Sánchez R S, Lynch V E, LaBombard B, Pedrosa M A, Hidalgo C, Gonçalves B and Balbín R 2005 Additional evidence for the universality of turbulent fluctuations and fluxes in the scrape-off layer region of fusion plasmas *Phys. Plasmas* **12** 052507
[19] Bramwell S T, Holdsworth P C W and Pinton J F 1998 Universality of rare fluctuations in turbulence and critical phenomena *Nature* **396** 552–4
[20] Bramwell S T *et al* 2000 universal fluctuations in correlated systems *Phys. Rev. Lett.* **84** 3744–7
[21] Hnat B, Dudson B D, Dendy R O, Counsell G F and Kirk A 2008 Characterization of edge turbulence in relation to edge magnetic field configuration in ohmic l-mode plasmas in the mega amp spherical tokamak *Nucl. Fusion* **48** 085009
[22] Graves J P, Horacek J, Pitts R A and Hopcraft K I 2005 Self-similar density turbulence in the TCV tokamak scrape-off layer *Plasma Phys. Control. Fusion* **47** L1–9
[23] Carreras B A and Sánchez R 2008 Statistical characterization of fluctuations: towards understanding turbulence-induced transport in magnetically-confined plasma *Anomalous Fluctuation Phenomena in Complex Systems: Plasmas, Fluids and Financial Markets* (Kerala: Research Signpost) p 1
[24] Melnikov A V *et al* 2013 Changes in plasma potential and turbulent particle flux in the core plasma measured by heavy ion beam probe during L–H transitions in the TJ-II stellarator *Nucl. Fusion* **53** 092002
[25] Carreras B A and Newman D E 1999 Long-range time dependence in the cross-correlation function *Phys. Plasmas* **6** 485
[26] Silva C, Goncalves B, Pedrosa M A, Hidalgo C, McCarthy K, Calderon E, Herranz J, Pastor I and Orozco O 2005 Transport and fluctuations during electrode biasing on TJ-II *Czech. J. Phys.* **55** 1589
[27] van Milligen B P, Hernández Nicolau J, Liu B, Grenfell G, Losada U, Carreras B, García L and Hidalgo C 2018 Filaments in the edge confinement region of TJ-II *Nucl. Fusion* **58** 026030
[28] Pedrosa M A, Silva C, Hidalgo C, Carreras B A, Orozco R O and Carralero D 2008 Evidence of long-distance correlation of fluctuations during edge transitions to improved-confinement regimes in the TJ-II stellarator *Phys. Rev. Lett.* **100** 215003
[29] Hidalgo C *et al* 2009 Multi-scale physics mechanisms and spontaneous edge transport bifurcations in fusion plasmas *Europhys. Lett.* **87** 55002
[30] Sánchez J *et al* 2009 Confinement transitions in TJ-II under Li-coated wall conditions *Nucl. Fusion* **49** 104018

[31] Diamond P H, Itoh S-I, Itoh K and Hahm T S 2005 Zonal flows in plasma—a review *Plasma Phys. Control. Fusion* **47** R35
[32] Walkden N R, Dudson B D and Fishpool G 2013 Characterization of 3d filament dynamics in a MAST SOL flux tube geometry *Plasma Phys. Control. Fusion* **55** 105005
[33] Kocsis G *et al* 2016 Investigation of edge filament dynamics in W7-X limiter plasmas *European Physical Society Conference on Plasma Physics, ECA* **40A** P4.003
[34] van Milligen B P, Lopez Fraguas A, Pedrosa M A, Hidalgo C, Martín de Aguilera A and Ascasíbar E 2013 Investigation of edge filament dynamics in W7-X limiter plasmas *Nucl. Fusion* **53** 093025
[35] Birkenmeier G, Ramisch M, Fuchert G, Köhn A, Nold B and Stroth U 2012 Spatial structure of drift-wave turbulence and transport in a stellarator *Plasma Phys. Control. Fusion* **55** 015003
[36] Teliban I, Block D, Piel A and Greiner F 2007 Improved conditional averaging technique for plasma fluctuation diagnostics *Plasma Phys. Control. Fusion* **49** 485
[37] Grenfell G *et al* 2020 Turbulent filament properties in L and H-mode regime in the RFX-mod operating as a tokamak *Nucl. Fusion* **60** 126006
[38] Ch P, Ritz E J and Powers T L 1988 Advanced plasma fluctuation analysis techniques and their impact on fusion research *Rev. Sci. Instrum.* **59** 1739
[39] Hidalgo C, Pedrosa M A, García L and Ware A 2004 Experimental evidence of coupling between sheared-flow development and an increase in the level of turbulence in the TJ–II stellarator *Phys. Rev.* E **70** 067402
[40] Frisch U 1995 *Turbulence* (Cambridge: Cambridge University Press)
[41] Carreras B A *et al* 1998 Self-similarity of the plasma edge fluctuations *Phys. Plasmas* **5** 3632
[42] Rhodes T L, Moyer R A, Groebner R J, Doyle E J, Lehmer R D, Peebles W A and Rettig C L 1999 Experimental evidence for self-organized criticality in tokamak plasma turbulence *Phys. Lett.* A **253** 181
[43] Pedrosa M A *et al* 1999 Empirical similarity of frequency spectra of the edge-plasma fluctuations in toroidal magnetic-confinement systems *Phys. Rev. Lett.* **82** 3621
[44] Carreras B A *et al* 1999 Characterization of the frequency ranges of the plasma edge fluctuation spectra *Phys. Plasmas* **6** 4615
[45] Happel T, Estrada T, Blanco E, Tribaldos V, Cappa A and Bustos A 2009 Doppler reflectometer system in the stellarator TJ-II *J-II. Rev. Sci. Instrum.* **80** 073502
[46] Estrada T *et al* 2012 Spatial, temporal and spectral structure of the turbulence-flow interaction at the L–H transition *Plasma Phys. Control. Fusion* **54** 124024
[47] Happel T *et al* 2017 Comparison of detailed experimental wavenumber spectra with gyrokinetic simulation aided by two-dimensional full-wave simulations *Plasma Phys. Control. Fusion* **59** 054009
[48] Dudok de Wit T, Pecquet A-L, Vallet J-C and Lima R 1994 The biorthogonal decomposition as a tool for investigating fluctuations in plasmas *Phys. Plasmas* **1** 3288
[49] Kerschen G, Golinval J-C, Vakakis A F and Bergman L A 2005 The method of proper orthogonal decomposition for dynamical characterization and order reduction of mechanical systems: an overview *Nonlinear Dyn.* **41** 147
[50] Higham D J and Nicholas 2016 *MATLAB Guide* vol 150 (Philadelphia, PA: SIAM)
[51] Alonso J A, Hidalgo C and Pedrosa M A 2012 Dynamic transport regulation by zonal flow-like structures in the TJ-II stellarator *Nucl. Fusion* **52** 063010

[52] van Milligen B P, Sánchez E, Alonso A, Pedrosa M A, Hidalgo C, Martin de Aguilera A and Lopez Fraguas A 2015 The use of the biorthogonal decomposition for the identification of zonal flows at TJ-II *Plasma Phys. Control. Fusion* **57** 025005
[53] Hahn S L 1996 *Hilbert Transforms in Signal Processing* (London: Artech House)
[54] Pedrosa M A *et al* 1999 Fast movable remotely controlled Langmuir probe system *Rev. Sci. Instrum.* **70** 415
[55] Pedrosa M A, Hidalgo C, Silva C, Carreras B A, Carralero D and Calvo I 2010 Long-range correlations during plasma transitions in the TJ-II stellarator *Contrib. Plasma Phys.* **50** 507
[56] Hidalgo C *et al* 2006 Physics of sheared flow development in the boundary of fusion plasmas *Plasma Phys. Control. Fusion* **48** S169
[57] Velasco J L, Alonso J A, Calvo I and Arévalo J 2012 Vanishing neoclassical viscosity and physics of the shear layer in stellarators *Phys. Rev. Lett.* **109** 135003
[58] Kim Y C and Powers E J 1979 Digital bispectral analysis and its applications to nonlinear wave interactions *IEEE Trans. Plasma Sci.* **7** 120
[59] Ullmann T, Schmid B, Manz P, van Milligen B, Tovar G E M and Ramisch M 2020 Experimental observation of resonance manifold shrinking under zonal flow shear *Phys. Rev. E* **102** 063201
[60] Pedrosa M A, Hidalgo C, Calderón E, Estrada T, Fernández A, Herranz J and Pastor I 2005 Threshold for sheared flow and turbulence development in the TJ-II stellarator *Plasma Phys. Control. Fusion* **47** 777
[61] Tynan G R, Moyer R A, Burin M J and Holland C 2001 On the nonlinear turbulent dynamics of shear-flow decorrelation and zonal flow generation *Phys. Plasmas* **8** 2691
[62] Carreras B A, García L, Pedrosa M A and Hidalgo C 2006 Critical transition for the edge shear layer formation: comparison of model and experiment *Phys. Plasmas* **13** 122509
[63] van Milligen B P, Kalhoff L, Pedrosa M A and Hidalgo C 2008 Bicoherence during confinement transitions in the TJ-II stellarator *Nucl. Fusion* **48** 115003
[64] Vargas V I, López-Bruna D, Herranz J and Castejón F 2007 Experimental electron heat diffusion in ECH plasmas of the TJ-II stellarator *Nucl. Fusion* **47** 1367
[65] Pedrosa M A, Hidalgo C, Calderón E, Alonso A, Orozco R O and de Pablos J L 2005 Spontaneous edge $E \times B$ sheared flow development studies in the TJ-II stellarator *Czech. J. Phys.* **55** 1579
[66] Pedrosa M A, Hidalgo C, López-Fraguas A, Ochando M A, Pastor I and Calderón E 2004 Interplay between parallel and perpendicular sheared flows and fluctuations in the plasma boundary region of the TJ-II stellarator *Plasma Phys. Control. Fusion* **46** 221
[67] Pedrosa M A, Hidalgo C, Alonso A, Calderón E, Orozco R O and de Pablos J L 2005 Experimental investigation of edge sheared flow development and configuration effects in the TJ-II stellarator *Proc. 15th International Stellarator Workshop* P3–17
[68] Nagashima Y *et al* 2006 Bispectral analysis applied to coherent floating potential fluctuations obtained in the edge plasmas on JFT-2M *Plasma Phys. Control. Fusion* **48** S1
[69] Fujisawa A *et al* 2007 Intermittent characteristics in coupling between turbulence and zonal flows *Plasma Phys. Control. Fusion* **49** 211
[70] van Milligen B P, Estrada T, Hidalgo C, Happel T and Ascasíbar E 2013 Spatiotemporal and wavenumber resolved bicoherence at the low to high confinement transition in the TJ-II stellarator *Nucl. Fusion* **53** 113034
[71] van Milligen B P, Hidalgo C and Sánchez E 1995 Nonlinear phenomena and intermittency in plasma turbulence *Phys. Rev. Lett.* **74** 395

[72] van den Berg J C (ed) 1999 *Wavelets in Physics* (Cambridge: Cambridge University Press)

[73] Hudgins L, Friehe C A and Mayer M E 1993 Wavelet transforms and atmospheric turbulence *Phys. Rev. Lett.* **71** 3279

[74] Lyon J F *et al* 1986 The advanced toroidal facility *Fusion Technol.* **10** 179–226

[75] Hidalgo C, Brañas B, Uckan T, Harris J H, Isler R, Ritz C P and Wootton A 1994 On the role of neutral particles on edge turbulence and electric fields in the advanced toroidal facility *Phys. Plasmas* **1** 3–5

[76] van Milligen B p, Sánchez E, Estrada T, Hidalgo C, Brañas B, Carreras B A and García L 1995 Wavelet bicoherence: a new turbulence analysis tool *Phys. Plasmas* **2** 3017

[77] Hartfuss H J, Estrada T, Hirsch M, Geist T and Sanchez J 1994 Broadband heterodyne reflectometer at the W7-AS stellarator *Rev. Sci. Instrum.* **65** 2284–90

[78] Renner H *et al* 1989 Initial operation of the Wendelstein 7AS Advanced Stellarator *Plasma Phys. Control. Fusion* **31** 1579

[79] Erckmann V *et al* 1993 H mode of the W 7-AS stellarator *Phys. Rev. Lett.* **70** 2086–9

[80] Diamond P H *et al* 2000 In search of the elusive zonal flow using cross-bicoherence analysis *Phys. Rev. Lett.* **84** 4842

[81] Taylor R J, Brown M L, Fried B D, Grote H, Liberati J R, Morales G J, Pribyl P, Darrow D and Ono M 1989 H-mode behavior induced by cross-field currents in a tokamak *Phys. Rev. Lett.* **63** 2365

[82] Tynan G R, Schmitz L, Conn R W, Doerner R and Lehmer R 1992 Steady-state convection and fluctuation-driven particle transport in the H-mode transition *Phys. Rev. Lett.* **68** 3032–5

[83] Pecseli H L and Trulsen J 1993 On the interpretation of experimental methods for investigating nonlinear wave phenomena *Plasma Phys. Control. Fusion* **35** 1701–15

[84] van Milligen B P, Hidalgo C, Sánchez E, Pedrosa M A, Balbín R, García-Cortés I and Tynan G R 1997 Statistically robust linear and nonlinear wavelet analysis applied to plasma edge turbulence *Rev. Sci. Instrum.* **68** 967

[85] Ch P, Ritz E J, Powers and Bengtson R D 1989 Experimental measurement of three-wave coupling and energy cascading *Phys. Fluids* B **1** 153

[86] Kim J S, Durst R D, Fonck R J, Fernandez E, Ware A and Terry P W 1996 Technique for the experimental estimation of nonlinear energy transfer in fully developed turbulence *Phys. Plasmas* **3** 3998

[87] Xu M *et al* 2012 Frequency-resolved nonlinear turbulent energy transfer into zonal flows in strongly heated L-mode plasmas in the HL-2A tokamak *Phys. Rev. Lett.* **108** 245001

[88] Per B 1996 *How Nature Works* (New York: Springer Science)

Chapter 3

Complex features of plasma turbulence

3.1 Introduction

The preceding chapter has presented some evidence that fusion plasmas are self-organized. How does this situation arise and what consequences does it have? A large amount of experimental evidence in both tokamak and stellarator plasmas shows that plasma profiles (of temperature, pressure, etc) tend to remain locally close to some specific thresholds, at least in certain regimes, and that these thresholds correspond to gradient values at which instabilities are triggered ('critical gradients') [1]. Some authors have suggested that this closeness to the trigger point, or marginality, could be the main cause for the observed divergence of radial turbulent transport from the classical diffusive paradigm[1] [2–4].

Indeed, the concept of near-marginality may provide a simple physical explanation for the existence of phenomena such as the superdiffusive propagation of perturbations [5], profile resilience, and the non-diffusive experimental scalings of the global energy confinement observed in many magnetic confinement devices [6, 7] (see chapter 1). Namely, these phenomena would be the result of relaxations of the temperature/pressure profiles, occurring when the local instability thresholds are overcome, which in turn may be triggered by similar relaxations occurring previously at nearby locations. This would lead to a chain of relaxation events, often referred to as 'avalanches' due to the analogy with sand piles [8]. In any confinement regime where such avalanches dominate the radial transport channel, one might expect perturbations to propagate super-diffusively[2].

[1] Recall that the diffusive paradigm holds that transport is the cumulative effect of small random steps (see section 1.5), and in particular that turbulent transport would be induced by stochastic motion around turbulent eddies. Assuming the eddies have a characteristic size and timescale, this hypothesis would lead to an effective turbulent diffusion (conduction and convection) that can be treated similarly as the background (neoclassical) diffusion.

[2] Faster than expected from diffusion: $\Delta x = (D\Delta t)^{1/2}$, where D is the diffusion coefficient.

doi:10.1088/978-0-7503-4856-0ch3

Another consequence of this situation is profile stiffness: in near-marginal conditions, plasma profiles will tend to be locally clamped to values just below the local critical threshold. Namely, whenever the local gradient exceeds the threshold, this will trigger avalanches on a fast (turbulent) timescale that will only stop once profiles are again below marginal: so, a feedback between turbulence and profiles is established.

Finally, because the radial extent of the avalanche is only limited by the radial size of the system, this would lead to the non-diffusive scaling of the global energy confinement, in which the system size enters as a parameter. Ordinary diffusion is always characterized by a (small) characteristic scale or 'step size' associated with the transport mechanism, i.e. it is independent of the system size. Contrariwise, the avalanche size and the associated transport are limited only by the radial extent of the system, making transport dependent on the system size [9].

The resulting slowly driven[3], open system containing local thresholds that separate transport channels with large disparity in transport efficiency is strongly reminiscent of self-organized criticality (SOC), a dynamical paradigm that was popularized in the 1990s [10, 11]. In SOC systems, profile gradients wander around the local critical threshold under the action of the external drive while staying below marginal everywhere, on average, due to the more efficient transport channel that becomes active when the threshold is exceeded. The redistribution of the profiles is carried out by avalanches, transport events that are triggered at locations that are sufficiently close to the threshold so as to be made unstable under the action of the external drive, and that are halted at locations that are well below the threshold. Some of these avalanches will eventually reach the edge of the system and exit it, thus providing an additional mechanism to balance the system drive (fuelling/heating) in steady state. Note, in particular, that transport in such a system will be 'bursty', particularly in the edge—in accordance with observations.

One-dimensional SOC systems have some interesting properties that make them behave as complex systems [12, 13]. First, avalanches lack any characteristic scale, being limited only by the system size. This lack of scales usually translates into the avalanches exhibiting scale-free statistics. If one assumes that the PDF of the spatial extent (l) of the avalanches behaves like

$$p(l) \propto l^{-(1+a)}, \quad 0 < a < 1, \tag{3.1}$$

then the corresponding mean linear extent becomes

$$\langle l \rangle = \int_0^L dl \cdot l \cdot p(l) \propto L^{1-a}, \tag{3.2}$$

showing that the mean extent depends on the system size, L. Since the avalanches obey power law statistics, at least over a certain range of scales, the dynamics of the associated transport are also characterized by scale-invariance or self-similarity.

[3] We call a system slowly driven when the drive is sufficiently weak to allow the system to self-regulate near the marginal state.

Self-similar, divergent fluctuations are characteristic of 'equilibrium critical systems', and occur routinely near phase transitions [14]. In SOC systems, however, a similar state is reached without the need for any careful tuning of system parameters, in contrast to their equilibrium counterparts that require carefully tuning a control parameter such as the temperature. For this reason, this type of critical dynamics is called 'self-organized'.

Due to their non-equilibrium nature, SOC systems also display temporal features typical of complex systems. One important aspect is that the triggering of avalanches, the most relevant dynamic player in the system, is not random but governed by long-range, scale-invariant memory. Memory is physically stored in the shape of the system profile, and remains rather insensitive to the details of the external drive [15]. Avalanches tend to be triggered at those locations that are closest to marginal since they are most likely to become unstable under the action of the external drive. On the other hand, avalanches will tend to stop at locations where the profiles are sufficiently below marginal, since these cannot become unstable even after neighbouring relaxations take place. The distribution of the two types of locations (triggering and stopping points) is determined by the preceding history of the system [16], and this is therefore the mechanism through which memory is stored in the system over timescales far exceeding the turbulent timescale.

Given that these general ideas possibly offer an explanation for some of the remarkable phenomena observed in fusion plasmas, as previously discussed in section 1.7, one would like to confirm that this description makes sense on the basis of experimental data from fusion devices. Do plasmas really display SOC characteristics, either generally or in some specific regimes? To address this issue, one would need to look for evidence of self-similarity (scale-invariance), critical exponents (associated with divergent scalings with the system size) and the kind of long-range, self-similar memory effects that are characteristic of SOC systems. We already noted that experimental accessibility in fusion plasmas is restricted due to the extreme conditions. Most of the available experimental data come in the form of time series, such as those obtained with Langmuir probes in the edge region of fusion devices, or ECE signals obtained at different radial locations.

In this chapter we will discuss in some detail how to look for evidence of complex features in time series, briefly reviewing the fundamentals first, then describing and discussing various methods to analyze experimental time series and interpret the results in this context, and concluding by briefly showcasing some of the actual applications of these methods to fusion plasmas that can be found in literature.

3.2 Self-similarity in random processes

The key mathematical concept in the context of the analysis of turbulent time series is that of a random or 'stochastic' process [17]. A stochastic process is a process that consists of a sequence of random values taken from a given distribution. Thus, each realization of this process will consist of a set of different values. A Gaussian random process, for instance, is a process in which the values are taken from a Gaussian distribution. Because each realization of the process consists of different values, the properties of the process can only be defined statistically. An experimental time series, on the other hand, is the discrete record obtained when sampling a continuous process at a certain rate. Due to the rapidly varying, almost unpredictable character of turbulence, turbulent data series look strikingly similar to stochastic processes, and many techniques for the analysis of stochastic processes are also useful for turbulent data series. In the following, we will therefore review a few of the fundamental ideas of stochastic processes.

3.2.1 Self-similar time processes

A random process $y(t)$ is said to be statistically self-similar, with self-similarity exponent H, if and only if

$$y(at) \equiv^d a^H y(t), \quad \forall a, t \geqslant 0. \tag{3.3}$$

Here, the symbol $\equiv^d$ means *equivalence in distribution*. Basically, it means that the statistical properties of the original process, $y(t)$, are the same as those of the rescaled process. One should also note that $y(at)$ does not mean 'multiplying the time t by a factor a', as that operation would not change the distribution. Rather, it means that one is examining the time series y at another time scale, much as we did in section 2.1.2, where the PDF of the particle flux was studied as a function of the smoothing scale, m.

One could naively think that this equivalence would hold when the statistical moments of the two processes (i.e. mean, variance, etc) are the same. However, the statistical moments do not carry any information about multi-point probability distributions that can be associated with the process. These multi-point probability distributions are the joint probability of the process taking certain values at several different time points t, and they therefore codify the memory contained in the process. Strict self-similarity would require any n-point probability distributions to be identical as well—noting there is an infinity of such multipoint distributions. However, in most practical situations, it is acceptable to restrict oneself to requiring that only the two-point probability distributions of the two processes be identical [13]. A two-point probability distribution specifies the probability of the process taking a value between y_1 and $y_1 + dy_1$ at time t_1 and between y_2 and $y_2 + dy_2$ at time t_2:

$$p_2(y_1, t_1; y_2, t_2)dy_1dy_2. \tag{3.4}$$

We will simplify the discussion by assuming also that the process is time-invariant, meaning that the dynamics of the process do not change as time progresses, so that only the elapsed time is meaningful. This allows rewriting p_2 in terms of a prescribed initial time, $t_0 = 0$, and the corresponding initial value of the process, y_0:

$$p_2(y_0, 0; y, t) = G(y, t|y_0)p(y_0). \tag{3.5}$$

Here, $G(y, t|y_0)$ is the *propagator* of the process, and it contains all the information about its two-point dynamics. $p(y_0)$, on the other hand, clearly depends only on how the system is initialized. The propagator describes the temporal evolution of the process. Due to the time invariance of the process, it must satisfy

$$G(y, t|y_0) = \int dy_1 G(y, t - t_1|y_1) \cdot G(y_1, t_1|y_0), \quad 0 < t_1 < t, \tag{3.6}$$

so that the probability of reaching a value y at time t after starting with a value of y_0 is the sum of all the possible intermediate paths that could connect y_0 with y.

The two-point joint PDF contains a lot of interesting information. One can calculate the one-point statistics at any time simply by integrating it over the initial condition y_0:

$$p_1(y_1, t) \equiv \int dy_0 p_2(y_0, 0; y, t) = \int dy_0 G(y, t|y_0) p(y_0). \tag{3.7}$$

This is the probability distribution of the values of the process at time t. One can also calculate the statistical q-moments of the process:

$$m_q(t) \equiv \left\langle |y(t) - y_0|^q \right\rangle = \int dy_0 p(y_0) \int dy |y - y_0|^q G(y, t|y_0), \quad \forall q > 0, \tag{3.8}$$

and the autocorrelation function of the process:

$$\begin{aligned} \mathrm{Ac}_{[y]}(t_2 - t_1) &\equiv \langle y(t_1) \cdot y(t_2) \rangle - \langle y(t_1) \rangle \cdot \langle y(t_2) \rangle = \\ &= \int dy_0 p(y_0) \int dy_1 y_1 \int dy_2 y_2 \cdot G(y_2, t_2 - t_1|y_1) \cdot G(y_1, t_1|y_0) \\ &\quad - \int dy_0 p(y_0) \int dy_1 y_1 \cdot G(y_1, t_1|y_0) \cdot \int dy_0 p(y_0) \int dy_2 y_2 G(y_2, t_2|y_0). \end{aligned} \tag{3.9}$$

3.2.1.1 Self-similar, time-invariant random process

A self-similar, time-invariant process with self-similarity exponent H is any random process whose propagator verifies

$$G(a^H y, at|a^H y_0) = a^{-H} G\left(y, t \mid y_0\right), \quad \forall a \geqslant 0, \tag{3.10}$$

where the symbol '=' now stands for true identity. This equation states that the propagator remains unchanged (up to a factor a^{-H}) when the time is scaled by a while the value y is scaled by a^H, as required by equation (3.3). More precisely, the propagator of a self-similar, time-invariant process is a *self-affine* function, since each of its two variables (y and t) is rescaled with a different exponent [18]. Typically, propagators for self-similar processes take the general form

$$G(y, t|y_0) = t^{-H} \Phi\left(\frac{y - y_0}{t^H}\right), \tag{3.11}$$

for some positive kernel $\Phi(x) \geqslant 0, \quad \forall x$, whose integral over the whole spatial domain equals 1. It is trivial to check that equation (3.10) is then verified.

The self-affinity of the time-invariant propagator leads to very interesting properties:

1. The one-point PDF $p_1(y, t)$, obtained by integrating p_2 over all possible y_0 values, is also self-affine:

$$p_1(a^H y, at) = a^{-H} p_1(y, t). \tag{3.12}$$

2. The q-moments of the process satisfy

$$m_q(t) \propto |t|^{qH}, \quad \forall q > 0. \tag{3.13}$$

We will be making use of this property in the following.

3.2.2 Fractional Brownian motion

In principle, there exist many self-similar, time-invariant random processes to which we might compare the experimental data. The central limit theorem [19], however, predicts that just a few processes act as attracting distributions for physical processes of additive nature. In particular, this theorem states that any sum of a sufficiently large number of random variables converges to the *Gaussian* distribution if one assumes that the variables are independent and have finite variance. Many processes in physics can be approximated by such sums, hence the ubiquitous role played by the Gaussian. Choosing the kernel Φ equal to a Gaussian with zero mean and variance σ^2 and setting $H = 1/2$, the usual *Brownian motion* [20] is obtained:

$$\Phi_{\mathrm{Bm}}(y - y_0|t) = \frac{1}{\sqrt{2\pi\sigma^2 t}} \exp\left(-\frac{(y - y_0)^2}{2\sigma^2 t}\right). \tag{3.14}$$

The well-known fact that the variance of Brownian motion satisfies $\langle|y(t) - y(t_0)|^2\rangle \propto t$ now follows naturally from equation (3.13).

Brownian motion is not the only possible process that may result from a Gaussian kernel. In the late 1960s, Mandelbrot introduced the so-called *fractional Brownian motion* (fBm) [21] which has the same Gaussian kernel but exponents $H \in (0, 1)$. Or, in other words:

$$\Phi_H^{\mathrm{fBm}}(y - y_0|t) = \frac{1}{\sqrt{2\pi\sigma^2 t^{2H}}} \exp\left(-\frac{(y - y_0)^2}{2\sigma^2 t^{2H}}\right). \tag{3.15}$$

It can be shown [21] that fBm corresponds to the random process defined by the integral

$$y_H^{\mathrm{fBm}}(t) = y_0 + \frac{1}{\Gamma(H + 1/2)} \int_0^t ds (t - s)^{H-1/2} \xi_2(s), \tag{3.16}$$

where ξ_2 is a random, uniformly distributed noise, which provides a neat way to generate it numerically (see figure 3.1). Incidentally, this integral shows that fBm differs from ordinary Brownian motion through a dependence on *history*. According to equation (3.13), the fBm moments satisfy (see figure 3.2)

$$\langle |y(t) - y(t_0)|^2 \rangle \propto t^{2H}. \tag{3.17}$$

In the following, we will also be making use of this expression.

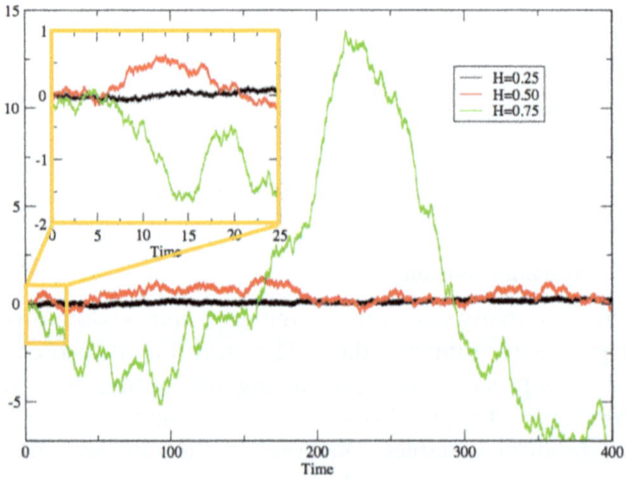

Figure 3.1. Examples of fBm realizations for different self-similarity exponents. Inset: zoom-in of the first part of the signals, highlighting the self-similarity.

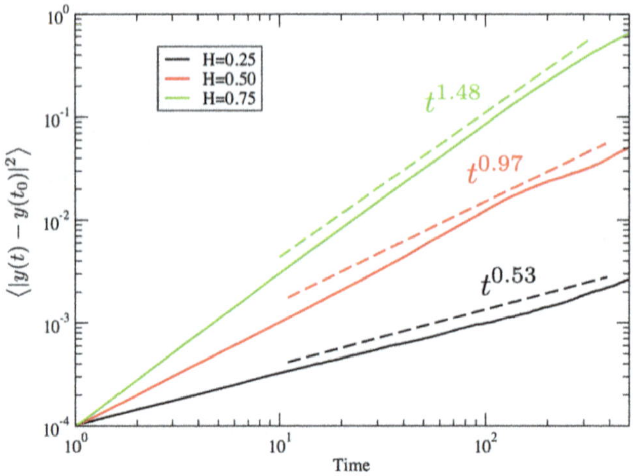

Figure 3.2. Time scaling of the variance for the same fBm realizations shown in figure 3.1. The best power law fits are shown as dashed lines.

3.2.3 Fractional Lévy motion

Most turbulent time series should have similar properties as the Bm and fBm processes due to the mentioned central limit theorem. Nevertheless, Gaussians are not the only limiting distributions allowed by the central limit theorem. Namely, if the restriction regarding finite variance is lifted, a new family of attracting distributions appear, known as *Lévy distributions* [18].

In contrast to the Gaussian distribution, Lévy PDFs do not have an analytical closed form. Yet their Fourier transform, $\hat{p}(k) = \int_{-\infty}^{\infty} dx\ e^{ikx}p(x)$, or *characteristic function*, does. For our purposes, we will focus on the symmetric, zero-mean members of the family, which can be characterized by two parameters: α, the *tail index*, and σ, the *scale factor*. Their characteristic function is [18]

$$L_{[\alpha,\sigma]}(k) = \exp(-\sigma^{\alpha}|k|^{\alpha}). \tag{3.18}$$

Even if its inverse Fourier transform does not have a general analytical form, it is possible to explain the meaning of the two parameters as follows. The tail index, α, defines the behaviour of the tail of the distribution (see figure 3.3), which falls off like

$$L^{\rm sym}_{[\alpha,\,\sigma]}(y) \propto |y|^{-(1+\alpha)}, \quad \text{for } |y| \to \infty. \tag{3.19}$$

Since the values of α lie in the range (0, 2), Lévy PDFs do not have a finite variance. The scale factor σ can be understood from

$$L^{\rm sym}_{[\alpha,\,\sigma]}(a \cdot y) = L^{\rm sym}_{[\alpha,\,|a|\sigma]}(y). \tag{3.20}$$

It also determines the value of the only finite moments of the Lévy PDF, since

$$\langle |y|^q \rangle \propto \sigma^q, \quad \text{if } q < \alpha. \tag{3.21}$$

Using these symmetric Lévy laws as kernels, self-similar, time-invariant processes akin to fBm can easily be defined. The resulting processes are referred to as *fractional Lévy motion (fLm)* [18]. Their propagator is

$$\Phi^{\rm fLm}_{\alpha,\,H}(y - y_0|t) = t^{-H} L^{\rm sym}_{[\alpha,\,\sigma]}\left(\frac{y - y_0}{t^H}\right). \tag{3.22}$$

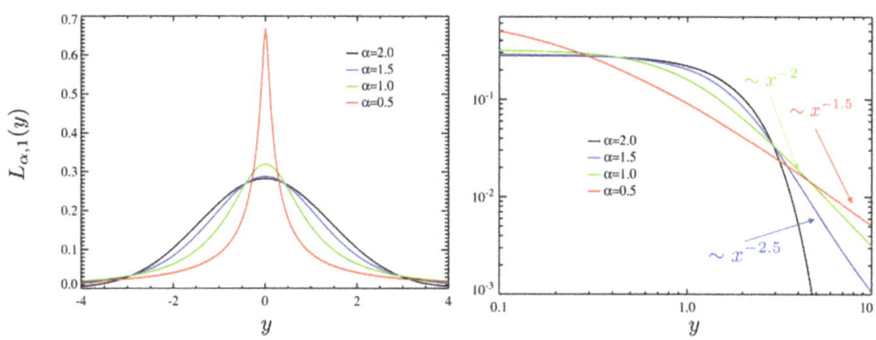

Figure 3.3. Linear–linear (left) and log–log plots (right) of a few symmetric Lévy PDFs with $\sigma = 1$ and various values of the tail index, α.

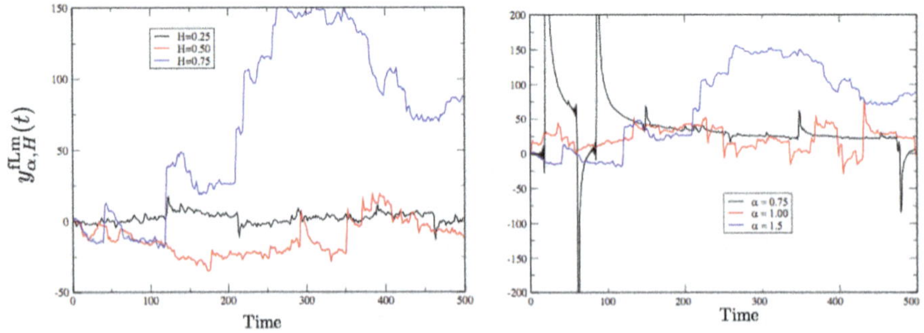

Figure 3.4. Various fLm processes generated using equation (3.23). Left: realizations with fixed $\alpha = 1.5$ and variable H. Right: realizations with fixed $H = 0.75$ and varying α.

Positivity of the propagator requires that the self-similarity exponent lie in the range $0 < H \leqslant \max(1, 1/\alpha)$. The most important parameter for fLm is α, the tail index. The smaller it is, the more significant the contribution of the largest values of y becomes to the overall process, such that large excursions may be dominant in spite of their rarity.

Similarly to fBm, it is also possible to write fLm processes explicitly in integral form [18]:

$$y^{\mathrm{fLm}}_{\alpha,\, H}(t) = y_0 + \frac{1}{\Gamma(H - 1/\alpha + 1)} \int_0^t ds(t - s)^{H-1/\alpha} \xi_\alpha(s), \tag{3.23}$$

where ξ_α corresponds to Lévy-distributed random noise with parameter α. This formula allows generating them easily using a computer (see figure 3.4). Similarly to Brownian motion, the case $H = 1/\alpha$ is special, since it eliminates the temporal kernel inside the integral in equation (3.23). This case is called a *Lévy flight*. Finally, equation (3.13) again guarantees that all fLm satisfy

$$m_q^{\mathrm{fLm};\, \alpha,\mathrm{H}}(t) = \left\langle |y(t) - y_0|^q \right\rangle \propto t^{qH}, \quad -1 < q < \alpha, \tag{3.24}$$

a property that will be of great importance for the applications mentioned below.

3.2.4 Stationary versus self-similar, time-invariant processes

Stationarity is another concept that is often encountered in the context of the analysis of time series and, in particular, of random processes. Loosely speaking, stationarity means that the statistical properties of the process are independent of time. To be more precise, stationary processes are those whose one-point PDF is time-independent:

$$p_1(y, t) \to p_1(y). \tag{3.25}$$

It follows that all q-moments of the process must be independent of time as well.

Stationarity is a nice property to have from a practical point of view. It justifies the equivalence of temporal and ensemble averages, which is convenient since the averaging over successive time windows usually replaces the latter in practice. However, self-similarity and stationarity are two mutually exclusive properties for a time series. Indeed, we just saw that self-similar processes have moments that increase as t^H! Does this mean that we need to abandon the advantages of stationarity when searching for evidence of self-similarity in practical situations? The answer is definitely no. The way around this apparent conundrum is to work with the *increments* of the process instead of the process itself.

Fractional Gaussian noise
Consider fBm. Its h-increments are defined as [21]

$$\Delta_h y_H^{\mathrm{fBm}}(t) = \frac{y_H^{\mathrm{fBm}}(t+h) - y_H^{\mathrm{fBm}}(t)}{h}. \tag{3.26}$$

Here h defines the step size of the increments. The resulting random process is known as *fractional Gaussian noise* (fGn). It can be shown that fGn is stationary in the above sense. More interestingly, however, it turns out that fBm is the *only* self-similar process generated by Gaussian noise that has stationary increments [21].

fGn has very interesting properties that will be quite useful when analyzing actual data. First, its one-point PDF is (see figure 3.5)

$$p_1(\Delta_h y_H^{\mathrm{fBm}}) = \frac{1}{\sqrt{2\pi\sigma_H^2}} \exp\left(-\frac{(\Delta_h y)^2}{2\sigma_H^2}\right), \tag{3.27}$$

i.e. it is a Gaussian with variance $\sigma_H^2 = h^{2H-2}\sigma^2$, where σ^2 is the variance of the underlying Gaussian noise, ξ_2. As a result, the one-point PDF of the increments is a rescaled version of the propagator of the integrated process and can therefore be used to identify it in practice!

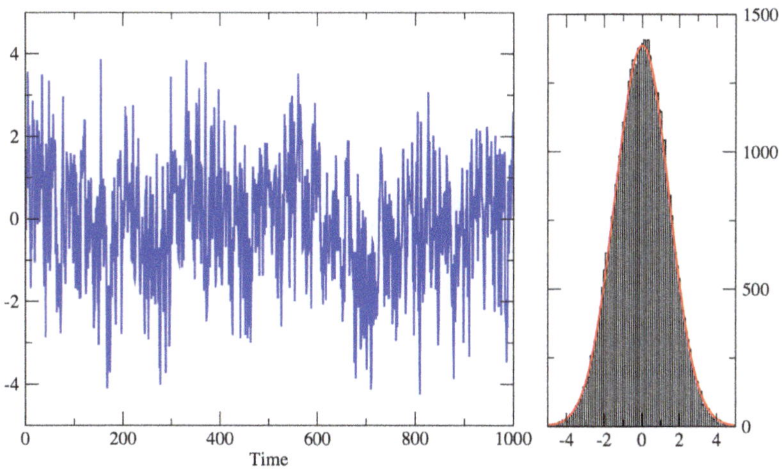

Figure 3.5. Left: fractional Gaussian noise with $H = 0.8$ generated from equation (3.16). Right: one-point PDF of the data, clearly conforming to a Gaussian shape, as expected.

Second, upon rescaling the stepping length from h to rh, its one-point PDF becomes

$$p_1(\Delta_{rh}y_H^{\text{fBm}}) = r^{H-1}p_1(\Delta_h y_H^{\text{fBm}}), \tag{3.28}$$

and its q-moments become

$$\left\langle \left| \Delta_{rh}y_H^{\text{fBm}} \right|^q \right\rangle = r^{q(H-1)}\left\langle \left| \Delta_h y_H^{\text{fBm}} \right|^q \right\rangle. \tag{3.29}$$

These two formulas are very useful when looking for self-similarity in stationary signals in practice.

Fractional Lévy noise

Fractional Lévy motion can also be shown to have stationary increments. The resulting process is known as *fractional Lévy noise* [18]. Similarly to fGn, the one-point PDF of fLn also turns out to be a rescaled version of the propagator of the integrated fLm process. Namely, it corresponds to the symmetric Lévy PDF

$$p_1\left(\Delta_h y_H^{\text{fLm}}\right) = L_{\alpha,\sigma_h}(\Delta_h y), \tag{3.30}$$

where the scale factor is $\sigma_h = h^{H-1}\sigma_\xi$, and α and σ_ξ are the parameters of the underlying Lévy noise, ξ_α.

Similarly, upon rescaling the stepping length from h to rh, its one-point PDF becomes

$$p_1(\Delta_{rh}y_H^{\text{fLm}}) = r^{H-1}p_1(\Delta_h y_H^{\text{fLm}}), \tag{3.31}$$

and its q-moments become

$$\left\langle \left| \Delta_{rh}\, y_H^{\text{fLm}} \right|^q \right\rangle = r^{q(H-1)}\left\langle \left| \Delta_h\, y_H^{\text{fLm}} \right|^q \right\rangle. \tag{3.32}$$

We note that this expression is restricted to the finite moments, $-1 < q < \alpha$.

3.2.5 Determining self-similarity from moments

The properties of scale-invariance in time processes discussed in the preceding sections constitute a useful toolbox to address the practical characterization of self-similarity in experimental time series [13]. Experimental data are sampled at a finite rate, resulting in a discrete temporal record:

$$Y = \{ Y_k,\ k = 1, 2, \ldots, N\}. \tag{3.33}$$

How does one test for self-similarity or scale-invariance in the measured signal?

Typically, this is done by comparing the statistical properties of the signal with those of a true self-similar time process, such as fBm or fLm. The motivation for this is the central limit theorem, which states that these processes act as attractors. Two different paths can be followed to make this comparison, depending on whether our experimental signal is stationary or not. As noted above, stationarity is incompatible with self-similarity, yet the increments of fBm/fLm are stationary. Therefore, if the experimental signal is stationary, it should be compared to fGn/fLn. If it is not, it can either be compared directly with fBm/fLm, or one could differentiate[4] the non-stationary data first, and compare the resulting series with fBn/fGn. Let us see how this works in practice.

Non-stationary data
Let us start with the case in which the data are not stationary. The best option is to compare the scaling of q-moments of the data with the expected scaling from self-similar fBm/fLm processes. To calculate the q-moments of the data up to time index k, it is convenient to start by subdividing the signal with a total length of N points in $M = \lfloor N/k \rfloor$ non-overlapping blocks of length k and calculate

$$m_k^q \equiv \langle | Y_k - Y_1 |^q \rangle = \frac{1}{M} \sum_{j=1}^{M} \left| Y_{k+(j-1)M} - Y_{1+(j-1)M} \right|^q, \quad k = 1, 2, \ldots, N \tag{3.34}$$

for any value of $q > -1$. As the reader will appreciate, this expression averages over as many blocks of length k as will fit in the available time series of length N. This way, all available information is used and statistical errors are reduced as much as possible.

Once the q-moments are computed, self-similarity would follow if (see equation (3.17))

$$m_k^q \propto k^{qH} \tag{3.35}$$

at least over a significant range of k values, since this is the behaviour expected for fBm (Figure 3.6) and fLm (figure 3.7). The obtained value of H would be an estimate of the self-similarity exponent of the process over the corresponding k-range. One must be careful to distinguish between Gaussian-like or Lévy-like processes, noting

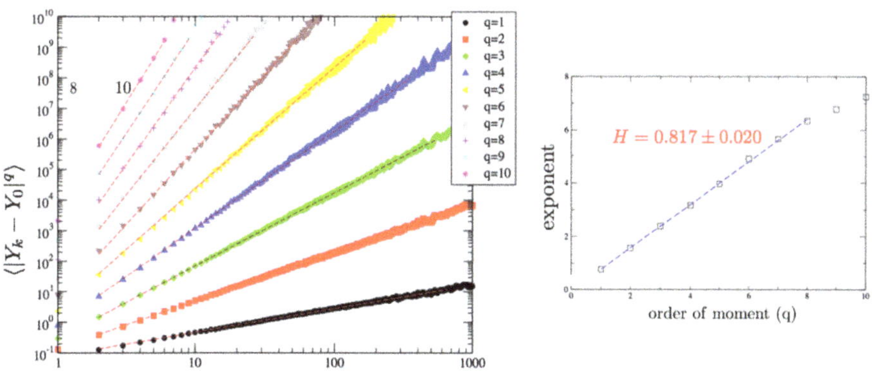

Figure 3.6. Left: first six moments computed using equation (3.34) for a fractional Brownian motion series with $H = 0.8$, generated from equation (3.16). Right: determination of the self-similarity exponent, H.

[4] That is, calculate the increments.

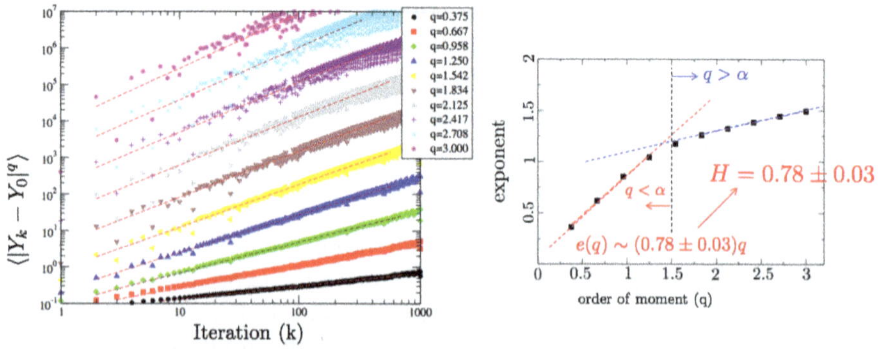

Figure 3.7. Left: assorted non-integer moments ($q < 3$) computed using equation (3.34) for a fractional Lévy motion series with $H = 0.8$ and $\alpha = 1.5$, generated from equation (3.23). Right: determination of the self-similarity exponent, H, from q-moments with $q < \alpha$. The break point of the scaling around $q \simeq \alpha$ is apparent.

that the latter will have diverging moments for $q \geqslant \alpha$ (see right frame in figure 3.7). Usually, this fact will appear as a change in scaling around $q \simeq \alpha$, which will in addition allow estimating the tail exponent α.

The moment method described here provides an estimate of the self-similarity exponent H. In addition, the shape of the propagator kernel, Φ, can also be determined from the data (see equations (3.15) and (3.22)). To do so, one simply needs to estimate the PDF of the displacements from the initial locations at any prescribed iteration k using all blocks available. Since k is arbitrary, one can improve statistics by including all possible values of k (in the range of k values for which self-similarity holds) to better estimate the PDF, after rescaling the associated increments in time by multiplying them by k^{-H}, where H is the value obtained from the moments. By estimating the propagator, one can determine whether the process is fBm or fLm, estimate α, and determine the maximum value of q for which moments are meaningful.

Stationary data

If the data record of interest is stationary, one should compare its statistical properties to those of fGn/fLn instead. Before discussing how to do this best, we recall that a non-stationary signal suspect of being self-similar can be converted to a stationary signal by calculating its increments:

$$\Delta Y^1_{k+1} = Y_{k+1} - Y_k, \quad k = 1, 2, \ldots \tag{3.36}$$

If the resulting increments are not stationary, the original signal is not similar to a fBm or fLm, and therefore not self-similar over any range of scales. In the case of experimental data, it may happen that a fluctuating, self-similar process is superposed on a slow trend (i.e. a low frequency oscillation, or a linear or polynomial trend). In this case, one could try to apply a trend removal technique before analyzing the data. Trend removal techniques, however, often introduce spurious frequencies[5], modifying oscillation phases, notably affecting the low frequencies (long times), which may interfere with the determination of H, so one should proceed very cautiously. Equation (3.36) defines increments for a time step that equals 1 sampling time unit. It is also convenient to define increments with a time spacing $M > 1$ using

$$\Delta Y_j^M = \frac{1}{M}\left(Y_{j\cdot M+1} - Y_{(j-1)\cdot M+1}\right), \quad j = 1, 2, ..., N_b, \tag{3.37}$$

where $N_b = \lfloor N/M \rfloor$ is the number of non-overlapping blocks of size M in the signal. Note that the normalization by $1/M$ is arbitrary. Here, it has been added for better consistency with the discussion of multifractality in section 3.2.6. M-spacing increments can also be expressed as

$$\Delta Y_j^M = \frac{1}{M} \sum_{k=(j-1)\cdot M+1}^{jM} \Delta Y_{k+1}^1, \quad j = 1, 2, ..., N_b. \tag{3.38}$$

Based on these considerations, we can test whether the stationary data record resembles a fGn/fLn process. To do so, we first check whether the one-point PDF of the $M = 1$ increments resembles a Gaussian or a Lévy distribution (see equations (3.27) and (3.30)). Recall that the PDF of the increments has the same shape as the propagator of the integrated fBm/fLm motion. If the process is of the Lévy type, this analysis will also provide an estimate of the tail index α.

After determining the nature of the data record, we can estimate the self-similarity exponent by comparing the scaling of the moments to the expected behaviour for fGn (equation (3.29)) or fLn (equation (3.32)). In other words, one verifies if the following relation holds,

$$\frac{\langle |\Delta Y^{(M)}|^q \rangle}{\langle |\Delta Y^{(1)}|^q \rangle} \propto M^{q(H-1)}, \tag{3.39}$$

for a sufficiently large range of M. Here, the angular brackets represent an average over all possible blocks, understood to be equivalent to an ensemble average in the case of stationary signals, allowing us to carry out these calculations with a single data record, provided it is sufficiently long. As with non-stationary data, Lévy processes generate diverging moments for $q \geqslant \alpha$, which will be apparent as a change in the scaling around $q \simeq \alpha$, thus providing a second estimate of the tail exponent, to be compared with the one obtained from the one-point PDF. As an illustration, we have performed this procedure for two synthetic series of fGn (see figure 3.8) and fLn (figure 3.9) with $H = 0.8$ and, in the Lévy case, $\alpha = 1.5$.

3.2.6 Multifractal analysis

A *multifractal* process is a process in which a single similarity exponent is not sufficient to describe the dynamics. Such systems may arise due to the interaction of

[5] Frequencies not present in the original data.

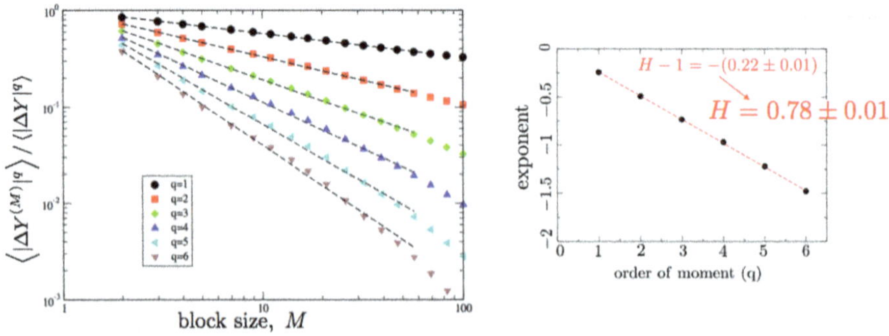

Figure 3.8. Left: first ten moments computed using equation (3.39) for a fractional Gaussian noise series obtained by calculating the M-increments of a fBm series generated with $H = 0.8$ using equation (3.16). Right: determination of the self-similarity exponent, H.

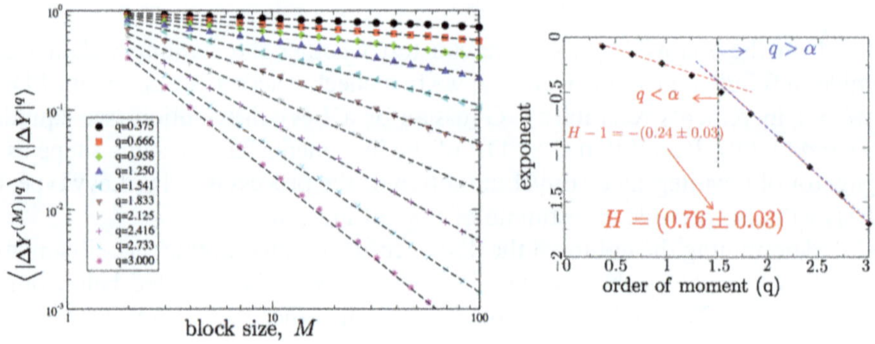

Figure 3.9. Left: assorted non-integer moments ($q < 3$) computed using equation (3.39) for a fractional Lévy noise series obtained by calculating the M-increments of a fractional Lévy motion series with $H = 0.8$ and $\alpha = 1.5$ generated from equation (3.23). Right: determination of the self-similarity exponent, H, from q-moments with $q < \alpha$. The break point of the scaling around $q \simeq \alpha$ is apparent.

various self-similar processes with different self-similarity exponents. If the value of these individual exponents are close together, the resulting process would behave very much like a single self-similar process. But if the individual contributing processes have rather different self-similarity exponents, the overall process will not be self-similar. At best, it might have quasi-self-similar features across a reduced range of scales over which a particular component dominates. The predominance of a component can only be detected if this range of scales is sufficiently large. This mixing of various types of processes may certainly occur in actual data, so that multifractal analysis is often used to test (limited) self-similar behaviour [22].

One way to determine multifractality in a data record with a length of N, $y(t_j)$, $j = 1, 2, \ldots, N$, is to divide the record in N_b non-overlapping blocks of size $T = \lfloor N/N_b \rfloor$. For each block i ($i = 1, 2, \ldots, N_b$), one defines the following measure for the time derivative at scale T:

$$w_i^T := \frac{|y(t_i^* + T) - y(t_i^*)|}{T} \equiv |\Delta_T y(t_i^*)| \propto T^{h_i^T - 1}, \tag{3.40}$$

where $t_i^* = t_{(i-1)N_b+1}$ is the time at the beginning of each block. One expects this measure to scale as a power of the block size, T. The exponent h_i^T is known as a *singularity exponent*. To understand this, consider that a smoothly differentiable function f would satisfy $|f(x + \delta) - f(x)| \propto \delta$, at least for small δ, implying $h \simeq 1$. For a self-similar process, however, the exponent could be less than one, quantifying how strongly non-differentiable or singular the self-similar process is. For a self-similar process, it can be shown that $h = H$, regardless of the choice of T, the scale at which the process is being explored.

In a multifractal signal, however, various singular exponents might be present in the signal and one should thus expect that the dominant self-similar contribution at a given scale T will determine the value of h. This idea suggests a method to quantify the exponents of the signal in the following way. One computes the so-called *partition function* or *structure function*:

$$S(T, q) = \frac{1}{N_b} \sum_{i=1}^{N_b} w_i^q \propto T^{\xi(q)}, \tag{3.41}$$

where the value of q is any real positive number. Accepting the hypothesis that the partition function should scale as a power of T, one obtains a family of *multifractal exponents* $\xi(q)$, simply by performing a direct least-squares fit to a straight line of $\log S(T, q)$ versus $\log T$. For a Gaussian self-similar process[6], equation (3.40) predicts that $\xi(q) = q(H - 1)$, so that the self-similarity exponent of the process is $H = 1 + \xi(q)/q$. In this case, H should not depend on q, since a self-similar process is characterized by a single exponent.

Note that the contribution of the time derivative measures w_i in the sum of equation (3.41) is dominated by small blocks (small T) for small positive q, and by large blocks (large T) for large positive q, respectively, so that the different values of q also explore different scales. In a multifractal signal, the contributions at different scales will be dominated by components with different scaling exponents. As a result, the *generalized self-similarity* exponent (also referred to as *generalized Hurst exponent*):

$$H(q) = 1 + \frac{\xi(q)}{q} \tag{3.42}$$

may depend (strongly) on q.

The theory of multifractality also allows quantifying the individual importance of each singularity exponent h to the partition function. The corresponding mathematics are a bit involved, and details can be found elsewhere. The bottom line is that the relevance of each value of h can be measured in terms of the *singularity spectrum*, $f(h)$, easily calculated for a time series [23]. This function can be computed from the

[6] For a Lévy self-similar process, this only holds for $q < \alpha$.

above family of multifractal exponents, $\xi(q)$. The dominant exponent contributing to the partition function for a specific value of q, $h^*(q)$, and the value of the corresponding singularity spectrum, $f(h^*(q))$, can be stated as [24]

$$h^*(q) = 1 + \frac{d\xi(q)}{dq} \tag{3.43}$$

$$f(h^*(q)) = q(h^*(q) - 1) + 1 - \xi(q). \tag{3.44}$$

By plotting $f(h^*)$ versus h^*, one can visualize the degree of self-similarity of the process; a self-similar process will result in a graph concentrated around a dominant value of h^*, whereas multifractal processes will lead to spread-out or scattered graphs.

3.3 The search for self-similarity in fusion plasmas

Illustration using probe data from the edge of the W7-AS stellarator

To illustrate the application of these techniques to experimental data, we will use a turbulent signal obtained by a Langmuir probe at a location in the edge of the plasma confined in the W7-AS stellarator (see appendix A.3), in the absence of significant poloidal flow. In particular, we will look at the ion saturation current, I_{sat}, collected at one of the probe pins, understood to be a proxy for the local plasma density, in discharge No. 35425. The raw data are shown in figure 3.10. Upon first inspection, the data look rather stationary. Therefore, we will compare their scaling properties with fGn/fLn to search for evidence of self-similarity, as explained above.

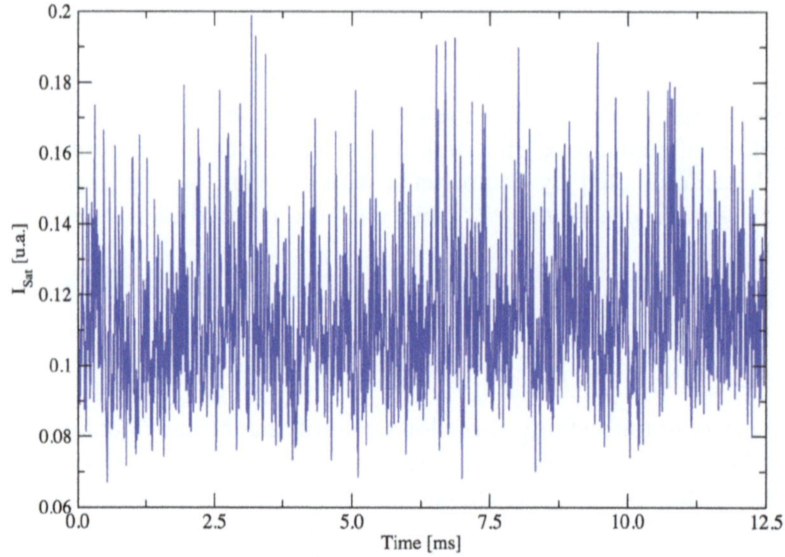

Figure 3.10. Ion saturation current data from a Langmuir probe in the edge of the plasma in W7-AS discharge No. 35425. These data will be used for illustration purposes in the rest of this chapter.

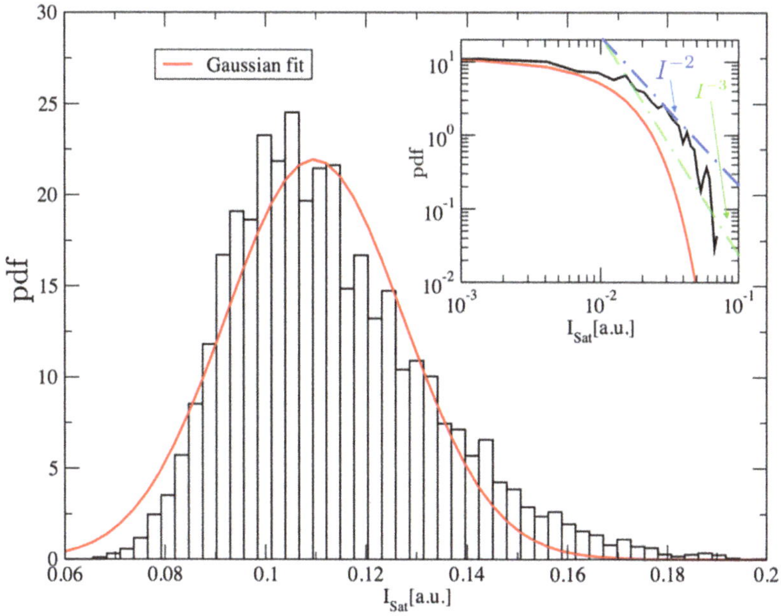

Figure 3.11. Left: 1-point PDF of the W7-AS ion saturation current on a double linear scale. Right: one-point PDF of the same data after subtracting the mean, shown on a double logarithmic scale. The best Gaussian fit is shown in red.

First, we calculated the one-point PDF of the signal, shown in figure 3.11 on both double linear and logarithmic scale (inset of the figure), the latter after subtracting the mean from the data. The best Gaussian fit to the data has also been included in both cases, and is shown in red. Clearly, the data deviates from the Gaussian shape, particularly in the tail region, which appears to be slightly asymmetric, but still, the Gaussian is not a bad fit. This suggests that a comparison of the signal to fGn might be adequate. The PDF of the data after subtracting the mean also provides some interesting information. As seen in the inset, the positive tail falls off slower than the Gaussian tail, but no power law behaviour is apparent. Still, there is a small range of scales, for I_{sat} between 10^{-2} and 4×10^{-2}, where the tail exponent could be close to -2, which could be consistent with a Lévy tail, but we consider this range to be too short to be truly significant.

We then proceed to examine the scaling properties of moments of the signal according to equation (3.34). The result is shown in figure 3.12, to be compared to its fGn counterpart, figure 3.8. In contrast to the case of fGn, the power law scaling does not hold over the full range of block sizes. However, there is a range of values, extending over almost two decades, from $M \sim (30 - 50)$ to $M \sim 2000$, where a clear power law scaling is apparent. Restricting the power law fits to this range, a self-similarity exponent $H \simeq 0.64$ can be inferred. Also, this exponent does not depend on q, so there is no need to compare with fLn. At block sizes $M < 50$ and $M > 2000$, the scaling is lost. This is a common situation in the analysis of experimental signals. Self-similarity is usually restricted to a limited range of scales due to finite-size effects

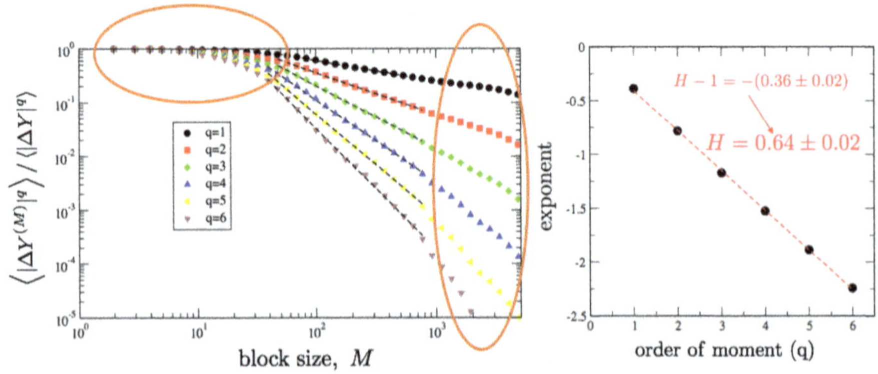

Figure 3.12. Left: first six integer moments computed using equation (3.39) from the W7-AS data. Right: determination of the self-similarity exponent, *H*.

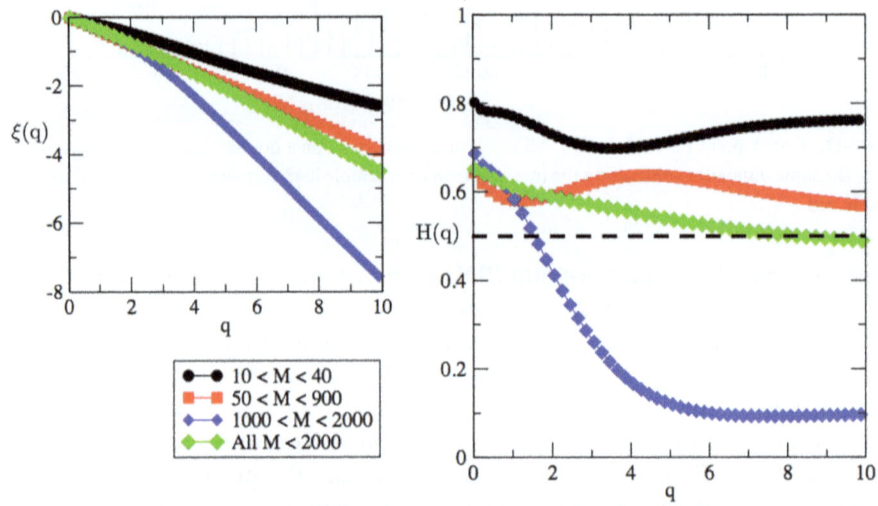

Figure 3.13. Multifractal exponent $\xi(q)$ (left) and generalized self-similarity exponent $H(q)$ (right) for the ion saturation current signal from W7-AS discharge No. 35425. Results corresponding to restricted ranges of block sizes are shown as well.

(related to the size of the system, which sets an upper limit) and the fact that the dynamics of the turbulence below the eddy scale (where dissipation occurs) is often disengaged from the complex, nonlinear dynamics which is the result of the interaction between instabilities and large scale plasma profiles. Indeed, the value $M \simeq 50$ corresponds to a timescale of roughly 25 μs since the signal is sampled at 2 MHz and this timescale is close to the typical local turbulence correlation time in these plasmas [25].

Next, we will illustrate the use of multifractal techniques with the same W7-AS data. Figure 3.13 shows both the multifractal exponents $\xi(q)$ and generalized self-similar exponents $H(q)$ obtained using equations (3.41) and (3.42), respectively.

As discussed above, strict self-similarity requires $H(q) \simeq H_0, \quad \forall q > 0$. However, the overall result for the W7-AS dataset (in green) is clearly not fully self-similar, yielding values of H varying between 0.7 for the smallest scales to about 0.45 for the largest scales. In addition, there is no range of values of q for which a well-defined (constant) value of H is apparent either.

This result is consistent with our previous analysis of this dataset using moments, presented above. There, it was shown that the dynamics are different for timescales shorter than a few tens of μs ($M < 50$, corresponding to 25 μs), where local turbulent fluctuation dynamics dominate, as compared with longer timescales, where behaviour was scale-invariant with $H \simeq 0.65$, up to times of the order of one millisecond (i.e. $M \simeq 2000$; see figure 3.12). To take account of this scale dependence, we therefore calculated the multifractal exponents and associated generalized self-similar exponents by restraining the range of block sizes to the interval $50 < M < 900$. The result (red) is clearly much more consistent with self-similar behaviour, characterized by a self-similarity exponent $H \simeq 0.6$, roughly consistent with the result from moment analysis. If we perform the analysis using block sizes in the turbulent time range, for block sizes in $10 < M < 40$ (black), a different self-similar region with a much larger exponent is found, $H \simeq 0.8$, which might be related with the autocorrelation dynamics of the turbulent fluctuations. Finally, when using only much longer block sizes, $1000 < M < 2000$, the behaviour (blue) is much less self-similar. One might be tempted to infer self-similarity at the largest block sizes, since a rather uniform exponent $H(q) \simeq 0.1$ is found for $q > 5$. However, one should bear in mind that the number of independent blocks is low for the largest scales, so the reliability of this inference is low and may suffer from statistical error and a saturation effect.

Further information can be gathered from the calculation of the singularity spectrum resulting from the multifractal exponents, $\xi(q)$. The results are shown in figure 3.14 for three different ranges of block sizes. In the intermediate range, with block sizes in the range $50 < M < 900$ (red), the dominant contribution comes from a narrow set of exponent values centered at around $H \simeq 0.6$, as expected. Likewise, in the small size range, with block sizes $10 < M < 40$, the spectrum is centered at $H \simeq 0.85$. The width of the regions with significant contributions to h^* are quite narrow in both cases. However, this is not the case for the larger block sizes, $1000 < M < 2000$, yielding clear multifractal behaviour, with significant contributions coming from a very broad range: $h^* \sim$ 0.15–0.7. It is interesting to note that the density of exponent values contributing to the partition function is larger around $h^* \simeq 0.65$ and $h^* \simeq 0.2$, probably reflecting the fact that the self-similar region at intermediate scales is still active for block sizes bigger than $M = 900$.

Applications in fusion plasmas

Several authors in the field of plasma physics have used the methods outlined above, or closely related variations, to study plasma turbulence [1]. For example, fluctuation data from a reflectometer at the DIII-D tokamak has been analyzed using a variation of equation (3.34) [26], describing the scaling of the moments of a

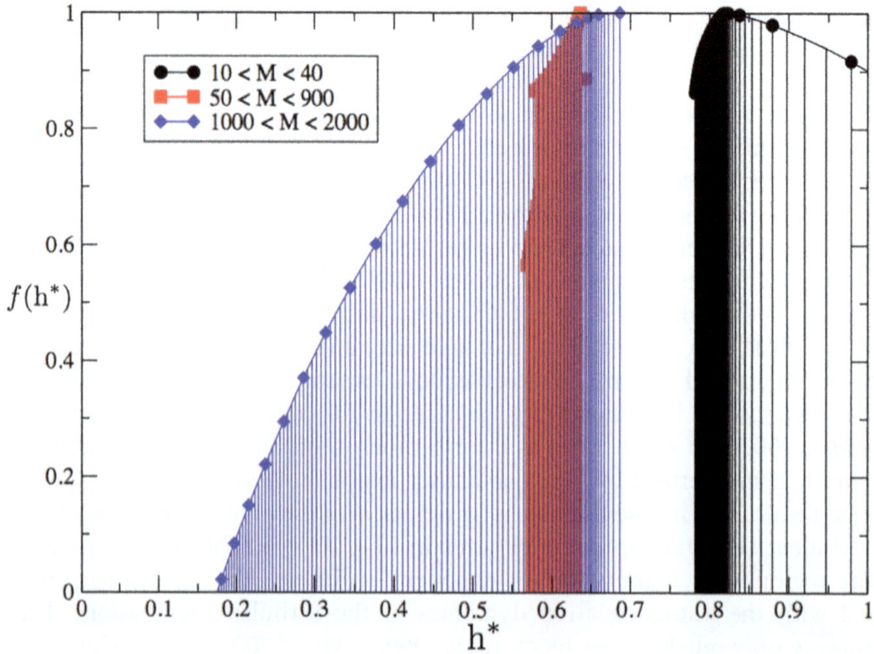

Figure 3.14. Singularity spectrum for the W7-AS ion saturation signal. Results are shown for the same ranges of block sizes shown in figure 3.13.

self-similar process. The reflectometry signal used was stationary, so the records had to be integrated first. Overlapping (instead of non-overlapping) blocks were used to calculate the moments (referred to as *structure functions* in these papers). That is, the sum in equation (3.34) was modified to

$$m_k^q \equiv \langle | Y_{k+i} - Y_i |^q \rangle_i = \frac{1}{N-k} \sum_{i=1}^{N-k} | Y_{k+i} - Y_i |^q, \quad k = 1, 2, ..., N, \tag{3.45}$$

which certainly improves statistics, at the price of not guaranteeing the statistical independence of successive realizations, which might bias the outcome. The autocorrelation time of the analyzed signal was 20 μs, but the investigation carried out by these authors showed that self-similarity, with a self-similarity exponent $H \simeq 0.77$, persisted up to times of several milliseconds; similar to what was found for the W7-AS in section 3.3. These studies were further elaborated in a second paper [27], in which a comparison with other methods to determine H were discussed, and where the radial dependence of H in the DIII-D tokamak was studied for several discharges, including L-mode and Ohmic regimes.

With the present knowledge in hand, it is worthwhile to revisit the results presented in section 2.1.2, since we are now in a position to shed some additional light on their deeper significance. That section examined the properties of the (one-point) PDF of edge turbulent fluxes measured at the edge of the W7-AS with Langmuir probes. The PDF of the m-point averaged flux signals was studied:

$$\Gamma_i^{(m)} = \frac{1}{m}\sum_{j=1}^{m}\Gamma_{m(i-1)+j}. \tag{3.46}$$

It was found (see figure 2.6) that the standard deviation (i.e. the squared root of the second moment) of the m-point averaged signal scaled as

$$\sigma(m) \propto m^{H-1}, \tag{3.47}$$

with $H \simeq 0.84$ for $m < 32$ and $H \simeq 0.57$ for $32 < m < 1024$. It was also found that the PDFs of the m-point averaged signal could be made to collapse onto a single shape if they were rescaled using (see equation (2.10))

$$p(\Gamma) = L^{-1}g(\Gamma/L), \quad \text{with } L = m^{H-1}, \tag{3.48}$$

in each of the two m ranges, with the corresponding value of the H exponent. At that point of the book, the full significance of these results might not have been clear. But now we are able to interpret these results in the light of the discussion of section 3.2.4.1. First, by comparing equation (3.48) with equation (3.28), one sees that the former corresponds to the scaling of the PDF of the h-increments of fBm (i.e. fractional Gaussian noise) upon rescaling $h \to r \cdot h$. Second, by comparing equation (3.47) to the square root of equation (3.29) after setting $q = 2$, one sees that the former corresponds to the scaling of the standard deviation of the h-increments of fBm.

Thus, the two pairs of equations are quite similar if one assumes that m, the block size, plays the role of the rescaling factor r. How does that arise? We first compute the integrated flux:

$$W_k = \sum_{j=1}^{k}\Gamma_j, \quad k = 1, 2, \ldots \tag{3.49}$$

If the process W_k were similar to a fBm, then Γ_k would be akin to a fGn process obtained with an increment $h = 1$. This immediately follows from (see equation (3.26))

$$\Delta_1 W_k = W_{k+1} - W_k = \Gamma_{k+1}. \tag{3.50}$$

Interestingly, the corresponding fGn with (integer) increment $h > 1$ would be given by

$$\Delta_h W_k = \frac{W_{k+h} - W_k}{h} = \frac{1}{h}\sum_{j=k+1}^{k+h}\Gamma_j, \tag{3.51}$$

which corresponds to a block-average of the flux using a block of size h, in line with the definition of equation (3.46). That is, the block averaging scheme used is a (non-overlapping) discrete version of the step-size rescaling for fGn. Therefore, the results of the study do indeed suggest that the W7-AS turbulent fluxes look very much like fGn, although with two different values of the self-similarity exponent H over the two identified m ranges, the fluctuation range for $m < 32$ and the mesoscale range, for $32 < m < 1024$. Very similar scaling behaviour has been also obtained in other fusion devices, as shown in section 2.1.2 as well as in other papers reporting similar results for the Italian reversed field pinch RFX [28], the Chinese tokamak CT-6B [29] or the Indian tokamak ADITYA [30]. However, not every author agrees on the significance of the existence of these two scaling regions. Some authors take this as evidence of multifractality [28]. In our opinion, in line with section 2.1.2, it is significant that the value $m = 32$ corresponds to a time scale of roughly 30 μs, of the same order as the local decorrelation time. This suggests that the low-m PDF shape is related to the dynamics of the individual, local turbulent fluctuations or eddies, necessarily below the range where complex dynamics might dominate, which would involve the nonlinear interaction between many individual fluctuations. On the other hand, the high-m PDF shape, unchanged over quite an extended range that lies well beyond the fluctuation scales, could signal the dominance of critical, self-similar dynamics.

We would like to conclude this part by making a few comments about multifractality studies in fusion plasmas, noting that the number of published studies in this area is not very high. Nevertheless, a few studies report the singularity spectrum for Langmuir probe data obtained in both tokamaks and stellarators [31–33]. The results show singularity spectra that are much broader than what would be expected for a purely monofractal signal. These studies, however, include all possible block sizes in the estimation of the spectrum, which may not be the best way to expose the relevant dynamics, as illustrated by the analysis of W7-AS data carried out here. We will return to the issue of multifractality in chapter 5.

3.4 Long-term memory in time-series

The preceding section has shown that plasma turbulence contains 'something' that lasts very much longer that the turbulence autocorrelation time (up to three orders of magnitude longer). The plasma therefore contains some type of *memory*. Memory is a term that is often mentioned in relation to complex dynamics, to express that the past history of the system influences its future evolution. In complex systems, this influence materializes in a scale-free fashion, only limited by finite-size effects. To understand what is meant by this idea, it is best to discuss it in comparison with the more familiar short-range memory. The common thread will be provided by the well-known autocorrelation function.

3.4.1 Memory and the autocorrelation function

We already introduced the correlation briefly in section 2.2. Here, we will review this important quantifier again in the framework of random processes and complex systems. The autocorrelation of a stationary random process, $y(t)$, is defined as the expected value,

$$\mathrm{Ac}_{[y]}(\tau) \equiv \langle y^*(t+\tau) \cdot y^*(t) \rangle, \tag{3.52}$$

where the random process $y^*(t)$ is obtained from the original process $y(t)$ by subtracting its mean and normalizing the result to the standard deviation:

$$y^*(t) \equiv \frac{y(t) - \langle y \rangle}{\sigma_y}, \quad \sigma_y = \langle y^2 - \langle y \rangle^2 \rangle^{1/2}. \tag{3.53}$$

Thus, the process $y^*(t)$ has zero mean, standard deviation equal to one, and no physical units. The autocorrelation depends only on the time lag τ due to the assumed stationarity. In addition, it satisfies $|\mathrm{Ac}_{[y]}(\tau)| \leqslant 1$, with equality occurring at least for time lapse $\tau = 0$. It is also a symmetric function with respect to the sign of the lag τ, so that only positive lags need to be considered. Finally, it is non-negative definite in the sense that [34]

$$\lim_{T\to\infty} \int_0^T \mathrm{Ac}_{[y]}(\tau) d\tau \geqslant 0. \tag{3.54}$$

This integral converges only if $\mathrm{Ac}_{[y]}(\tau)$ decays faster than τ^{-1}, which is not always the case. This is an important detail, since the value of this integral carries significant physical information about the process.

In practical situations, however, often one has access to just one realization of the process, so ensemble averaging must be replaced by time averaging. In these situations, equation (3.52) can be replaced by

$$\mathrm{Ac}_{[y]}(\tau) = \frac{1}{T} \int_0^T dt\; y^*(t) \cdot y^*(t + \tau), \tag{3.55}$$

where T must be sufficiently large.

Discrete autocorrelation

In case of a discretely sampled time series, $\{Y_k,\ k = 1, 2, \ldots, N\}$, the definition of the autocorrelation can be easily reformulated as a sum:

$$\mathrm{Ac}_{[y]}(k) = \frac{1}{(N-k)} \sum_{i=1}^{N-k} Y_i^* \cdot Y_{i+k}^*, \quad Y_k^* \equiv \frac{Y_k - \bar{Y}}{\sigma_Y}, \tag{3.56}$$

where $\bar{Y}$ and σ_Y^2 are the usual discrete mean and variance,

$$\bar{Y} = \frac{1}{N} \sum_{i=1}^{N} Y_k, \quad \sigma_Y^2 = \frac{1}{N} \sum_{i=1}^{N} (Y_k - \bar{Y})^2. \tag{3.57}$$

Interpretation

The autocorrelation function of a time-series is used to determine the scales (lag ranges) over which positive or negative correlation exists. Positive correlation means that the signal tends to keep the same sign (on average), while negative correlation means that it is more likely that the signal reverses sign after a given lag τ. The autocorrelation function can also be used to determine the characteristic correlation time of the process. This is the time lag beyond which correlations are small. Correspondingly, one could simply determine the time lag τ_c such that $\mathrm{Ac}_{[y]}(\tau) \sim 0$ for any $\tau > \tau_c$. If this condition is met, the process has *short range memory*, and τ gives the characteristic decorrelation timescale of the process. By assuming an exponential decay of the autocorrelation function, $\mathrm{Ac} \simeq \exp(-\tau/\tau_c)$, one easily obtains

$$\tau_c \equiv \lim_{T\to\infty} \int_0^T \mathrm{Ac}_{[y]}(\tau)d\tau. \tag{3.58}$$

This generally turns out to be a rather good estimate of the decorrelation time[7]. Note that this integral is identical to the one mentioned earlier (equation (3.54)), which would suggest that *all processes yielding a finite, positive value for this integral exhibit short-range memory*. A limiting case occurs when the integral is exactly zero, which would point to a complete lack of memory. *Long-term memory*, on the other hand, would correspond to a diverging value of this integral as T increases.

In the context of complex systems, the interpretation is slightly more subtle [13]. The *long-term, scale-free memory* occurring in complex systems should be reflected by a decay of the autocorrelation function in proportion to a power law, $\mathrm{Ac}_{[y]}(\tau) \sim \tau^{-a}$, with $a > 0$ (see figure 3.15; green curve). When the value of this 'memory exponent' lies in the range $0 < a < 1$, the integral diverges:

$$\mathrm{Ac}_{[y]}(\tau) \sim B|\tau|^{-a}, \quad 0 < a < 1, \quad \tau \to \infty. \tag{3.59}$$

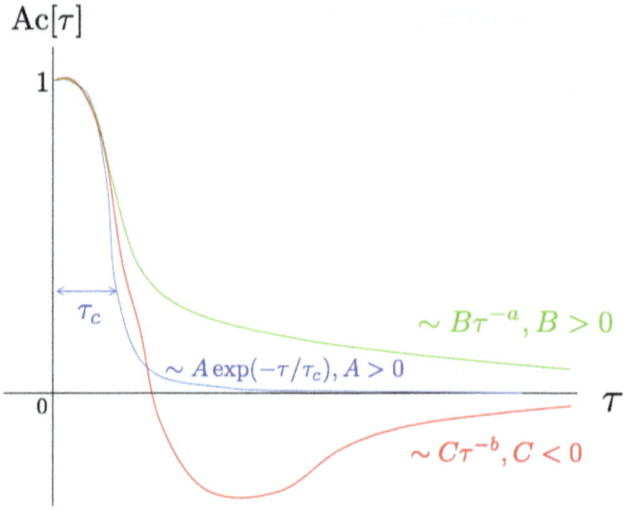

Figure 3.15. Classification of autocorrelation curves: exponential decay (blue), positive correlation with power law decay (green), and negative correlation with power law decay (blue).

[7] It is by no means the only one. It is also common to define the autocorrelation time as the value of the time lag, τ, at which the autocorrelation first reaches the value $1/e$. Other authors take the first lag at which the autocorrelation crosses zero. For most time-series, all these criteria lead to comparable values of the decorrelation time, at least to the same order of magnitude.

The constant B should be positive, to guarantee the presence of *positive* correlations at long time lags.

Negative long-term correlations can also be accommodated. These occurs for processes yielding (see figure 3.15; red curve)

$$\mathrm{Ac}_{[y]}(\tau) \sim C|\tau|^{-b}, \quad b > 1, \quad \tau \to \infty, \tag{3.60}$$

with $C < 0$ and

$$\lim_{T\to\infty} \int_0^T \mathrm{Ac}_{[y]}(\tau)d\tau = 0. \tag{3.61}$$

Here, the condition on the memory exponent is $b > 1$, that guarantees convergence of the integral, while $C < 0$ allows compensating the positive contribution of the autocorrelation around zero lag. This is an example of a process that possesses long-term memory in spite of a vanishing autocorrelation integral. It is evident that this type of memory must necessarily be *antipersistent*.

3.4.2 Memory and the power spectrum

The preceding discussion might suggest that the autocorrelation is an optimal tool to detect the presence of long-range temporal correlations in time series. The practical application of these methods is however limited by the presence of noise in measurement time series, such that the determination of the tail behaviour of the autocorrelation, with the precision required to estimate the integral in equation (3.54), is quite challenging. An alternative method is provided by the *power spectrum* of the signal, based on the Fourier transform.

Statisticians often express the Fourier transform as follows:

$$\hat{f}(\omega) = \int_{-\infty}^{\infty} dt e^{i\omega t} f(t). \tag{3.62}$$

Its inverse is given by

$$f(t) = \frac{1}{2\pi} \int_{-\infty}^{\infty} d\omega e^{-i\omega t} \hat{f}(\omega). \tag{3.63}$$

Converting to other definitions of the Fourier transform is rather straightforward [35]. Some important results concerning Fourier transforms are *Parseval's theorem*,

$$\int_{-\infty}^{\infty} dt|f(t)|^2 = \frac{1}{2\pi} \int_{-\infty}^{\infty} d\omega|\hat{f}(\omega)|^2, \tag{3.64}$$

and the *convolution theorem*,

$$\mathrm{FT}\left[\int_{-\infty}^{\infty} f(t)g(\tau - t)dt\right] = \hat{f}(\omega)\hat{g}(\omega). \tag{3.65}$$

The power spectrum of a stationary random process is defined as the *Fourier transform of its autocorrelation function*:

$$\hat{S}_{[y]}(\omega) = \int_{-\infty}^{\infty} d\tau e^{i\omega\tau} \mathrm{Ac}_{[y]}(\tau). \tag{3.66}$$

Therefore, all the information contained in the autocorrelation function is also present in the spectrum. Furthermore, the spectrum of the autocorrelation function can be directly expressed in terms of the Fourier coefficients of the process, so that there is no actual need to deal with the autocorrelation function. This is easily shown as follows:

$$\begin{aligned}
\hat{S}_{[y]}(\omega) &= \int_{-\infty}^{\infty} d\tau e^{i\omega\tau} \mathrm{Ac}_{[y]}(\tau) \\
&= \lim_{T\to\infty} \frac{1}{T} \left[\int_{-\infty}^{\infty} d\tau e^{i\omega\tau} \left(\int_{-\infty}^{\infty} dt y^*_{[0,\,T]}(t) \cdot y^*_{[0,\,T]}(t+\tau) \right) \right] \\
&= \lim_{T\to\infty} \frac{1}{T} \left[\int_{-\infty}^{\infty} d\tau e^{i\omega\tau} \left(\int_{-\infty}^{\infty} du y^*_{[0,\,T]}(u-\tau) \cdot y^*_{[0,\,T]}(u) \right) \right] \\
&= \lim_{T\to\infty} \frac{1}{T} \left[\int_{-\infty}^{\infty} ds \int_{-\infty}^{\infty} du e^{i\omega(u+s)} \cdot y^*_{[0,\,T]}(s) \cdot y^*_{[0,\,T]}(u) \right] \\
&= \lim_{T\to\infty} \frac{1}{T} \left[\int_{-\infty}^{\infty} ds\, e^{i\omega s} \cdot y^*_{[0,\,T]}(s) \right] \cdot \left[\int_{-\infty}^{\infty} du\, e^{i\omega u} \cdot y^*_{[-T,\,T]}(u) \right] \\
&= \lim_{T\to\infty} \frac{|\hat{y}^*_{[0,\,T]}(\omega)|^2}{T},
\end{aligned} \tag{3.67}$$

where $y^*[a, b]$ represents the *truncation of the data to the finite interval* $[a, b]$ and $\hat{y}_{[a,b]}(\omega)$ is its Fourier transform. In the third line, the change of variables $u = t + \tau$, $du = dt$ was performed. In the fourth line[8], the change of variables $u - \tau = s$, $-d\tau = ds$ was performed. Finally, note that one can estimate the autocorrelation time directly from the power spectrum using

$$\tau_c \equiv \lim_{T\to\infty} \int_0^T \mathrm{Ac}_{[y]}(\tau) d\tau = \frac{1}{2} \lim_{\omega\to 0} \hat{S}_{[y]}(\omega). \tag{3.68}$$

Discrete Fourier transforms

The discrete version of the Fourier transform was already introduced in section 2.4. It is, however, worthwhile to discuss how it relates to the material presented in the previous section. We repeat the corresponding definitions (equation (2.22)) for convenience:

$$\tilde{Y}_k = \sum_{n=0}^{N_1} y_n e^{-2\pi i k(n/N)}, \quad y_n = \frac{1}{N} \sum_{k=0}^{N_1} \tilde{Y}_k e^{2\pi i k(n/N)}. \tag{3.69}$$

In contrast to equation (3.62), the direct transform is defined with a negative exponential, and the inverse transform with a positive one. It turns out that this is

[8] This is basically the proof of the convolution theorem!

irrelevant for the calculation of the power spectrum, since the autocorrelation function is symmetric with respect to the lag. The second aspect to consider refers to the determination of τ_c using equation (3.68). In terms of the discrete Fourier transform, it is written as

$$\tau_c = \frac{1}{2}\lim_{\omega\to 0} \hat{S}_{[y]}(w) = \frac{|\tilde{Y}_0|^2}{2N}. \tag{3.70}$$

In the above equations, τ_c is expressed in the same units as the inverse sampling rate of the time series.

Interpretation
The spectrum reveals the presence of long-term memory effects directly. Namely, if $f(t) \propto t^{-s}$ with $0 < s < 1$ when $t \to \infty$, then $\hat{f}(\omega) \propto \omega^{-(1-s)}$ when $\omega \to 0$ [35]. In the case of positive long-term memory, the autocorrelation behaves like $\mathrm{Ac}(\tau) \sim B\tau^{-a}$, with $0 < a < 1$ and $B > 0$. Therefore, the spectrum will behave like

$$\hat{S}_{[y]}(\omega) \sim \omega^{-(1-a)}, \quad \omega \to 0. \tag{3.71}$$

This limit diverges when $\omega \to 0$, which is a reflection of the fact that there are no characteristic timescales. But, more interestingly, we can now relate any frequency region in the power spectrum that behaves as a decaying power law with exponent between 0 and −1 with a corresponding range of temporal scales over which the action of positive, scale-free positive correlations dominates.

Similarly, for cases with negative long-term memory, the autocorrelation behaves like $\mathrm{Ac}(\tau) \sim C\tau^{-b}$, with $b > 1$ and $C < 0$. Therefore, the spectrum will behave like:

$$\hat{S}_{[y]}(\omega) \sim \omega^{(b-1)}, \quad \omega \to 0. \tag{3.72}$$

This allows one to identify any region in the power spectrum that follows an increasing power law scaling as indicative of the presence of negative, scale-free correlations over the corresponding range of scales.

3.4.3 Memory in self-similar random processes

As noted above, the detection of self-similarity in time series is based on a comparison with self-similar random processes with stationary increments: fBm and fLm. It is therefore useful to discuss the autocorrelation and power spectra of these processes.

In the case of fBn, one has [21]

$$\mathrm{Ac}_{\Delta_h y^{fGn;H}}(\tau) \sim H(2H-1)|\tau|^{-2(1-H)}, \quad \tau \gg h, \tag{3.73}$$

where h is the spacing we introduced in section 3.2.4.1. That is, it has a power law tail, as expected. The memory exponent is $2(1 - H)$, where H is the self-similarity exponent of the process. In the special case $H = 1/2$ the complete tail vanishes due to the factor $(2H - 1)$. This is a consequence of the fact that all increments are uncorrelated. For $1/2 < H \leqslant 1$ the right-hand side is positive and the memory exponent lies between 0 and 1, revealing that positive long-term correlations are at play. For $0 \leqslant H < 1/2$, the right-hand side becomes negative while the memory exponent is larger than 1, implying that negative, long-term correlations become dominant. Its power spectrum, on the other hand, can be shown to scale as [21]

$$\hat{f}_h^{\mathrm{fgn;H}} \sim \frac{K(H)}{|\omega|^{2H-1}}, \quad \omega \ll h^{-1}, \tag{3.74}$$

with $K(H) > 0$. The spectrum is flat for $H = 1/2$, which corresponds to 'white' or uncorrelated noise. For $1/2 < H \leqslant 1$, the spectrum follows a decreasing power law, approaching -1 as $H \to 1$. For $0 \leqslant H < 1/2$, the power spectrum is an increasing power law, whose exponent approaches 1 as $H \to 0$. Interestingly, the tail exponent of the power spectrum, $S(\omega) \propto \omega^c$, and the tail exponent of the autocorrelation function, $\mathrm{Ac}(\tau) \propto \tau^{-a}$, are related by $c = a - 1$ for fGn. However, this relation only holds because the self-similarity of fGn is exact, over all scales. It often fails for processes that only exhibit self-similarity over finite, although significant, ranges of scales, which has led to some misinterpretations in the literature.

We have discussed the shape of the autocorrelation function and the power spectrum in relation to the memory of the process, and some studies indeed use the shape of the power spectrum (i.e. equation (3.74)) to estimate the self-similarity exponent H of the underlying process, instead of the more robust moment methods discussed in previous sections. The autocorrelation function (equation (3.73)) is rarely used for this task due to its usually very noisy tail.

Unfortunately, neither the autocorrelation nor the power spectrum can be computed for fLn processes. The reason for this is that the autocorrelation is a second moment and, as such, diverges for all Lévy laws, as does the power spectrum. Thus, neither is a useful quantifier for memory (or self-similarity) in the case of fLn. This is even true for finite data sets, where numerical estimates can always be made, yet the scaling exponents will diverge with the length of the data series and are therefore not meaningful. Some authors have proposed alternative measures to address this limitation (such as the autodifference [18]), but these schemes are rather involved. Below, we will discuss a practical method to circumvent this difficulty.

3.4.4 The rescaled range method

When dealing with experimental data, several methods can be used to look for long-term memory. Some of these are based on the autocorrelation or spectrum, as discussed above. However, these techniques may fail due to the presence of noise in the measurements which, combined with the finite length of experimental time series, can often make it hard or impossible to determine tail exponents [36]. In view of this and the mentioned limitations of these methods for Lévy processes, here we discuss a more robust technique for determining the tail exponents that does not rely on second order moments: the rescaled range method.

The rescaled range (or R/S) method was originally developed by H E Hurst in an attempt to determine the optimum capacity of a dam in the River Nile on the basis of historic river flow records [37], which requires knowing the range of water flow values occurring in a given period of time. In its original formulation, the R/S method was applied to a stationary time-series with Gaussian statistics, $\{y_k,\ k = 1, 2, \ldots, N\}$. The time series must be first demeaned,

$$\tilde{y}_k = y_k - \bar{y}, \quad \bar{y} = \frac{1}{N}\sum_{i=1}^{N} y_i, \tag{3.75}$$

and then integrated to form the cumulative variable

$$z_k = \sum_{i=1}^{k} \tilde{y}_i, \quad k = 1, 2, \ldots N. \tag{3.76}$$

Then, the so-called *rescaled range* is computed:

$$(R/S)_k = \frac{\max\{z_i,\ i = 1, \ldots, k\} - \min\{z_i,\ i = 1, \ldots, k\}}{\sigma_k}, \tag{3.77}$$

where σ_k^2 is the running variance:

$$\sigma_k^2 = \frac{1}{k}\sum_{i=1}^{k} \tilde{y}_i^2 - \left(\frac{1}{k}\sum_{i=1}^{k} \tilde{y}_i\right)^2. \tag{3.78}$$

The rescaled range measures the range of values explored by the integrated process during the first k iterations, normalized to the running standard deviation up to that time (see figure 3.16). If the original stationary series lacks any long-term dependence, its demeaned version should behave like a random Gaussian process so that the integrated process would essentially be a Gaussian random walk. The range of such a random walk increases with time (k) as $\sqrt{k}$. The presence of scale-free positive long-term correlations should lead to a scaling

$$(R/S)_k \propto k^{H_{R/S}}, \tag{3.79}$$

with $H_{R/S} > 1/2$. This occurs because any positive temporal correlation would translate into increments that tend to keep their sign more often than not (called 'persistence'), leading to a range that increases faster than would be the case for Gaussian noise. Likewise, one will obtain $H_{R/S} < 1/2$ if negative long-term correlations are present in the signal, since sign reversal of increments would then be more likely ('antipersistence').

Indeed, if this method is applied to computer-generated fGn with a self-similarity exponent H, the R/S analysis yields a power law scaling with $H_{R/S} = H$, confirming the usefulness of this method to test for self-similarity and to quantify the kind and extent of long-term memory of the underlying process [21]. In addition, the rescaled range method is quite resilient to noise, as has been shown elsewhere [25]. This makes it into a very powerful tool for the analysis of noisy experimental data.

Interestingly, the R/S technique can be modified to be applicable to a large family of Lévy processes, thus replacing the need to introduce more complicated concepts

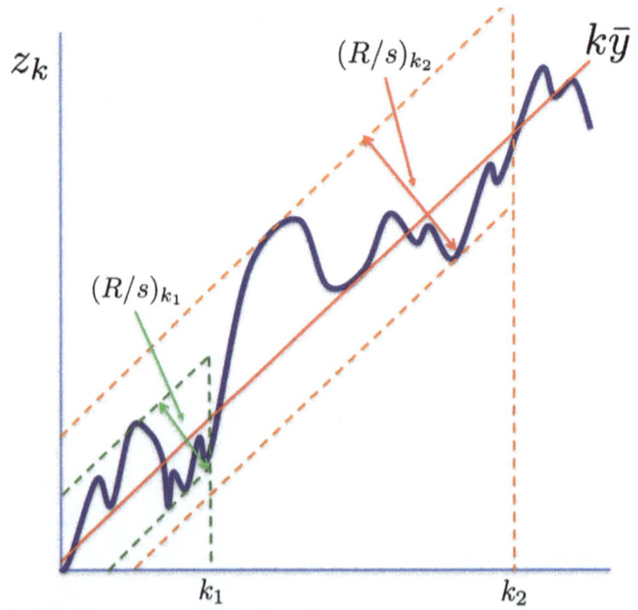

Figure 3.16. Illustration of the R/S method. The rescaled range is shown at two values of k.

such as the autodifference, referred to above. The range is a first order quantity, and as such, it converges for Lévy processes with $1 < \alpha < 2$. The variance used in the normalization of the range in the original R/S formulation does not, since it is a second-order moment. However, the variance is just a normalization that can easily be replaced by any other moment of order lower than α. Indeed, by redefining the rescaled range as [13]

$$(R/S)_k^s = \frac{\max\{z_i,\ i = 1,\dots,k\} - \min\{z_i,\ i = 1,\dots,k\}}{\left(\frac{1}{k}\sum_{i=1}^{k}|\bar{y}_i|^s\right)^{1/s}}, \quad s \ll \alpha, \tag{3.80}$$

a new Hurst exponent $H_{R/S;s}$ can be determined from

$$(R/S)_k^s \propto k^{H_{R/S;s}}. \tag{3.81}$$

It can be shown that fLn with $1 < \alpha < 2$ yields $H_{R/S;s} = H$, the self-similarity exponent of the associated fLm. Note, however, that the interpretation in terms of memory differs from above. Random Lévy motion has a self-similarity exponent of $1/\alpha$. Therefore, $H_{R/S;s} > 1/\alpha$ implies the presence of positive long-term memory, while $H_{R/S;s} < 1/\alpha$ implies antipersistence.

3.5 The search for long-term memory in fusion plasmas

First, we will illustrate the application of these techniques using the same turbulent signal used earlier in section 3.3 of this chapter: a measurement made by a Langmuir probe at the edge of the confined plasma in discharge No. 35427 of the W7-AS stellarator. Then, we will briefly discuss other interesting applications of these methods to fusion plasmas reported in the literature.

Illustration using probe data measured at the edge of the W7-AS stellarator

As discussed in some detail in section 3.3, the PDF of the W7-AS data is sufficiently close to a Gaussian to validate the use of the autocorrelation and power spectra. Starting with the autocorrelation function, one would like to establish whether its tail drops off like a power law or follows an exponential decay. Figure 3.17 illustrates the difficulties of answering this question directly. As shown in the inset, the tail oscillates with an amplitude too small to be significant but completely obscuring the decay of the tail. This behaviour is typical of noisy experimental data. On the other hand, the analysis of the low-lag part provides estimates for the autocorrelation time of the signal in the range of (12–36) μs, depending on the method used. These estimates lie in the same ballpark, as expected.

The power spectrum of this time series was already discussed in section 2.4.1 and shown in figure 2.25. As noted in the figure caption, an averaging procedure was used to improve the reliability of the spectral estimate. Here, we focus exclusively on the dynamical interpretation of the three power law scaling regions that are visible in figure 2.25. Assuming the spectrum behaves like $\hat{S}(f) \sim f^{-\alpha}$ in each region, we obtained: $\alpha \sim 0$ for $f < 10$ kHz; $\alpha \sim 1$ for $10 < f < 100$ kHz; and $\alpha \sim 4$ for $f > 100$ kHz. The scaling exponent in the intermediate region is consistent with what is expected for a transport process that exhibits long-term, self-similar, persistent correlations.

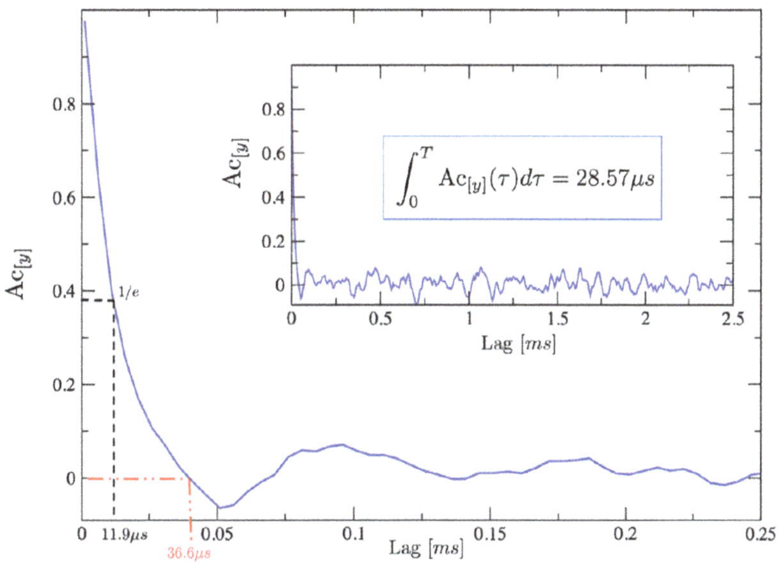

Figure 3.17. Autocorrelation function of the ion saturation current signal from W7-AS discharge No. 35425. The inset shows the same function for much longer time lags.

The intermediate frequency range corresponds to timescales that are well above the local turbulence correlation time, suggesting that this spectral feature arises and is maintained at scales exceeding local turbulent dynamics. In addition, we found that the cross spectrum of the ion saturation signal (a proxy for density fluctuations) and poloidal electric field, measured simultaneously by the same Langmuir probe, was found to have the largest magnitude in this range (see figure 2.28). This shows that the overall turbulent flux is dominated by contributions from this intermediate spectral range, something that was not evident from the spectrum of the ion saturation current by itself, and indicates that the self-organization of the plasma has a significant impact on the turbulent flux [38].

The values of the exponents found in the other two scaling regions can also be understood. The exponent corresponding to the high frequency range ($\alpha \sim 4$) is determined by the average shape of the individual transport bursts present in the signal, with durations of a few microseconds. In fact, it can be shown [15] that the value of this exponent can be changed at will by modifying the characteristic shape of these bursts, but keeping their temporal order (i.e. preserving temporal correlations). The flat spectral shape ($\alpha \sim 0$) of the low frequency range ('white' noise), on the other hand, suggests the presence of decorrelation. It is impossible, in a finite system, to maintain correlations over all scales. At some point, finite-size effects will enter and terminate the correlation range; this region is probably a reflection of this fact.

To continue our search for scale-invariant, long-term memory, we turn to the R/S analysis. The rescaled range graph for the W7-AS ion saturation signal is shown in figure 3.18. Three regions are apparent in the graph of the rescaled range, as shown in the top panel. First, there is a region where $R/S \propto \tau$ for time lags smaller than $\tau \sim (50 - 60)$ μs, followed by a second scaling region with $R/S \propto \tau^{(0.63-0.64)}$ for lags in the range of 100 μs–2 ms, and a third, rather short, region extending for $\tau > 3$ ms, for which it is difficult to determine any exponent. The extension and value of the Hurst exponent in each region can be clarified by calculating the instantaneous Hurst exponent, defined as

$$H^{\text{inst}}_{[R/S]}(\tau) = \frac{\tau}{R/S}\frac{d}{d\tau}(R/S), \tag{3.82}$$

shown in the bottom panel of figure 3.18. Even though the derivative amplifies fluctuations, this graph is very useful since it provides a clear visualization of the constancy of H over an extended range of scales, which occurs when $R/S \propto \tau^H$. This constancy is particularly clear for the second region, but fails for the largest time lags where the oscillations of the value of $H(\tau)$ are too large to make a decision.

The dynamical interpretation of these regions follows our previous discussions on R/S analysis. The region with $H \simeq 0.63$ for lags between 100 μs and 2×10^3 μs is consistent with the existence of long-term, scale-invariant positive correlations in the signal, noting that these timescales are well above the correlation time of the signal. Interestingly, this range overlaps very nicely with the [1 – 10] kHz frequency range in the power spectrum, where we found a decaying power law ω^{-1}, thus reinforcing the interpretation of previous paragraphs. In addition, the region with $H \simeq 1$ for

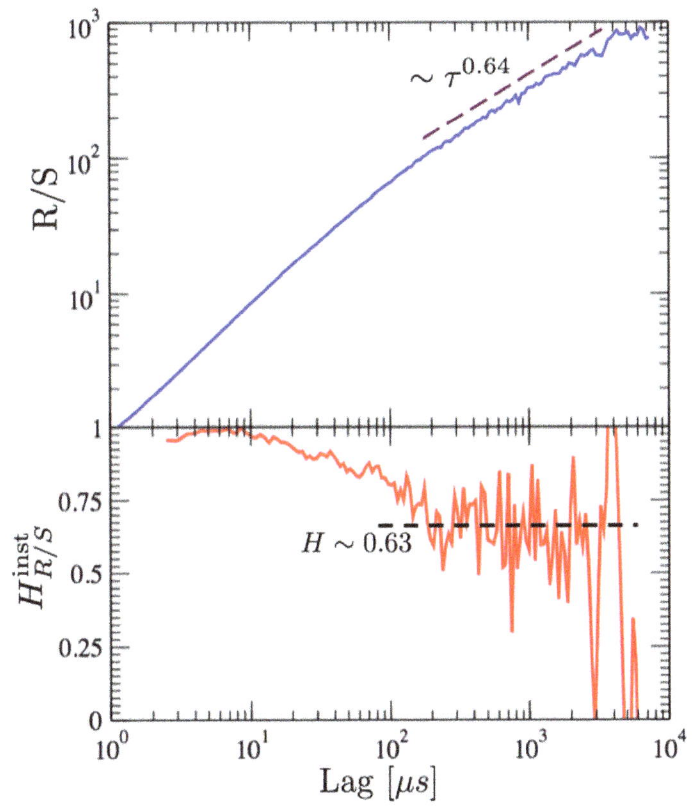

Figure 3.18. Rescaled range analysis of the ion saturation current signal from W7-AS discharge No. 35425. The top panel shows the rescaled range as a function of lag; the bottom panel shows the instantaneous Hurst exponent defined in the main text.

$\tau < (50 - 60)\ \mu s$ is also consistent with the range of frequencies where the spectrum falls off like ω^{-4}, dominated by single-burst dynamics.

Selected investigations on fusion plasmas

The features just described for the W7-AS data from discharge No. 35425 are not unique, but rather turn out to be typical for turbulence measurements in many fusion devices [25, 29, 39–41]. Consequently, the R/S analysis has become a common and useful tool for the characterization of memory in the edge of fusion plasmas[9]. The Hurst exponent has been determined from Langmuir probe data in the edge and scrape-off layer (SOL) of seven different magnetic confinement devices (tokamaks, stellarators, a reversed field pinch and a linear device), as reported in [25]. The observed mesorange value for the Hurst exponent was very similar for floating

[9] Furthermore, it has also been used extensively as a tool to estimate the self-similarity exponent H, due to the fact that it coincides with the Hurst exponent for fBn [21]. Several papers follow this approach and even compare its performance with some of the other methods discussed previously [26, 27, 36].

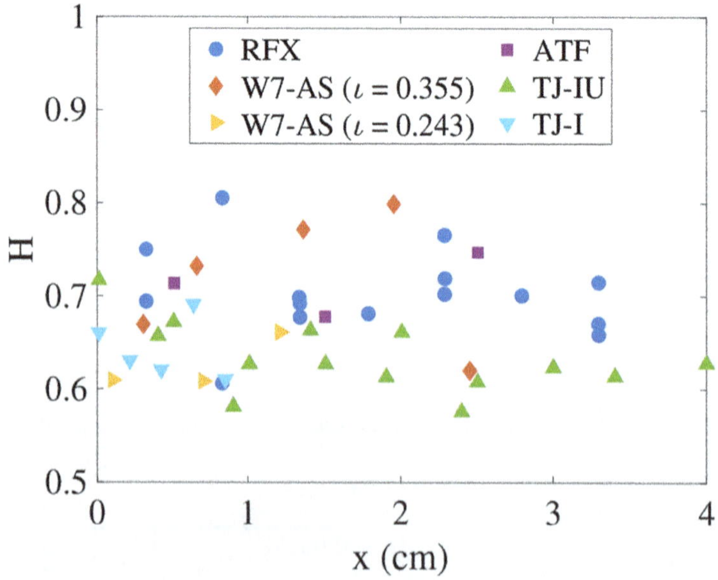

Figure 3.19. The radial profile of the Hurst coefficient H inside the last flux surface for several devices. (Reproduced after [25], with the permission of AIP Publishing.)

potential and ion saturation current measurements. It was also found that the mesorange Hurst exponent varied over a large range in the SOL, $0.5 < H < 1$, but tended to be concentrated in a narrow range in the plasma edge, $0.6 \leqslant H \leqslant 0.72$ (not considering the linear device). A few example results are shown in figure 3.19. The comparison with the linear device is particularly interesting, as it has no rotational transform and confinement times are very small. In this case, H was close to 0.5. For that reason, the high asymptotic values of H were linked to the existence of significant confinement, or, in other words, to a clear separation of the turbulent timescale and the confinement timescale.

Another interesting study that uses the R/S method to calculate the Hurst exponent is illustrated in figure 3.20. This figure shows the autocorrelation function for ion saturation current measurements taken at the W7-AS edge, calculated over small 1 ms time windows in order to be able to follow its evolution in time. The autocorrelation function broadens intermittently, implying the tail decays slower for specific time windows. In the same time windows, the H exponent is also estimated. It can be seen that the times of broadening of the autocorrelation function tend to coincide with peaks in H. This is experimental confirmation of several statements made in section 3.4.1 that associated a value of $H \neq 0.5$ to the presence of power law (i.e. fat) tails of the autocorrelation function, as required for the establishment of long-term memory. The physical phenomena underlying these observations could involve large, intermittent structures.

Yet another study involving the use of the Hurst exponent and R/S analysis was performed at the TJ-II stellarator. This study involved experiments to characterize the behaviour of turbulence in the vicinity of a magnetic island in ECRH heated

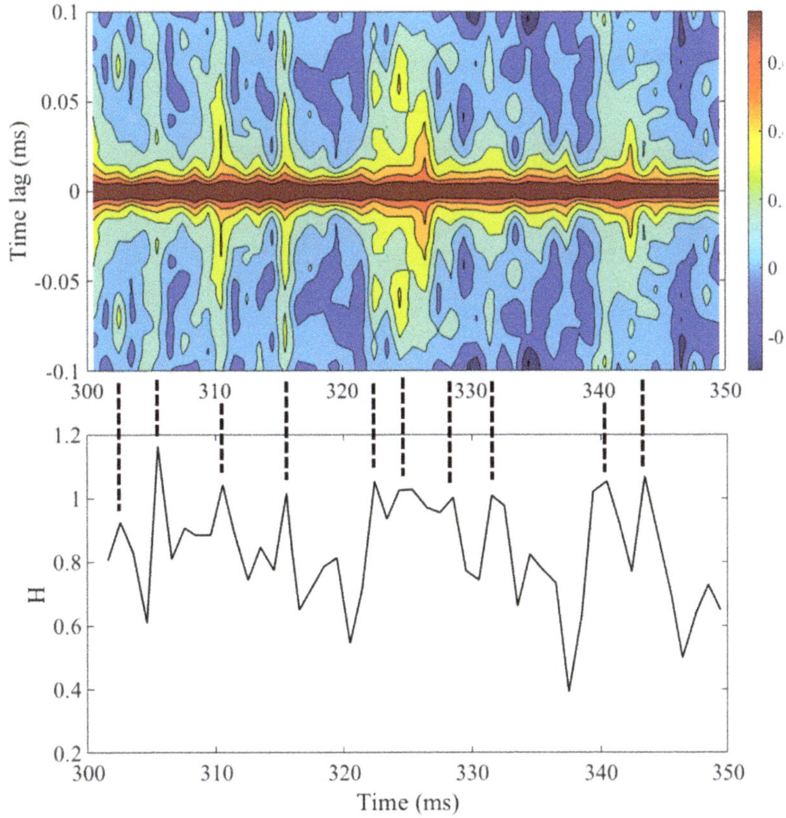

Figure 3.20. The autocorrelation function for edge fluctuation data from W7-AS calculated in intervals of 1 ms (top). The tail of the autocorrelation function broadens intermittently, and this is correlated with increases of the *H* coefficient, shown in the lower plot, as indicated by the dashed lines. (Reproduced after [25], with the permission of AIP Publishing.)

plasmas. For this purpose, the position of a large magnetic island, associated with the rational surface $\iota\!\!\!-= 3/2$, was slowly scanned outwards by modifying the rotational transform via the external induction of Ohmic current [42]. By performing Doppler reflectometry (DR) measurements with an approximately constant radial position of the reflection layer, the effective radial variation of various quantities in the vicinity of the island could be explored. The two-channel reflectometer detected a rather strong variation of the perpendicular flow velocity $v_\perp$ (and hence radial electric field) as the magnetic island was swept past the observation point of the reflectometer, as shown in figure 3.21 [43]. This allowed establishing the width of the island at about 3 cm—considerable in view of the minor radius of about 20 cm.

The Hurst exponent *H* was then determined by applying the R/S analysis to the fluctuating amplitude of the reflectometry signal. To improve statistics, results were averaged over several discharges, taking the time of maximum variation of $v_\perp$ as a reference time. The resulting average evolution of *H* versus Δt, the time minus the reference time, is shown in figure 3.22, for the two Doppler reflectometry channels,

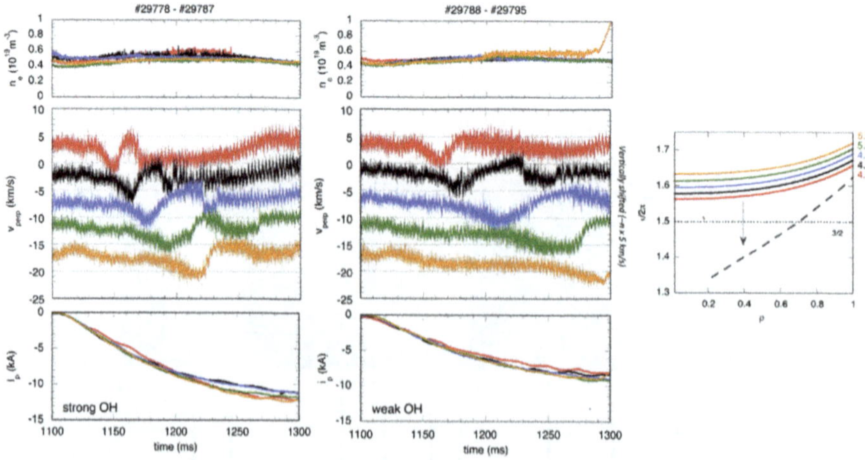

Figure 3.21. The time evolution of the line averaged plasma density (top), perpendicular rotation velocity (middle), and net plasma current (bottom) measured in five different magnetic configurations whose rotational transform profiles are shown in the right panel for strong (left) and weak (centre) OH current intensity. Each magnetic configuration represented with a different colour is characterized by the current in the helical coil; it changes from 4.4 kA (in red) to 5.2 kA (in yellow) with the subsequent change in the rotational transform (from 1.66 to 1.72 at the plasma edge). Note that the velocity traces have been shifted in the vertical direction in steps of 5 km s^{-1} for a clearer representation. (Reproduced with permission from [43]. © IOP Publishing. All rights reserved.)

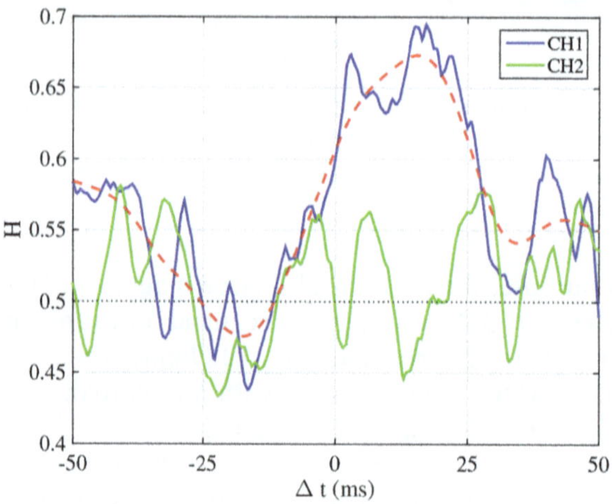

Figure 3.22. The mean Hurst coefficient (calculated from $|\tilde{n}|$ for eight discharges and averaged) for Doppler reflectometry channels 1 and 2. The red dashed line indicates the smoothed trend (channel 1). (Reproduced with permission from [44]. © IOP Publishing. All rights reserved.)

located at about $\rho \simeq 0.75$ (channel 1) and $\rho \simeq 0.68$ (channel 2). This means that channel 1 is located roughly at the position of the steepest density gradient, whereas channel 2 is located further inward, where the gradient is less steep. Interestingly,

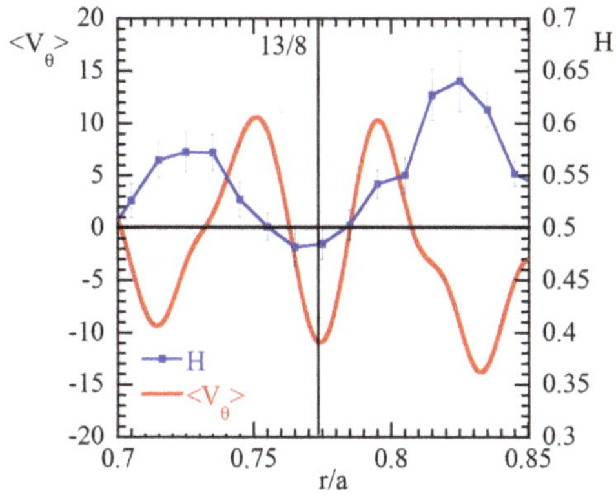

Figure 3.23. Modelling results: mean poloidal flow and Hurst exponent near an important rational (13/8, indicated by the vertical line). (Reproduced with permission from [44]. © IOP Publishing. All rights reserved.)

channel 1 shows a very significant response of H at the trailing edge of the island as it sweeps past the observation point (i.e. inward from the island), whereas the response of channel 2 is not impressive.

To better understand these observations, a simulation was performed of magnetohydrodynamic turbulence induced by resistive interchange modes in a periodic cylinder (see appendix B). Rather than attempting to model the experimental situation in detail, the profile of poloidal flow and Hurst coefficient around a few major low-order rational surfaces was studied. A typical result is shown in figure 3.23. The averaged poloidal flow $\langle V_\theta \rangle$ has a minimum at the rational surface, while the flow has a strong shear in the region of the magnetic island associated with the resonant surface. H was observed to exhibit a very clear structure associated with the rational surface. Specifically, one observes that the H coefficient has a minimum near the rational surface, with minimum values even dropping below 0.5. This is consistent with the idea that the region immediately surrounding a singular surface constitutes a trapping zone (or attractor) for particle motion, leading to antipersistent behaviour. The H coefficient increases significantly above 0.5 as one moves away from the resonant surface. This behaviour was seen to occur fairly systematically at several low-order resonant surfaces in the model [44].

When comparing this result to the experimental data, one should be very cautious, as the rationals considered in the model do not correspond to the rational of the experiment ($\iota\!\!\!-$ = 3/2)—among other differences. One important such difference is the in–out symmetry of the $\langle V_\theta \rangle$ profile around the rational. While the experimental velocity profile appears to be asymmetric (Figure 3.21), the model velocity profile is symmetric (figure 3.23). This symmetry depends on the eigenfunction of the instability corresponding to the rational surface [45]. In addition, one should consider that the mean poloidal flow $\langle V_\theta \rangle$, averaged over the poloidal and toroidal angles, is not the same as the local perpendicular flow, $v_\perp$, measured by the reflectometer, which may depend on the particular location of measurement.

Nevertheless, the decrease of H below 0.5 could be consistent with the experimental result of figure 3.22, which shows H values below 0.5 at $\Delta t \simeq -20$ ms, and relatively high values of H on both sides (the peak for positive Δt being very clear, although the peak for negative Δt is not well resolved).

The large value of H in the region surrounding the rational surface is interesting, as $H > 0.5$ is associated with plasma self-organization and avalanches. In [44] we also looked at heat transport in these experiments (based on ECE electron temperature measurements with modulated ECR heating), and found indications that the region of large H is also a region with relatively enhanced radial heat transport. Thus, a tantalizing link between the self-similarity of density fluctuations and enhanced heat transport is suggested, possibly providing an experimental handle on the turbulent mechanism of heat transport (avalanches?). Clearly, further work is needed to clarify these issues.

3.6 Characterization of self-similarity and memory via event burst sampling

So far in this chapter we have characterized the self-similarity and memory of time series by comparison to (one-dimensional) random processes. However, turbulence is not a one-dimensional random process, but rather a spatiotemporal fluctuating 'field'. One of the ways that this fact expresses itself is through 'burstiness': we saw a hint of this in section 2.1.1, and we will come back to this important issue in chapter 5. An example of the complex spatiotemporal dynamics of turbulence is provided by the ion saturation current measured by a Langmuir probe (as shown earlier in figure 2.2): it fluctuates rapidly in time, but the fluctuations are much larger over the specific time intervals corresponding to significant transport events (i.e. an 'avalanche') passing by the probe position. These 'bursts' can be identified and quantified by choosing a proper amplitude threshold to filter out small fluctuations. There are various ways to do this, but a well-accepted method is to consider that a large event occurs whenever the signal exceeds a few[10] times the standard deviation of the signal (see figure 3.24). The *duration* of the event then corresponds to the lapse of time during which the signal remains above that threshold. The *energy* of the event is defined as the time integral of the signal (minus the threshold) over that same lapse of time. Finally, the *waiting* or *quiet time* is defined as the lapse of time between two successive events.

3.6.1 Self-similarity of events and criticality

If the complex process that drives transport in the system is evolving according to self-similar, critical dynamics, one of its main features will be the lack of any characteristic scale. Mathematically, this translates into the divergence of the first

[10] Two or three.

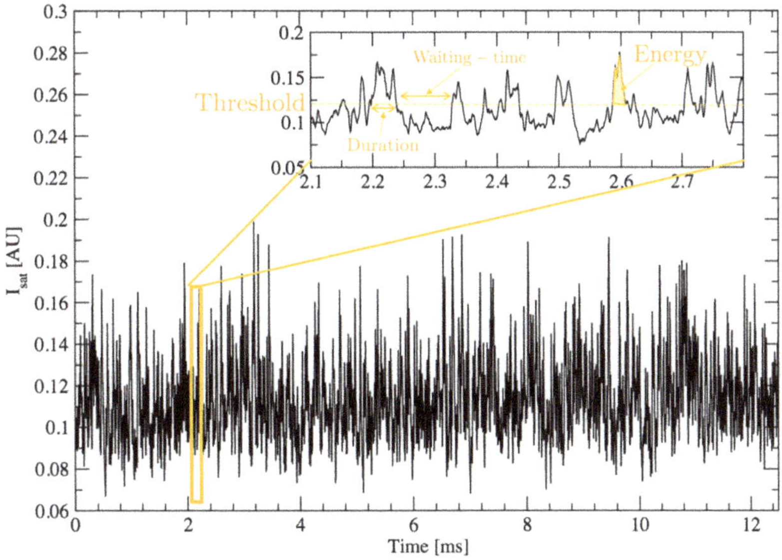

Figure 3.24. Conditional sampling of the ion saturation current signal from W7-AS discharge No. 35425. The concepts of *threshold, event duration, waiting time,* and *energy* of an event are illustrated.

non-zero moment (the mean) associated with any positive quantity that characterizes the transport events, such as their duration or their energy. As noted, a divergent mean implies that the PDFs of both durations and energies follow a power law:

$$p(d) \propto d^{-(1+b)}, \quad p(e) \propto \mathrm{e}^{-(1+b')}, \tag{3.83}$$

where the exponents b and b' must be both positive and lie between 0 and 1.

3.6.2 Event correlation: waiting times

Any memory present in the identified set of bursts can be revealed by studying the statistics of waiting times. Indeed, the triggering of any burst event is akin to a point process—i.e. a collection of (possibly) random points distributed over the temporal axis. In the absence of correlations (i.e. memory), all point processes become Poisson processes [46], with a waiting time distribution that follows an exponential PDF,

$$p(w) = \tau_0^{-1}\exp(-w/\tau_0), \tag{3.84}$$

where τ_0 corresponds to the average time between event triggers (and τ_0^{-1} is the average triggering rate). Long-range, self-similar correlations between events, however, would generate a waiting time distribution with the following asymptotic form [47, 48],

$$p(w) \propto w^{-(1+\lambda)}, \quad \lambda \in (0, 1), \tag{3.85}$$

where the value of the exponent again ensures that no mean triggering time (characteristic timescale) can be defined.

3.7 Burst sampling in fusion plasmas

We will illustrate this methodology once again using the ion saturation current signal measured at the edge of the plasma in discharge No. 35427 from the W7-AS stellarator. The PDFs for the durations and energies associated with individual bursts, as identified in the ion saturation current signal after choosing a threshold value (in arbitrary units) of 0.118 (roughly, three times the standard deviation of the data), are shown in figure 3.25. For the durations (top frame), an extended decaying power law region is found, roughly scaling as $p(d) \propto d^{-1.6}$ for a range of durations between (10–70) μs. The value of the exponent is sufficiently small, according to our previous discussion in section 3.6.1, to suggest that a characteristic event duration is absent, which is consistent with critical behaviour. The distribution of energies (bottom, left frame) also reveals a power law scaling region, with a scaling $p(E) \propto E^{-1.20}$, again in the range of values that is consistent with critical and self-similar behaviour. In this case, the range of the scales of self-similar scaling is quite a bit longer, spanning well over a decade.

Finally, the PDF of the waiting times between successive bursts, identified using the same threshold, is shown in the bottom right frame of figure 3.25. The log–log scale used in the plot shows that the scaling is not exponential, but closer to power law. It is found that $p(w) \propto w^{-1.22}$ over more than a decade, with waiting times in the

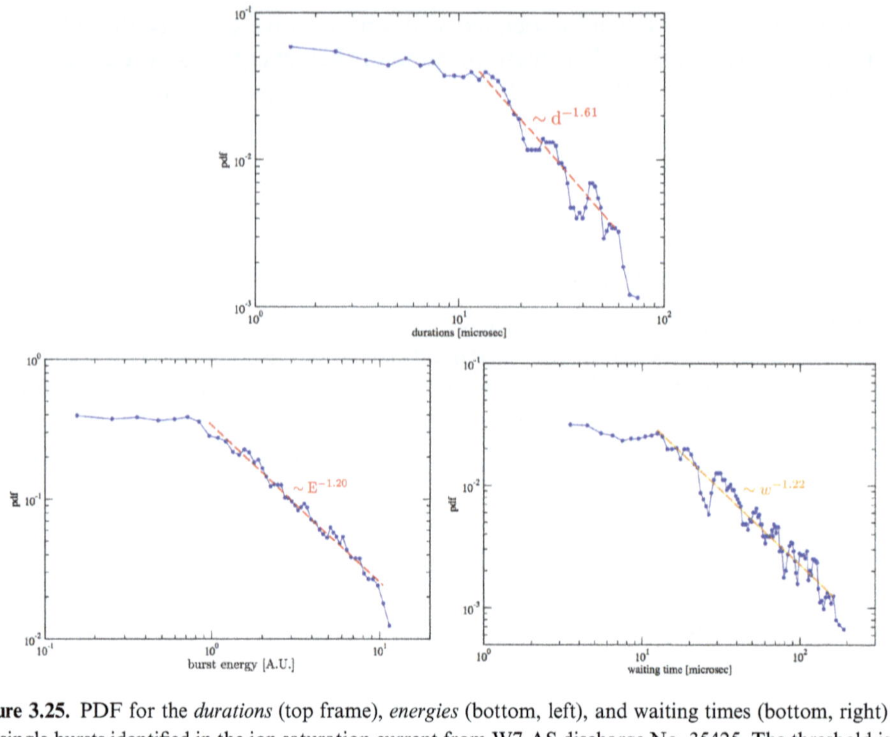

Figure 3.25. PDF for the *durations* (top frame), *energies* (bottom, left), and waiting times (bottom, right) for the single bursts identified in the ion saturation current from W7-AS discharge No. 35425. The threshold is set at three times the standard deviation.

range of $(10 - 200)$ μs. This means that the events above the threshold are not being triggered randomly, but correlated in time. That is, the energy and duration of events is somehow conditioned by the duration and energies of the bursts that preceded them in time. The small value of the exponent again points to critical dynamics.

Selected studies in fusion plasmas

Curiously, the identification of bursts using thresholding has been used mainly to carry out waiting time studies in the context of fusion research. Probably the first such investigation was performed at the Italian RFX reversed field pinch [49], using ion saturation current data obtained with Langmuir probes, as well as data from microwave reflectometry. Bursts were identified using a threshold a value set at 2σ, and the waiting times between them scaled as $p(w) \propto w^{-1.68}$ for probe data, and $p(w) \propto w^{-1.93}$ for the reflectometry data, both in the range of $(1 - 250)$ μs.

Other studies looked at probe data from the edge of the JET tokamak and the W7-AS and TJ-II stellarators. These investigations also found power law scalings for the waiting times, roughly as $p(w) \propto w^{-2}$, in a range of $(0.1 - 10)$ ms, using a similar threshold [48]. The dependence of these scalings on the choice of threshold was also studied. Interestingly, it was found that an exponential scaling was obtained when no thresholding was applied, or when the threshold was chosen too small, reflecting the fact that the noisy nature of the signals dominates the shortest timescales in the process. The minimum threshold needed to reveal power law behaviour was found to be the value for which the majority of identified bursts had durations (and carried energies) in the scale-free region of the duration (or energy) PDF [48]. Further studies, with similar results, were carried out at the TEXTOR tokamak [39] and the LHD stellarator [50], again confirming that exponential waiting time scalings changed into power laws when a sufficiently large threshold was used.

3.8 Summary

As noted earlier in this book, a turbulent time series cannot be equated with a random variable, as turbulence has an essential spatio-temporal aspect that a random variable lacks. Nevertheless, in this chapter we have relied heavily on the theory of the statistics of random variables in order to quantify significant aspects of turbulence: self-similarity, memory effects, and the shapes of probability distributions. There is no contradiction here, as the statistical theory of random numbers is only used to quantify these specific aspects of turbulent time series; at no point is turbulence equated to a random time series.

Using a series of methods, we have been able to confirm, in a wide range of devices and conditions, that fluctuating time series from the edge region of fusion plasmas show significant memory effects, extending up to three (!) orders of magnitude beyond the turbulence correlation time. We speculate that this long-term memory is stored in the plasma profiles, so that the detected memory is a clear reflection of the interaction between turbulence (fast timescales) and transport (establishing the plasma profiles on slow timescales, of the order of the confinement time). It follows, from these observations and many others reported in the literature, that any numerical simulation that intends to reproduce this type of spatio-temporal dynamics, even if only qualitatively, should therefore include the simultaneous evolution of and interaction between both fast and slow timescales, and not assume a strict separation from the start as is often done. In the same vein, these findings also cast some doubt on the validity of diffusive models to capture the fundamental properties of turbulent transport in these plasma regimes, since diffusion is local and lacks memory by definition.

References

[1] Sánchez R and Newman D E 2015 Self-organized criticality and the dynamics of near-marginal turbulent transport in magnetically confined fusion plasmas *Plasma Phys. Control. Fusion* **57** 123002

[2] Diamond P H and Hahm T S 1995 On the dynamics of turbulent transport near marginal stability *Phys. Plasmas* **2** 3640

[3] Carreras B A, Newman D, Lynch V E and Diamond P H 1996 A model realization of self-organized criticality for plasma confinement *Phys. Plasmas* **3** 2903

[4] Newman D E, Carreras B A, Diamond P H and Hahm T S 1996 The dynamics of marginality and self-organized criticality as a paradigm for turbulent transport *Phys. Plasmas* **3** 1858

[5] Gentle K W *et al* 1995 An experimental counter-example to the local transport paradigm *Phys. Plasmas* **2** 2292

[6] Perkins F W *et al* 1993 Nondimensional transport scaling in the Tokamak Fusion Test Reactor: is tokamak transport Bohm or gyro-Bohm? *Phys. Plasmas* **5** 477

[7] Petty C C *et al* 1995 Nondimensional transport scaling in DIII-D: Bohm versus gyro-Bohm resolved *Phys. Plasmas* **2** 2342–8

[8] Politzer P A 2000 Observation of avalanchelike phenomena in a magnetically confined plasma *Phys. Rev. Lett.* **84** 1192

[9] van Milligen B P, Sánchez R and Carreras B A 2004 Probabilistic finite-size transport models for fusion: anomalous transport and scaling laws *Phys. Plasmas* **11** 2272

[10] Bak P, Tang C and Wiesenfeld K 1987 Self-organized criticality: an explanation of the 1/*f* noise *Phys. Rev. Lett.* **59** 381

[11] Per B 1996 *How Nature Works* (New York: Springer Science)

[12] Jensen H J 1998 *Self-organized Criticality: Emergent Complex Behaviour in Physical and Biological Systems* (Cambridge: Cambridge University Press)

[13] Sanchez R and Newman D E 2018 A primer on complex systems: with applications to astrophysical and laboratory plasmas *Lecture Notes in Physics* vol 943 (Berlin: Springer)

[14] Huang K 2009 *Phase Transitions* (London: Taylor and Francis)

[15] Hwa T and Kardar K 1992 Avalanches, hydrodynamics and discharge events in models of sandpiles *Phys. Rev.* A **45** 7002

[16] Mier J A, Sanchez R and Newman D E 2016 Characterization of a transition in the transport dynamics of a diffusive sandpile by means of recurrence quantification analysis *Phys. Rev.* E **94** 022128

[17] Ross S M and Processes S II 1995 *Stochastic Processes* II edn (New York: Wiley)

[18] Samorodnitsky G and Taqqu M S 1994 *Stable Non-Gaussian Processes* (London: Chapman and Hall)
[19] Feller W 1968 *Introduction to Probability Theory and Applications* (New York: Wiley)
[20] Einstein A 1905 Über die von der molekulartheoretischen Theorie der Wärme geforderte Bewegung von in ruhenden Flüssigkeiten suspendierten Teilchen *Ann. Phys., Lpz.* **17** 549
[21] Mandelbrot B B 1968 Fractional Brownian motions, fractional noises and applications *SIAM Reviews* **10** 422
[22] Meneveau C and Sreenivasan K R 1991 The multifractal nature of turbulent energy dissipation *J. Fluid Mechanics* **224** 429
[23] Halsey T C, Jensen M H, Kadanoff L P, Procaccia I and Sraiman B I 1986 Fractal measures and their singularities *Phys. Rev.* A **33** 1141
[24] Feder J 1988 *Fractals* (New York: Plenum)
[25] Carreras B A 1998 Self-similarity of the plasma edge fluctuations *Phys. Plasmas* **5** 3632
[26] Gilmore M, Yu C X, Rhodes T L and Peebles W A 2002 Investigation of rescaled range analysis, the Hurst exponent, and long-time correlations in plasma turbulence *Phys. Plasmas* **9** 1312
[27] Yu C X, Gilmore M, Peebles W A and Rhodes T L 2003 Structure function analysis of long-range correlations in plasma turbulence *Phys. Plasmas* **10** 2772
[28] Antoni V, Carbone V, Martines E, Regnoli G, Serianni G, Vianello N and Veltri P 2001 Electrostatic turbulence intermittency and MHD relaxation phenomena in a RFP plasma *Europhys. Lett.* **54** 51
[29] Wang W H, Yu C X, Wen Y Z, Wang L, Yang X Z and Feng C H 2004 Effect of poloidal sheared flow on the long-range correlation characters of edge plasma turbulent transport *Phys. Plasmas* **11** 2075
[30] Jha R, Kaw P K, Kulkarni D R and Parikh J C 2003 Evidence of Lévy stable process in tokamak edge turbulence *Phys. Plasmas* **10** 699
[31] Ohno N, Budaev V P, Takamura S and Masuzaki S 2006 Superdiffusion and multifractal statistics of edge plasma turbulence in fusion devices *Nucl. Fusion* **46** S181–91
[32] Neto C R, Guimaraes-Filho Z O, Caldas I L, Nascimento I C and Kuznetsov Y K 2008 Multifractality in plasma edge electrostatic turbulence *Phys. Plasmas* **15** 082311
[33] Rajkovic M, Skoric M, Solna K and Antar G 2008 Characterization of local turbulence in magnetic confinement devices *Nucl. Fusion* **48** 024016
[34] Fuller W A 1986 *Introduction to Statistical Time Series Analysis* (New York: Wiley)
[35] Bracewell R N 2000 *The Fourier Transform and Its Applications* (New York: McGraw-Hill)
[36] Carreras B A and Newman D E 1999 Long-range time dependence in the cross-correlation function *Phys. Plasmas* **6** 485
[37] Hurst H E 1951 Long term storage capacity of reservoirs *Trans. Am. Soc. Civ. Eng.* **116** 770
[38] Carreras B A *et al* 1999 Characterization of the frequency ranges of the plasma edge fluctuation spectra *Phys. Plasmas* **6** 4615
[39] Xu Y H, Jachmich S and Weynants R R 2005 On the properties of turbulence intermittency in the boundary of the TEXTOR tokamak *Plasma Phys. Control. Fusion* **47** 1841
[40] Xu G S, Wan B N, Zhang W, Yang Q W and L Wang 2006 Multiscale coherent structures in tokamak plasma turbulence *Phys. Plasmas* **13** 102509
[41] dos Santos Lima G Z, Iarosz K C, Batista A M, Caldas I L, Guimarães-Filho Z O, Viana R L, Lopes S R, Nascimento I C and Kuznetsov Y K 2012 Self-organized criticality in MHD driven plasma edge turbulence *Phys. Lett.* A **376** 753–7

[42] Romero J A, López-Bruna D, López-Fraguas A and Ascasíbar E 2003 Controlling confinement with induced toroidal current in the flexible heliac TJ-II *Nucl. Fusion* **43** 387
[43] Estrada T, Ascasíbar E, Blanco E, Cappa A, Hidalgo C, Ida K and López-Fraguas A 2016 Plasma flow, turbulence and magnetic islands in TJ-II *Nucl. Fusion* **56** 026011
[44] van Milligen B P *et al* 2016 The role of magnetic islands in modifying long range temporal correlations of density fluctuations and local heat transport *Nucl. Fusion* **56** 016013
[45] van Milligen B P, Carreras B A, de la Luna E and Solano E R 2019 Radial variation of heat transport in L-mode JET discharges *Nucl. Fusion* **59** 056006
[46] Cox D R and Isham V 1980 *Point Processes* (London: Chapman and Hall)
[47] Sánchez R, Newman D E and Carreras B A 2002 Waiting-time statistics of self-organized-criticality systems *Phys. Rev. Lett.* **88** 068302
[48] Sánchez R 2003 Quiet-time statistics of electrostatic turbulent fluxes from the JET tokamak and the W7-AS and TJ-II stellarators *Phys. Rev. Lett.* **90** 185005
[49] Spada E *et al* 2001 Search of self-organized criticality processes in magnetically confined plasmas: hints from the reversed field pinch configuration *Phys. Rev. Lett.* **86** 3032–5
[50] Carralero D, Calvo I, Shoji M, Carreras B A, Ida K, Ohdachi S, Sakakibara S, Yamada H and Hidalgo C 2011 Influence of β on the self-similarity properties of LHD edge fluctuations *Plasma Phys. Control. Fusion* **53** 095010

IOP Publishing

Analysis of Turbulence in Fusion Plasmas

Boudewijn Ph van Milligen and Raul Sanchez

Chapter 4

Causality

As seen in the preceding chapters, the complex systems in which turbulence occurs involve various variables or 'fields' that interact in multiple feed-forward or feedback loops. For example, such a set of fields could consist of a gradient, the amplitude of the turbulence driven by this gradient, and a sheared flow, driven by the turbulence amplitude, that would suppress turbulence. Typically, such interactions would be described by differential equations that may be strongly non-linear.

To disentangle and understand these interactions and to be able to compare such non-linear interactions between experimental systems and the corresponding numerical models, appropriate tools are needed. The idea of feed-forward/feedback loops immediately brings one to consider the impact of one variable on another in terms of cause and effect. Which methods are available to study the causal relations between fluctuating variables?

If one can intervene in the system under study and modulate the value of one variable, the observation of the (delayed) reaction of other variables to this modulation sometimes allows establishing a probable causal relationship. A basic prior understanding of the underlying mechanics of the system is usually needed to interpret the results [1]. However, some systems do not permit such external intervention, or perhaps there is no practical way to modify a relevant variable (no control 'knob' is available), or the system could be too complex to allow a straightforward interpretation of the observations. How would one proceed to study causality in this case?

If a (rudimentary or partial) model is available, one may attempt to match the (local or instantaneous) system evolution to the predicted evolution, or to quantify parameters related to system evolution (growth rates, damping rates, etc).

Another possibility would be to simply observe the temporal evolution of the variables, and see whether there are systematic time delays between the variations of different variables. The standard procedure for detecting a systematic relation between two variables is to calculate the linear correlation coefficient $C_{X,Y}$

doi:10.1088/978-0-7503-4856-0ch4

(see section 2.2). However, if turbulence is strong, non-linear effects may blur and obliterate any linear correlation, if it exists, because waveforms may change significantly in shape as they propagate from one measurement location (of a variable X, say) to the next (of Y). In this case, the correlation coefficient may completely fail to identify the relation between the two concerned variables or measurements and lead to confusing or erroneous interpretations. This failure of the correlation coefficient for the study of strongly turbulent systems highlights the need for alternative techniques to detect causal relations between variables.

4.1 Antecedents

The starting point of considerations involving causality is the traditional philosophical concept of deterministic causation. It refers to the relation between two repeatedly occurring events, labelled X and Y. One considers that X causes Y when two conditions are met:

1. *Necessity*: if X occurs, then Y must occur.
2. *Sufficiency*: if Y occurs, then X must have occurred.

Clearly, causation involves a temporal ordering in the sense that the effect Y cannot occur before the cause X.

The definition of deterministic causation relies on the precise identification of the events X and Y, which is not a practical proposition for data analysis. Instead, a statistical approach based on probabilities is needed.

Correspondingly, Wiener introduced the concept of 'quantifiable causality' [2]: 'For two simultaneously measured signals, if we can predict the first signal [Y] better by using the past information from the second one [X] than by using the information without it, then we call the second signal causal to the first one.' Note that this definition involves the concept of 'information' in a fundamental way. Thus, causality is understood to involve not only a temporal ordering, but also a flow of information, which is directional in view of the temporal ordering.

This definition is amenable to algorithmic analysis and quantification. It was Granger[1] who first formulated an algorithm to evaluate and quantify causality on the basis of this idea [3]. His algorithm is based on a linear prediction of the evolution of a time series involving multivariate minimization, and comparing the result to the actual evolution. This approach does not appear to be very adequate for the analysis of turbulence due to the linear nature of the prediction. Although non-linear generalizations are possible and have indeed been elaborated [4], here we will turn to a non-parametric procedure for causality detection originating in the field of information theory. Instead of trying to predict the evolution of the time series, it relies on the amount of information we dispose of that *could* allow us to make such a prediction (without actually making it).

[1] Clive W J Granger, Nobel Prize in Economic Sciences, 2003.

4.2 The quantification of information

The information content of a random variable $X = \{x_i, i = 1, \dots, N\}$, with probability distribution $p_X(x)$ is given in terms of the well-known Shannon entropy

$$H(x) = -\sum_{x \in X} p_X(x) \log_2 p_X(x) \tag{4.1}$$

and is expressed as a number of bits of information (hence the logarithm base 2). Intuitively, one may understand this as a measure of uncertainty about the random variable X. The Shannon entropy tells one something about the statistical distribution of X, but not about its temporal evolution.

The mutual information of two discrete random variables, X and Y, is given by the following expression:

$$I(X;\, Y) = \sum_{y \in Y} \sum_{x \in X} p_{X,Y}(x,\, y) \log_2\left(\frac{p_{X,Y}(x,\, y)}{p_X(x) p_Y(y)}\right). \tag{4.2}$$

Here $p_{X,Y}(x,\, y)$ is the joint probability distribution function (PDF)—the probability of obtaining both $X = x$ and $Y = y$. The quantity $I(X;\, Y)$ measures the amount of information obtained on one variable by observing the other variable, or the amount of information that is shared between the two variables. If the two variables were statistically independent, then $p_{X,Y}(x,\, y) = p_X(x)p_Y(y)$, so that $I(X;\, Y) = 0$.

The mutual information is often considered a non-linear generalisation of the cross correlation. It behaves similarly in many ways, although unlike the cross correlation, it is nonnegative. For example, like the cross correlation, the mutual information is symmetric under the interchange of variables: $I(X;\, Y) = I(Y;\, X)$. Also, if X and Y are two discretely sampled time series, $X = \{x(t_i),\, i = 1, \dots, N\}$ and similarly for Y, the mutual information between X and a time-shifted version of Y, $Y_{\Delta i} = \{y(t_{i+\Delta i})\}$, will peak at a specific value of $\Delta i_{\max}$ and typically fall off (often exponentially) with $|\Delta i - \Delta i_{\max}|$, very similar to the behaviour of the cross correlation function. Therefore, a 'correlation time' can be defined on the basis of the rate of decay of the mutual information from its peak value, similar to what is common practice for the cross correlation (see section 2.2). The resulting correlation time may of course differ somewhat in value between the two methods, as the mutual information takes non-linear interactions into account, whereas the cross correlation does not.

To emphasize the relation of the mutual information with the concept of entropy, note that it can be expressed explicitly in terms of the Shannon entropy:

$$I(X;\, Y) = H(X) + H(Y) - H(X,\, Y). \tag{4.3}$$

Figure 4.1 shows this expression graphically in the form of a Venn diagram. In words, equation (4.3) expresses that the mutual (shared) information equals the information about X plus the information about Y minus the joint information about X and Y. From the diagram, it is clear that the mutual information will consist exclusively of the information corresponding to the overlaid part of the two circles.

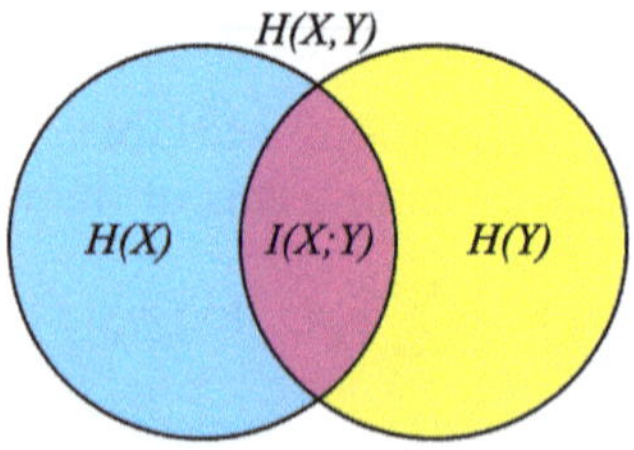

Figure 4.1. Venn diagram illustrating the relation between the mutual information and the Shannon entropy.

4.3 The transfer entropy

As noted in section 4.1, to determine a causal influence the temporal ordering comes into play, and neither the Shannon entropy nor the mutual information as such provide this information directly. However, the analysis can be enhanced by considering conditional probability distributions that measure the probability of obtaining a value (of X) given the value of another quantity (Y). Consider the *conditional* entropy

$$H(X|Y) = -\sum_{x\in X}\sum_{y\in Y} p_{X,Y}(x, y)\log_2\frac{p_{X,Y}(x, y)}{p_Y(y)}, \tag{4.4}$$

where $p_{X,Y}(x, y)$ is the joint probability that both $X = x$ and $Y = y$. $H(X|Y)$ specifies the amount of information or uncertainty regarding X if it is known that $Y = y$.

In order to implement the above concept of 'quantifiable causality', one needs to quantify the 'prediction capacity' for the next sample ($i + 1$) of signal Y, based on its history. This can be coded into a conditional version of the Shannon entropy, namely $H(y_{i+1}|y_{i-k})$, which is the information (or uncertainty) regarding y_{i+1} given the historic value y_{i-k}. Here, $k \geqslant 0$, and it can be understood to be a vector of values rather than a single value. The used notation requires some clarification for proper understanding. Although k takes a specific value, i should be understood to run over all possible values, so that only the statistical relation between y_{i+1} and y_{i-k} (averaged over all i) is considered in the expression $H(y_{i+1}|y_{i-k})$.

One then compares this information to the information given the history of both Y and X, which can be coded as $H(y_{i+1}|y_{i-k}, x_{i-k})$. The tenet of 'quantifiable causality' is that if our knowledge of y_{i+1} improves by considering the history of X as well as the history of Y, there is an information flow from X to Y.

Thus, the transfer entropy is defined as a difference between two conditional entropies,

$$T_{X\to Y} = H(y_{i+1}|y_{i-k}) - H(y_{i+1}|y_{i-k}, x_{i-k}), \tag{4.5}$$

and it measures the increase of information achieved by considering the history of X in addition to the history of Y when predicting the next value of Y. In terms of probability distribution functions, this can be written as

$$T_{X\to Y} = \sum p(y_{i+1}, y_{i-k}, x_{i-k})\log_2\frac{p(y_{i+1}|y_{i-k}, x_{i-k})}{p(y_{i+1}|y_{i-k})}. \tag{4.6}$$

One immediately recognizes that $T_{X \to Y} = 0$ when the enumerator and denominator of the fraction inside the logarithm are equal, i.e. when $p(y_{i+1}|y_{i-k}, x_{i-k})$ does not depend on x_{i-k}. In this case, the historic values of X do not add anything to our predictive capacity (information) regarding Y.

The numerical evaluation of the transfer entropy can be performed in various ways. One straightforward method consists in constructing the relevant probability distribution functions based on the data X, Y using binning. These multidimensional PDFs would need a large number of data to fill the bins with a statistically significant number of points per bin if the bin sizes are chosen small. Specifically, the main joint PDF $p(y_{i+1}, y_{i-k}, x_{i-k})$ depends on $2K + 1$ variables if k is a vector with length K. If the data ranges of X and Y are subdivided into a number m of bins, this means that the total number of bins in the $(2K + 1)$-dimensional PDF is m^{2K+1}. To fill the bins of this PDF with a significant number of points, the number of data points N required would be $N \gg m^{2K+1}$.

In plasma physics, the length of the available data arrays under stationary conditions is usually rather short ($N \sim 10^3$–10^5). Inevitably, therefore, m and K must be chosen small. Choosing m small (typically, 3) is called 'coarse graining' [5], and is sufficient to allow the detection of significant changes in the system state. In addition, we will set $K = 1$ (using only a single historic value to predict the next sample), which has the additional benefit that a unique time lag can be associated with this value, which, as we will see, facilitates interpretation. On the other hand, it then becomes a significant issue which specific value of k is used and typically, one will need to scan a range of values of k in order to explore the most informative choices.

From the above explanations, one may deduce that the transfer entropy is a quantifier of the causal impact of one variable (X) on the other (Y). Nevertheless, one should never forget that there may be a third variable (Z) that may be hidden or undetected, affecting both X and Y. In this case, there need not be a direct causal effect $X \to Y$ for the transfer entropy to become significant; if Z affects both X and Y, but has a slightly delayed effect on Y, there may appear to be a flow of information $X \to Y$ that is actually absent. Therefore, the transfer entropy only provides an indication of causality, rather than proof of it. While this may be disappointing, we also note that there exist no other fail-safe methods to determine the causality between two fluctuating variables, and one could even affirm that this goal is impossible to achieve. Given this general consideration, we will simply have to see whether the transfer entropy is useful for our purposes, and always bear in mind that a high value of the transfer entropy may indicate an interesting (non-linear) relation between fluctuating variables, but never constitutes a proof of causality.

Once we have calculated a value of the transfer entropy T, how do we know whether it is significant? First, the value of T, expressed in bits, can be compared with the total bit range, $T_{\max} = \log_2 m$, the maximum possible value of T. If T is not very small with respect to $T_{\max}$, it may be considered significant. Second, we can estimate the statistical significance of T by calculating it for two random (noise)

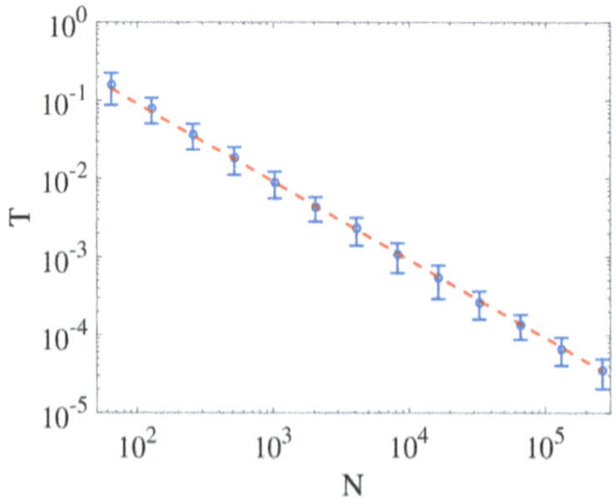

Figure 4.2. Transfer entropy for two random Gaussian signals, as a function of the number of samples of the signals, N (with $m = 3$). Each point is calculated as the average over 100 independent realizations, and the error bar indicates the variation of the result. The red dashed line is proportional to $1/N$. (Reproduced from [6], with the permission of AIP Publishing.)

signals that are statistically independent by construction. Figure 4.2 shows the transfer entropy calculated for two such random signals, with a Gaussian distribution, each with a number of samples equal to N. We repeated the calculation of each point 100 times with different random sequences for X and Y, and then determined the mean value of T (dots) and their standard deviation (indicated by the error bars). We can use this value as an estimate of the minimum statistical significance level of the transfer entropy: any values of T determined from experimental signals should be significantly larger than this level to be considered significant. One sees that the value of the transfer entropy for these uncorrelated random variables drops rapidly in proportion to $1/N$, attesting to the robustness of this quantity [6].

Finally, we note some other potential advantages of the transfer entropy. As it is based on probability distributions, it is fully insensitive to confounding details such as data amplitude or sign (and therefore, data calibration), as well as the shape or amplitude of propagating perturbations: namely, only the relative time delay (k) matters. And since it is directional, it allows distinguishing clearly between influences operating in one direction and in the opposite direction, even if these influences occur simultaneously. This is a rather amazing and powerful new contribution of the transfer entropy to this type of data analysis; an example of how this feature can be exploited will be given in section 4.6.3.

4.4 Tests of the transfer entropy

In this section, we will use several relatively simple model systems to test the capacity of the transfer entropy to detect the causal relation between fluctuating variables. Of course, we are well aware that these systems are not comparable to turbulence in complexity. However, the simplicity of the chosen models will allow us to understand the causal flow on the basis of the underlying equations, so that we can verify that the analysis based on the transfer entropy makes sense and is useful.

4.4.1 Van der Pol oscillators

Our first model system is a set of coupled Van der Pol oscillators. A system of M coupled Van der Pol oscillators may exhibit chaos without external driving for specific parameter choices. The system is described by the set of equations

$$\frac{\partial x_i}{\partial t} = y_i, \tag{4.7}$$

$$\frac{\partial y_i}{\partial t} = \left[\epsilon_i - \left(x_i + \sum_j \kappa_{ij} x_j\right)^2\right] y_i - \left(x_i + \sum_j \kappa_{ij} x_j\right). \tag{4.8}$$

The parameter ϵ_i determines the limit cycle of oscillator i (for $\kappa = 0$), while κ_{ij} specifies the non-linear coupling between oscillator i and oscillator j [7].

One of the simplest possible such systems is obtained for $M = 2$ oscillators. We set $\epsilon = (1,\ 1.1)$ so that the limit cycles of the two oscillators are slightly different, and choose the following coupling matrix:

$$\kappa = \begin{pmatrix} 0 & 1 \\ 0 & 0 \end{pmatrix}, \tag{4.9}$$

with only one off-diagonal term, such that oscillator 2 drives oscillator 1, but not vice versa. This establishes a clear direction of the mutual interaction, and the test will consist of determining whether the transfer entropy can recover this direction.

With this choice of parameters, the system is in a quasi-periodic state. Time was integrated from $t = 0$ to $t = 1000$, and 10 000 equally spaced data points were saved for analysis. Only the time interval from $t = 100$ to $t = 1000$ was used for analysis in order to remove the initial transient phase. A section of data is shown in figure 4.3.

From a spectral analysis, the mean oscillation period of the signals was about 6.72, corresponding to about 67 samples per period. To prevent the transfer entropy

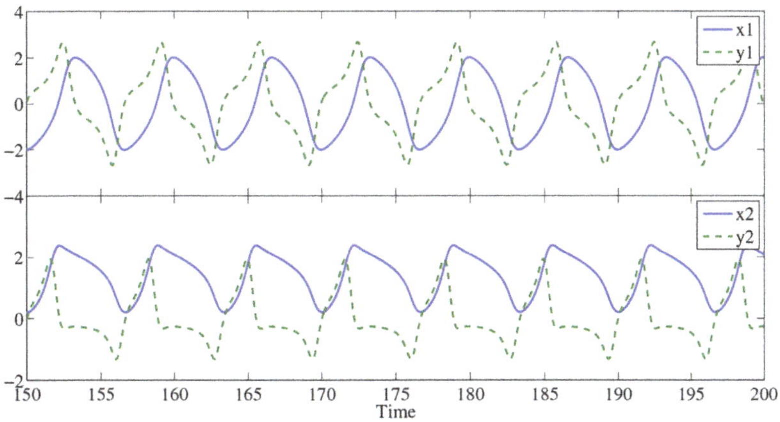

Figure 4.3. Time traces obtained from a system of $M = 2$ Van der Pol oscillators. (Reproduced with permission from [8].

from responding to information pertaining to different oscillation cycles, the time delay index k should be chosen less than about one fourth of the full cycle, i.e. less than 17 ($\simeq$67/4). The latter value (17) also roughly corresponds to the first minimum of the self-mutual information of x_1 (not shown).

We make the choice $k = 8$ for the delay index (about half the maximum value of 17), and calculate the transfer entropy using coarse graining with $m = 3$. Net information flow from signal i to signal j is computed as $T_{ij}^{\mathrm{net}} = T_{ij} - T_{ji}$, where the indices 1, … , 4 correspond to the signals x_1, y_1, x_2, y_2, respectively. The following net transfer entropy matrix is obtained (only the part of the matrix above the diagonal is shown; the rest follows from antisymmetry):

$$T^{\mathrm{net}} = \begin{pmatrix} 0 & -0.036 & -0.29 & 0.15 \\ - & 0 & 0.0007 & -0.20 \\ - & - & 0 & 0.21 \\ - & - & - & 0 \end{pmatrix}. \tag{4.10}$$

This can be represented graphically by representing the four signals in a plane as dots, with arrows connecting the dots, such that the direction of the arrow indicates the direction of net information flow, while the width of the arrow is proportional to the value of T_{ij}^{net}.

The result is shown in figure 4.4. There is a strong flow from x_2 to x_1 (corresponding to T_{13}^{net}). This non-trivial component corresponds to the fact that κ is set up in such a way that oscillator 2 affects oscillator 1 (but not vice versa). Another strong flow is from y_2 to y_1 (T_{24}^{net}), for the same reason. Arrows linking components of the same oscillator (e.g. x_2 and y_2) evidently do not reveal anything about couplings between distinct oscillators and can be ignored. The remaining arrows are smaller, so that it is clear that the information flow is dominantly from oscillator 2 to oscillator 1. Therefore, the analysis technique correctly recovers the direction of coupling among the components of the system.

The stability of the analysis method was tested by computing the transfer entropy for a varying length of the data arrays, N. Figure 4.5 shows, among others, T_{13} (the transfer entropy from signal 1, x_1, to signal 3, x_2) and T_{31} (the transfer entropy from signal 3, x_2, to signal 1, x_1) versus N, with the same analysis settings as above.

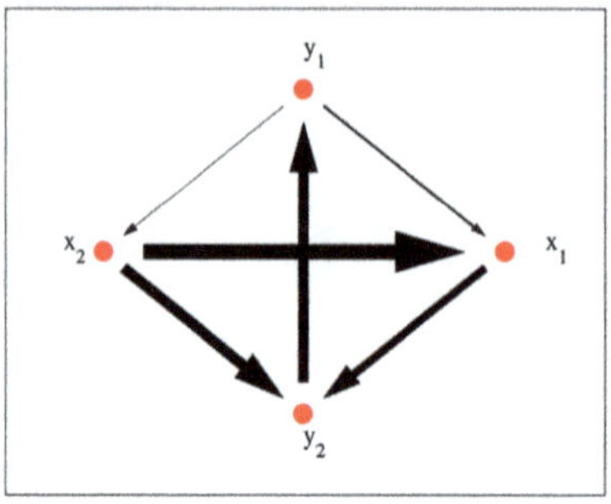

Figure 4.4. Information flow between the four signals of the system of $M = 2$ Van der Pol oscillators: the arrows indicate the direction of flow and the width is proportional to the amount of information transfer. (Reproduced with permission from [8]. © IOP Publishing. All rights reserved.)

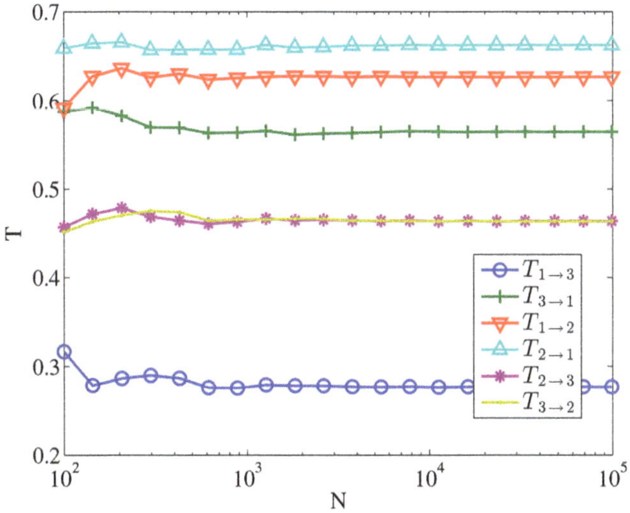

Figure 4.5. Numerical convergence of the transfer entropy versus length of the data arrays, N, for the system of $M = 2$ coupled Van der Pol oscillators ($m = 3$, $k = 8$). (Reproduced with permission from [8].)

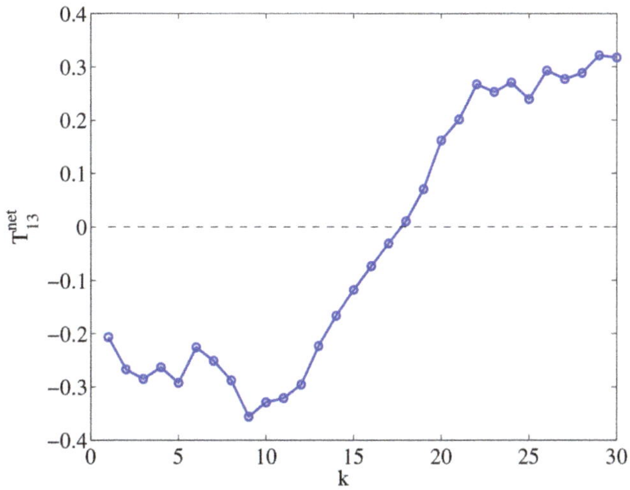

Figure 4.6. Net information flow between the two coordinates x_1 and x_2 of the system of $M = 2$ Van der Pol oscillators (flow from x_1 to x_2 being positive), for different values of k. (Reproduced with permission from [8].)

It is seen that T converges to a stable value for $N \gtrsim 10^3$, a rather modest number. Note that in this case, the total number of bins of the main PDF is $m^{2k+1} = 3^3 = 27$. The value of the transfer entropy T, expressed in bits, can be calibrated against the total bit range, $\log_2 m = 1.58$, implying that the coupling strength is quite significant.

To understand the evolution of the transfer entropy with the time delay index k, we calculated T_{13}^{net} (which specifies the net flow from signal 1, x_1, to signal 3, x_2) for a range of k-values (figure 4.6). The graph shows that the net information transfer from x_1 to x_2 is negative for $k \leqslant 17$ (as it should be, for we know that the information transfer should go from oscillator 2 to oscillator 1). For higher values of k, the net flow changes sign. This occurs when crossing the quarter-period value (17) or minimum self-mutual information value (17) mentioned above, and is caused by the fact that this system is quasi-periodic. It seems important, therefore, to keep k well below the mentioned reference values in the case of a quasi-periodic system.

4.4.2 Simplified predator–prey model

In the context of fusion plasmas, spontaneous confinement transitions are of prime interest. In recent years, models have been developed to describe such transitions, involving the non-linear interaction between various fields. In this subsection, we will use the model of [9] to generate signals for analysis using the transfer entropy technique. The model equations are

$$\begin{aligned}\frac{dE}{dt} &= \left(\frac{1}{1 + V'^2 + U'^2} - E\right)E,\\ \frac{dV'}{dt} &= (aE^2 + cU'^2 - b)V',\\ \frac{dU'}{dt} &= \left(\frac{aE^2}{1 + V'^2} - b\right)U' + dE^2V'.\end{aligned} \tag{4.11}$$

Here, E represents the turbulence amplitude, U' represents the zonal flow shear, and V' the sheared flow.

We performed a simulation run with $a = 0.204$, $b = 0.16$, $c = 0.714$, and $d = 0.5$ (see figure 5 of the cited paper), generating a set of 10 000 time points (at sampling rate $\Delta t = 1$). A short section of the model output is shown in figure 4.7. The mean period of the quasi-periodic oscillations is 40.09, corresponding to about 40 samples.

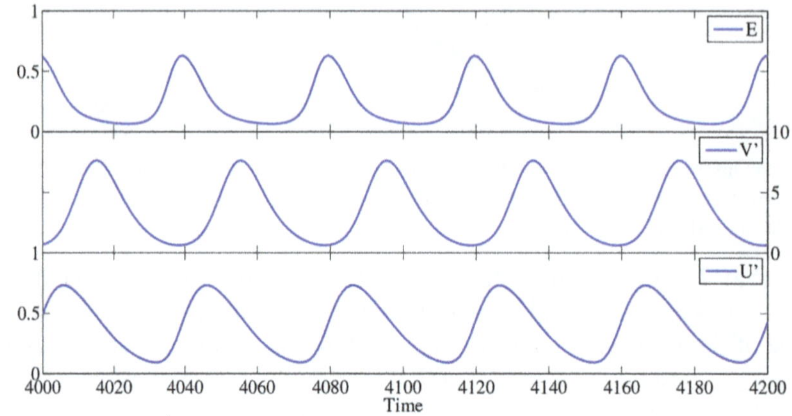

Figure 4.7. Signals from a run of the model equation (4.11) with parameters $a = 0.204$, $b = 0.16$, $c = 0.714$, and $d = 0.5$. (Reproduced with permission from [8]. © IOP Publishing. All rights reserved.)

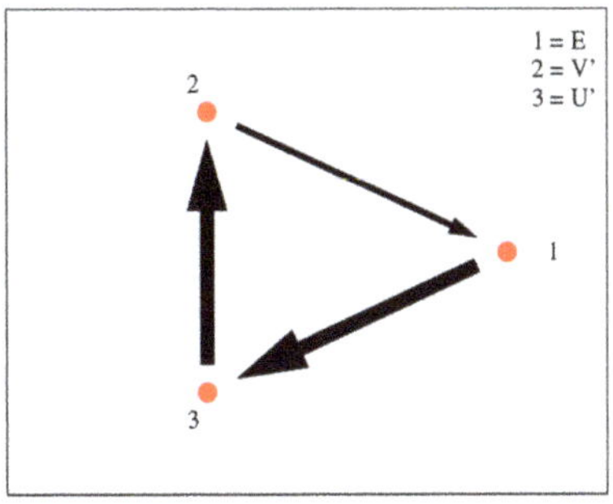

Figure 4.8. Information flow between the three signals of the predator–prey model: the arrows indicate the direction of flow and the width is proportional to the net amount of information transfer. (Reproduced with permission from [8]. © IOP Publishing. All rights reserved.)

The first minimum of the self-mutual information occurs at 8 (for U'), 10 (for V'), or 14 (for E) samples. Thus, the time lag k should be chosen well below 8.

The transfer entropy is computed for all nine possible combinations of signals ($1 = E$, $2 = V'$, $3 = U'$). The settings chosen are $k = 5$, $m = 5$. The following transfer entropy matrix is obtained:

$$T = \begin{pmatrix} 0 & 0.49 & 1.08 \\ 0.63 & 0 & 0.85 \\ 0.71 & 1.08 & 0 \end{pmatrix}. \tag{4.12}$$

Comparing the values of T with $\log_2 m = 2.32$, one concludes that the interactions are quite strong.

Net information flow from signal i to signal j is computed as $T_{ij}^{\text{net}} = T_{ij}T_{ji}$. This is again represented graphically by drawing three dots representing the three signals in a plane. The three dots are connected by arrows, such that the direction of the arrow indicates the direction of net information flow, while the width of the arrow is proportional to the value T_{ij}^{net}, see figure 4.8.

The resulting diagram makes eminent physical sense: E drives U' (turbulence drives zonal flow); U' drives V' (zonal flow drives sheared flow); and V' drives E (sheared flow controls turbulence). This is precisely the order of interaction that one would expect in this type of model. Of course, in this simplified case this sequence can also be obtained by simple inspection of the data (figure 4.7), as the mutual time delays are in clear accord with this result. On the other hand, the present method is generally applicable and hopefully allows determining the interactions even in less evident cases.

4.5 Unravelling relations between turbulent variables

In this section, we will analyze turbulence data from various magnetic confinement devices.

4.5.1 Zonal flows in a low field stellarator

The TJ-K stellarator (see appendix A.1) is a small magnetic confinement device operated at low magnetic field ($B = 72$ mT) and low normalised plasma pressure, β. The discharge analysed in the following corresponds to a helium plasma, heated by microwaves, with a central density of $n = 2.3 \times 10^{17}$ m^{-3} and an electron temperature of $T_e = 8$ eV and cold ions, as reported in detail elsewhere [10]. In this experiment, turbulence was dominated by electrostatic drift wave turbulence, and the total particle transport and zonal potential were found to be linked in a predator–prey cycle. Among other diagnostics, the device disposes of a set of 64 Langmuir probes, distributed over a poloidal circumference of the device. The probes are configured to measure floating potential and ion saturation current in an alternating fashion, at a sampling rate of 1 MHz. From these signals, we compute the zonal potential $\Phi_z(t)$ as the mean poloidal value of the floating potential, and the global radial particle flux $\Gamma_{tot}(t)$ as the poloidal mean of the local radial particle flux, proportional to the fluctuating ion saturation current times the local poloidal electric field, as described in more detail in the cited reference.

We quantify the 'global turbulence level' by computing the root mean square (RMS) deviation of the 32 poloidally distributed ion saturation current measurements (I_{sat}), thus obtaining a quantifier of the turbulence level with the same time resolution as $\Phi_z(t)$ and $\Gamma_{tot}(t)$.

A short section of data is shown in figure 4.9; clearly, these data are much less regular than the model data shown in the preceding section, making it considerably more difficult to understand the non-linear relationship between the signals. We emphasize that the three quantities thus obtained are essential ingredients of zonal flow models: the zonal potential, the global turbulent particle flux, and the turbulence amplitude.

We again apply the analysis described above, setting $m = 3$ (coarse graining). Figure 4.10 shows the transfer entropy between the two signals $\Phi_z(t)$ and $\Gamma_{tot}(t)$ a function of the delay k. The amplitude of the transfer entropy is rather small, namely

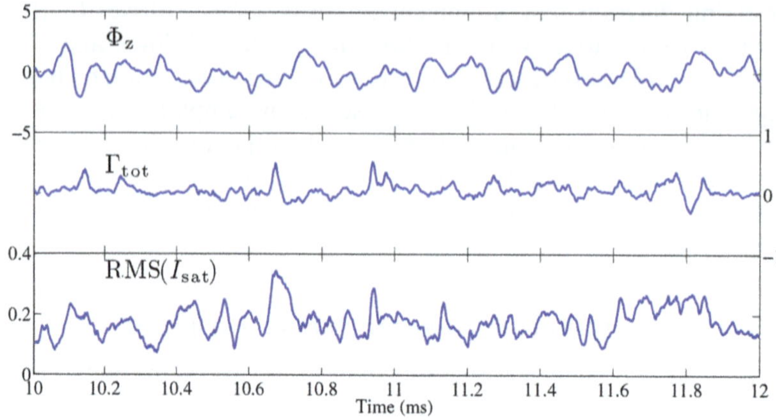

Figure 4.9. A short section of data from TJ-K (arbitrary units). (Reproduced with permission from [8]. © IOP Publishing. All rights reserved.)

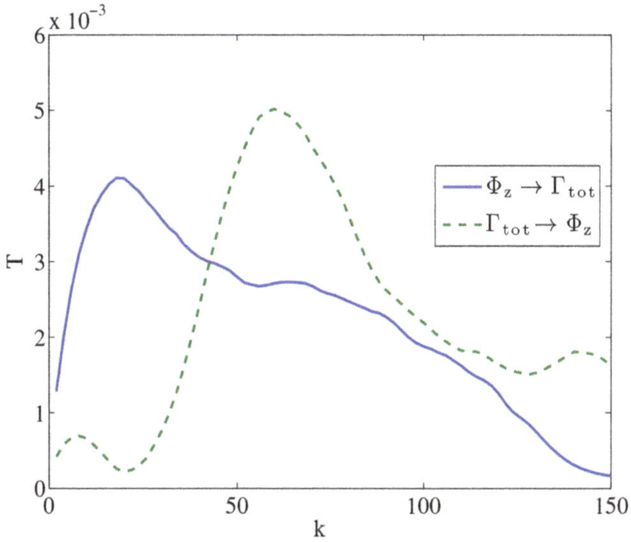

Figure 4.10. Transfer entropy between the zonal potential $\Phi_z(t)$ and the global particle transport $\Gamma_{tot}(t)$ at TJ-K versus time delay k (in sampling units, i.e. μs). (Reproduced with permission from [8].)

below 5×10^{-3}, compared with the full bit range $\log_2 m = 1.59$, which indicates that the causal link between these variables is not very strong. Nevertheless, it is unexpected and interesting to observe that $T_{\Phi\to\Gamma}$ and $T_{\Gamma\to\Phi}$ peak at different values of k. $T_{\Phi\to\Gamma}$ peaks at about 20 μs, while $T_{\Gamma\to\Phi}$ peaks at about 60 μs. Thus, zonal potential Φ_z has a rather fast impact on the total particle flux Γ_{tot}, while the latter acts back on the zonal potential Φ_z on a much longer timescale. The net information transfer flows from Φ_z to Γ_{tot} for timescales less than about 40 μs, and in the opposite direction for longer timescales. The two distinct timescales for mutual interaction would immediately give rise to oscillatory behaviour, as indeed observed. This seems coherent with the usual predator–prey models for the interaction between zonal flow and turbulence.

In the standard zonal flow model, the zonal flow has an impact on the global turbulence level. Figure 4.11 shows the interaction diagram between all three signals at the two most significant values of k. On a short timescale (20 μs), the zonal flow affects the transport, and the transport in turn affects the turbulence level (presumably, by modifying the driving gradients). On a longer timescale (60 μs), the transport affects the zonal flow, but the interaction with the turbulence level is insignificant.

The short timescale result is interesting, as it confirms the analysis of [10], where it was observed that the zonal flow does not affect the turbulence amplitude strongly, but rather it affects the transport (as noted in the cited paper, by modifying the phase relation between density and potential fluctuations). The modification of the transport then affects the turbulence amplitude. Although there is an arrow showing that the zonal potential also affects the turbulence level directly, its intensity is much less than the indirect route via the turbulent transport. The long timescale result is

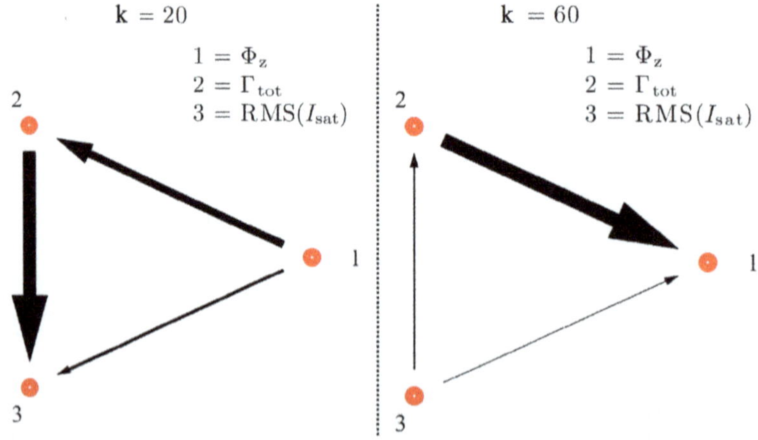

Figure 4.11. Graphical representation of the net transfer entropy between the zonal potential Φ_z, the global particle transport Γ_{tot}, and the global turbulence level RMS (I_{sat}) at TJ-K for two values of the delay k. (Reproduced with permission from [8].)

presumably simply due to a restoration of ambipolarity: a modification of transport must eventually lead to a modification of potential. We note that this longer timescale result is novel and reveals the potential significance of the technique to uncover new relationships [8].

4.5.2 Magnetic fluctuations and L–H transitions

A topic of great interest to the fusion community is the L–H transition, in which the plasma spontaneously transits from a low (L) to high (H) confinement state [11] via a self-regulation mechanism that involves turbulence and the generation of zonal flows. It has been suggested that the magnetic topology (rational magnetic surfaces) and MHD activity could facilitate the generation of zonal flows and hence confinement transitions [12–14]. Under specific circumstances, MHD activity associated with rational surfaces might in fact stimulate the development of zonal flows via the magnetic Reynolds stress, leading to the formation of transport barriers [15].

Experimentally, this issue is most easily addressed in stellarators, as these devices have a better control over the magnetic configuration than, e.g. tokamaks. Indeed, access to the H mode was found to depend on the magnetic configuration in the W7-AS stellarator [16].

In the TJ-II stellarator, the low to high confinement transition is a 'soft' transition in the sense that the confinement improvement factor is small, yet it possesses the typical features of any H mode (rapid drop of H_α emission, reduction of fluctuation amplitudes, formation of a sheared flow layer, ELM-like bursts) [17]. This currentless, flexible, low magnetic shear stellarator is ideally suited to study this issue due to nearly complete external control of the magnetic configuration (see appendix A.2) and, in particular, the capacity to modify the rotational transform profile significantly.

The experiments described below have been carried out in pure NBI heated plasmas (line averaged plasma density $\bar{n}_e = 2\text{–}4 \times 10^{19}\ \text{m}^{-3}$ and central electron temperature $T_e = 300\text{–}400$ eV) [18]. The NBI input heating power is kept constant at about 500 kW during the discharge, but the fraction of NBI absorbed power increases from 55% to 70% as the plasma density rises. The L–H transition time is characterised by several changes in experimental signals, but here we determine this time point as the time at which the amplitude of the density fluctuations measured by the Doppler reflectometer drops sharply. As shown in [19], this time is more precise than, e.g. the time needed for the formation of the shear layer, which may take several milliseconds. This procedure typically allows identifying the L–H transition time with a precision of about 1 ms. Figure 4.12(a) shows the mean line average density at the L–H transition as a function of the magnetic configuration, parametrized by the rotational transform value at $\rho = 2/3$. Figure 4.12(b) shows the position of the main low-order rational surfaces. The figure shows that the line average density at which the L–H transition occurs depends on the configuration.

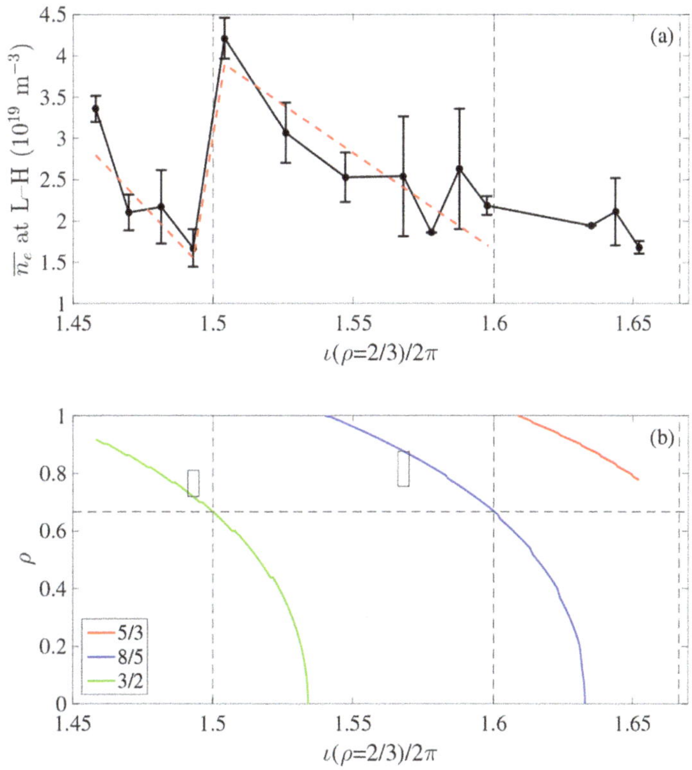

Figure 4.12. Magnetic configuration scan at TJ-II. The configurations identified via $\iota\!\!\!-(\rho = 2/3)$. (a) Line average density at L–H transition. The red dashed line is provided to guide the eye. The bars indicate the shot-to-shot variation (not the measurement error). (b) Position of the main low-order rational surfaces in vacuum. The horizontal dashed line indicates $\rho = 2/3$. The vertical dashed lines correspond to $\iota\!\!\!-(\rho = 2/3) = 3/2$, 8/5 and 5/3. (Reproduced from [18], with the permission of AIP Publishing.)

One can surmise that the transition is 'easier' (occurs at lower density) when there is a major low-order rational surface in the outer third of the plasma, which would suggest that a low-order rational surface in the edge region facilitates the L–H transition by making it easier for zonal flows to form.

To study the transition, we will focus on two specific magnetic configurations. These two configurations have slightly different L–H threshold densities but exhibit very different temporal behaviour:

- In configuration 100_35_61, with $\iota\!\!\!-(\rho = 2/3) = 1.493$, the radial position of the $\iota\!\!\!- = 2/3$ rational surface is located at $\rho \simeq 0.73$ in a vacuum. This configuration is characterized by a 'slow' transition. The transition is not straight into the H phase, but rather into an I phase, characterized by limit cycle oscillations (LCOs) [20, 21] and similar to LCOs reported at other devices [22].
- In configuration 100_42_64, with $\iota\!\!\!-(\rho = 2/3) = 1.568$, the radial position of the $\iota\!\!\!- = 8/5$ rational surface is located at $\rho \simeq 0.86$ in a vacuum. In this case, the transition is 'fast' and enters directly into the H phase.

The different behaviour during the transition in these two configurations may be helpful to understand the transition mechanism. Figure 4.13 shows the evolution of the magnetic activity, as measured by a Mirnov pick-up coil, for the two mentioned configurations. The shown results are averaged over a large number of discharges, so that the mean behaviour is reflected. Both the RMS (variance) of the coil signal and the spectrum versus time are shown. In the case of the 'slow' transitions, one observes a gradual decrease of MHD activity prior to the L–H transition time, the

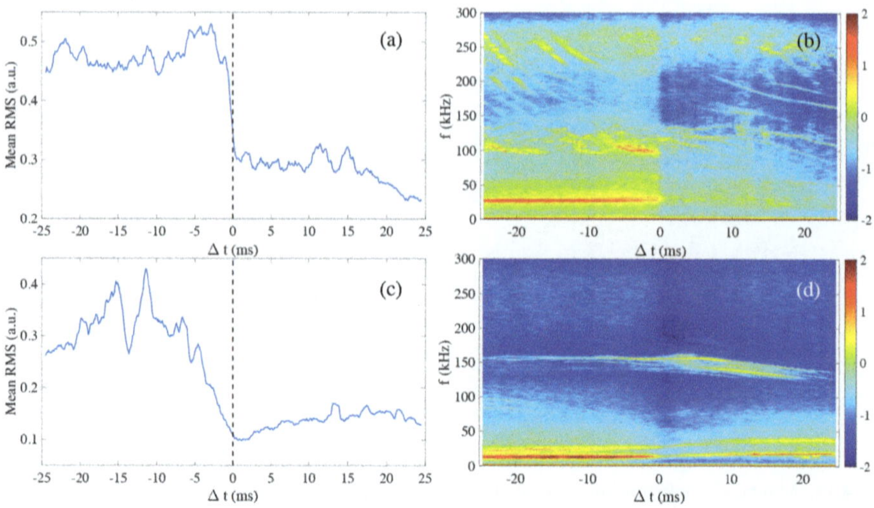

Figure 4.13. Mean evolution of magnetic activity across confinement transitions. (a), (b) Configuration 100_42_64 ('fast' L–H transitions, 90 discharges). (c), (d) Configuration 100_35_61 ('slow' L–I transitions, 36 discharges). (a), (c) Mean RMS of a Mirnov coil. (b), (d) Mean spectrogram of a Mirnov coil. (Reproduced from [18], with the permission of AIP Publishing.)

decay starting some 10 ms beforehand. In the case of the 'fast' transitions, the drop of RMS is rather sharp and lasts only a few ms, although one could still argue that the decay starts before the transition time. The RMS of the Mirnov coil signal drops by a factor of about two from the pre-transition peak to the transition time point. Afterwards, in I or H mode, the RMS amplitude remains low. The spectra show that the drop of RMS is mainly associated with the suppression of low frequency ($f < 80$ kHz) MHD modes, and not so much with that of high frequency ($f > 80$ kHz) Alfvén modes.

The decay of magnetic activity associated with confinement transitions is rather striking. *A priori*, the origin of this phenomenon is unclear. One might be inclined to think that the decay is related to a gradual evolution of profiles on the transport timescale (of the order of 5 ms) associated with the rising density. However, in the case of the 'fast' L–H transition (configuration 100_42_64), the decay is sufficiently fast to make it doubtful that transport effects are playing a significant role.

It is known that zonal flows play a major role in confinement transitions in general and this is also the case at TJ-II [23]. To study the interaction between magnetic fluctuations and zonal flows, we will use the signal from a magnetic poloidal field pick-up coil ($\dot{B}$) and signals from the Doppler reflectometry system installed at TJ-II. Although a Mirnov coil measures global magnetic activity, its signal is often dominated by low-order modes associated with low-order rational surfaces, and we exploit this property here. The Doppler reflectometer allows measuring the perpendicular rotation $v_\perp(\rho, t)$ (within the flux surface) of fluctuations at two specific radial locations [24], with high temporal resolution. We analyze Doppler reflectometry data from a series of discharges in which the radial position was varied systematically [25, 26]; the Doppler reflectometer measurement positions (in the two configurations studied) are roughly indicated in figure 4.12 by two small rectangles. Due to the expansion of the density profile across the transition, the radial position of the Doppler reflectometer channels prior to the transition (in the L phase) is somewhat further inward than indicated in the figures below [25]. Figure 4.14(a) shows the transfer entropy $T_{v_\perp \to \sigma(|\tilde{n}|)}$ ($m = 5$, lag $k = 500$, or 50 μs), reflecting the interaction between the perpendicular flow velocity and the density fluctuation amplitude, both measured by the reflectometer. The statistical error level of the reported T values is 0.01. $\sigma(|\tilde{n}|)$ is the running RMS of the density fluctuation amplitude signal (calculated using a running time window of 2 μs), considered a measure of the turbulence amplitude envelope. The radii reported on the ordinate of the graphs were calculated for density profiles corresponding to $\Delta t > 0$. The figure shows that the perpendicular velocity mainly has a causal impact (in the restricted sense explained above) on the density fluctuations in a period of about 30 ms after the L–I transition, in a specific radial range. In vacuum, the radial position of the $\iota\!\!\!-= 3/2$ rational is located at $\rho \simeq 0.73$. Probably, the rational surface is shifted outward somewhat in the presence of the plasma with a small negative net current, and possibly coincides with the radial position at which the transfer entropy is showing a strong response ($\rho \simeq 0.74 - 0.79$). We note that the locations and times of high transfer entropy coincide with the observation of LCOs in these same discharges [20], i.e. mainly after the L–I transition. The reverse interaction, $T_{\sigma(|\tilde{n}|) \to v_\perp}$

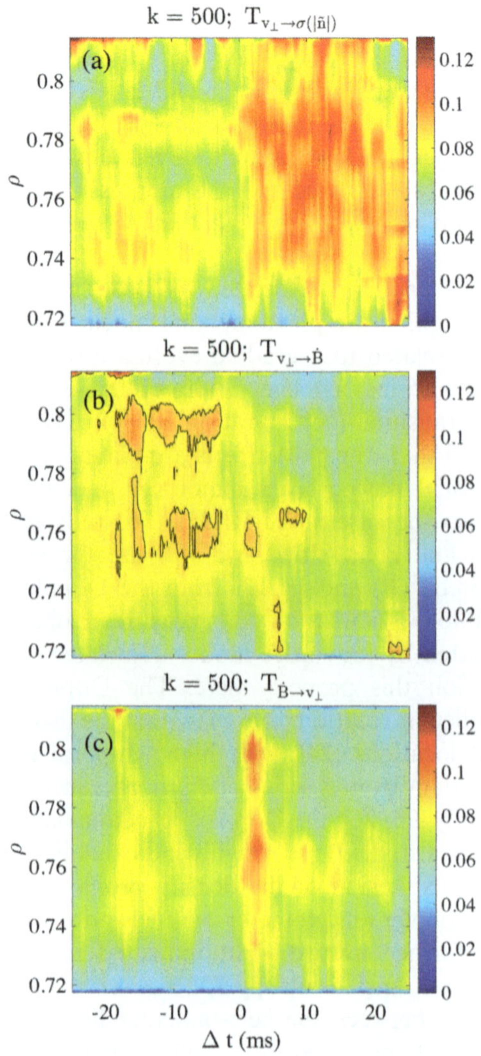

Figure 4.14. Configuration 100_35_61 ($\iota\!\!\!{-}(\rho = 2/3) = 1.493$). (a) Transfer entropy $T_{v_\perp \to \sigma(|\tilde{n}|)}$. (b) Transfer entropy $T_{v_\perp \to \dot{B}}$. (c) Transfer entropy $T_{\dot{B} \to v_\perp}$. (Reproduced from [18], with the permission of AIP Publishing.)

(not shown), is much smaller, in line with the expectations based on a similar analysis for a simplified predator–prey model [8], in which the dominant direction of the interaction between turbulence amplitude and sheared flow was found to be from sheared flow to turbulence amplitude. Thus, these results constitute a validation of the analysis method in the framework of confinement transitions and LCOs.

Figure 4.14(b) shows the transfer entropy $T_{v_\perp \to \dot{B}}$ ($m = 5$, lag $k = 500$, or 50 μs), to study the interaction between the perpendicular flow velocity and the magnetic fluctuations, measured by a pick-up coil. The transfer entropy is now only large in a time period of about 25 ms preceding the transition time. Thus, one may presume that the interaction between $v_\perp$ and $\dot{B}$ is associated with the gradual reduction of RMS ($\dot{B}$) near the transition, as shown in figure 4.13.

We draw attention to the fact that there seem to be two predominant zones of interaction: one at $\rho \simeq 0.76$ and one at $\rho \simeq 0.8$. To emphasize this fact, a single level contour has been drawn in the figure (black). It should be noted that the radii in the L phase ($\Delta t < 0$) are somewhat smaller than indicated on the ordinate axes of the graphs (valid for the I phase, $\Delta t > 0$), as mentioned above: $\rho_L \simeq -0.656 + 1.77\rho_I$, so the actual radii of these features are, respectively, $\rho_L \simeq 0.69$ and 0.76, bracketing the theoretical vacuum position of the rational surface (0.73). This is consistent with results from resistive MHD turbulence simulations, performed for similar parameters [27]. In these simulations, the mean poloidal flow was typically found to be symmetric around a rational surface, showing a double peak inside and outside from the rational surface, at distances compatible with the two radial zones of interaction found here.

Further evidence of the interaction between velocity oscillations and magnetic fluctuations is shown in figure 4.14(c), showing $T_{\dot{B}\to v_\perp}$ ($m = 5$, lag $k = 500$, or 50 μs). The transfer entropy exhibits a short burst at the L–I transition time, at the same two radial locations as in figure 4.14(b). This phenomenon corresponds to the short-lived interaction described in the preceding section. We note that the direction of the interaction is reversed; here, the magnetic fluctuations are 'influencing' the perpendicular fluctuating velocity, precisely at the transition. The results shown seem to indicate that magnetic fluctuations play an important role in confinement transitions.

In the case of configuration 100_42_64, the plasma makes a fast transition from the L to the H state.

Figure 4.15(a) shows the transfer entropy $T_{v_\perp\to\dot{B}}$ ($m = 5$, lag $k = 500$, or 50 μs), while figure 4.15(b) shows the transfer entropy $T_{\dot{B}\to v_\perp}$ ($m = 5$, lag $k = 500$, or 50 μs). The statistical error level of the reported T values is 0.01. The radii shown correspond to the H phase. The transfer entropy is sharply concentrated at the L–H transition. The temporal sharpness of the interaction between perpendicular velocity and magnetic fluctuations is likely related to the sharp decay of RMS($\dot{B}$) shown in figure 4.13 for this configuration. In vacuum, the radial position of the $\iota\!\!\!- = 8/5$ rational is located at $\rho \simeq 0.86$. Probably, the rational surface is shifted inward somewhat in the presence of the plasma with a small positive net current, and possibly coincides with the radial position at which the transfer entropy is showing a response ($\rho \simeq 0.76 - 0.84$).

Due to the fact that this transition is 'fast', the sequence of interactions at the transition is not fully resolved, and no precursor activity is detected.

It is useful to summarize these results, as they appear to provide evidence that the confinement transitions are affected by the presence of rational surfaces.

First, the line average density at which the L–H (or L–I) transition is triggered tends to be lower when a low-order rational surface is present in the edge (gradient) region (approximately, $0.6 < \rho < 0.9$), although this effect is only clear for the lowest order rationals studied (figure 4.12). Thus, the presence of a low-order rational in the plasma edge appears to facilitate the transition.

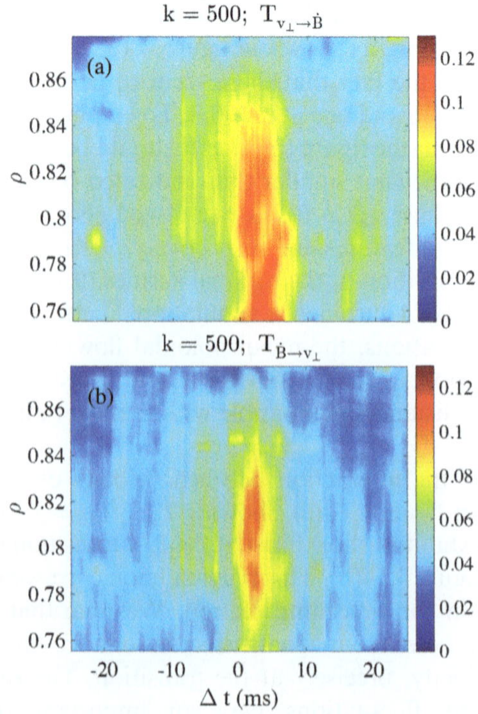

Figure 4.15. Configuration 100_42_64 ($\iota(\rho = 2/3) = 1.568$). (a) Transfer entropy $T_{v_\perp \to \dot{B}}$. (b) Transfer entropy $T_{\dot{B} \to v_\perp}$. (Reproduced from [18], with the permission of AIP Publishing.)

Second, we observed that low frequency MHD activity is strongly suppressed near the L–H (or L–I) transition, see figure 4.13. The suppression is either fast (within a few ms) or slow (starting around 10 ms before the transition time, as defined above), depending on the magnetic configuration.

Third, we used a causality detection technique, based on the transfer entropy (T), to study the interaction between magnetic fluctuations, $\dot{B}$ (measured by a poloidal field pick-up coil), and the local perpendicular plasma rotation velocity, $v_\perp$ (measured by Doppler reflectometry). We obtain a clear and significant interaction between magnetic fluctuations and the fluctuating velocity. Making a distinction between two magnetic configurations (corresponding to different low-order rational surfaces in the edge region), we find the following.

Configuration 100_35_61 ($\iota(\rho = 2/3) = 1.493$, rational 3/2 in the edge region) is characterized by 'slow' transitions and LCOs following the L–I transition. The transfer entropy $T_{v_\perp \to \sigma(|\tilde{n}|)}$ measures the causal impact of the poloidal velocity (zonal flow) on the turbulence amplitude envelope, and is large after the transition, during the LCOs reported in earlier work [20], see figure 4.14(a). The direction of the dominant interaction was in line with expectations obtained from a simplified predator–prey model [8].

To clarify the interaction between zonal flows and magnetic oscillations, we calculated the transfer entropy $T_{v_\perp \to \dot{B}}$, and found that it is large at two radial positions associated with the rational surface and during several tens of ms before the transition, see figure 4.14(b). Thus, it seems that the gradual reduction of magnetic fluctuation amplitude prior to the transition, reported in figure 4.13, can be explained by the gradual development of zonal flows associated with the rational surface. At the transition itself, however, there is a short burst of interactions in the opposite direction such that $T_{\dot{B}\to v_\perp}$ is large, see figure 4.14(c), suggesting that the magnetic fluctuations play an important role during the L–I transition, when the mean sheared flow is formed.

In configuration 100_42_64 ($\iota(\rho = 2/3) = 1.568$, rational 8/5 in the edge region), characterized by 'fast' transitions without LCOs, $T_{v_\perp\to\dot{B}}$ and $T_{\dot{B}\to v_\perp}$ are both large in a relatively narrow time window around the transition time, and near the radial position of the rational surface, see figure 4.15, although the precise sequence of events is hard to follow for these 'fast' transitions.

In any case, it seems the magnetic fluctuations, associated with rational surfaces (or magnetic islands) definitely play an essential role in the transition.

4.5.3 The flux–gradient relation

As noted in chapter 1, the basic quantity defining transport is the flux–gradient relation. The relation between fluctuating radial particle flux and density gradient values can be studied by computing the expectation value of the flux, conditioned on the gradient. This analysis technique is instantaneous (meaning the turbulent flux and gradient are taken at the same time instant). Based on the analysis of Langmuir probe data obtained in the edge region of magnetic confinement devices, it was found that the mean size of turbulent flux events was minimal in the proximity of the most probable (density) gradient, and that the mean size of the flux events increased for both negative and positive excursions from the most probable gradient value, so that the resulting flux versus gradient curve attains a 'U'-shape [28].

This observation suggests that the plasma somehow 'self-regulates' transport to keep the gradient at some preferred value, which would provide a possible explanation for 'profile stiffness', systematically observed in many devices [29, 30]. Numerical studies have shown that the plasma may be able to self-organize the pressure profile in order to maximize local stability [31]. This 'profile stiffness', in turn, is related to the important issue of 'power degradation', i.e. the detrimental scaling of plasma confinement with input power, observed in all magnetic confinement devices [32] and an obstacle to the achievement of economic power production via fusion.

At this point, we need to emphasize that the above-mentioned studies refer to the *mean fluctuating* fluxes and gradients (i.e. excluding the slowly varying mean values—slowly referring to transport timescales), rather than the *mean total* fluxes and gradients. The latter usually exhibit a monotonic relation: mean total transport fluxes tend to increase as a function of the negative mean total gradients, albeit not linearly [33–35]. The dichotomy between the fluctuating (fast timescales) and total mean (slow timescales) parameters in 'critical gradient' models, characterized by a non-linear increase of the total mean flux as a function of the negative total mean gradient, has been pointed out in earlier work [36].

The 'U'-shape of the conditional expectation value of the fluctuating flux, as a function of the gradient, implies that not only (negative) gradient values *above* its mean value, but also those *below* it lead to an increased mean fluctuating flux. As a possible explanation, it was suggested that some flux events may be the consequence of gradient excursions at a different location (non-locality) [37].

To study the statistical relation between the radial flux and gradient, one can compute the two-dimensional or joint PDF

$$p(\Gamma, \nabla n). \tag{4.13}$$

This joint PDF contains information about the *instantaneous* statistical relation between the fluctuating flux and gradient.

The conditional expectation value of the flux Γ at a given value of ∇n is

$$E(\Gamma|\nabla n) = \int \Gamma \; p(\Gamma|\nabla n) \; d\Gamma = \int \Gamma \; \frac{p(\Gamma, \nabla n)}{p(\nabla n)} \; d\Gamma, \tag{4.14}$$

where $p(\nabla n)$ is the total probability of obtaining ∇n: $p(\nabla n) = \int p(\Gamma, \nabla n) d\Gamma$. This quantity typically has an approximately Gaussian shape, peaking at the most likely gradient value. Physically, this means that the gradient fluctuates around the mean or most probable gradient with a probability distribution that decays exponentially fast for excursions from the mean gradient. Similarly, the total probability of the flux, $p(\Gamma) = \int p(\Gamma, \nabla n) d\nabla n$ is a peaked distribution with 'fat tails' (i.e. decaying algebraically). While the cause of the central peak of $p(\Gamma)$ is understood (see section 2.1.1), the origin of the 'fat tails', carrying a significant amount of the total turbulent flux and associated with anomalous transport, remains somewhat of a mystery (but see chapter 5).

Figure 4.16 shows an example of a flux–gradient plot, $E(\Gamma|\nabla n)$, calculated from 50 ms of Langmuir probe data obtained in the TJ-II stellarator in NBI heated plasmas ($1150 < t < 1200$ ms). The fluctuating flux is $\Gamma = \tilde{n}\tilde{E}_\theta/B$, where E_θ is obtained from the difference of two poloidally displaced probe pins measuring the floating potential

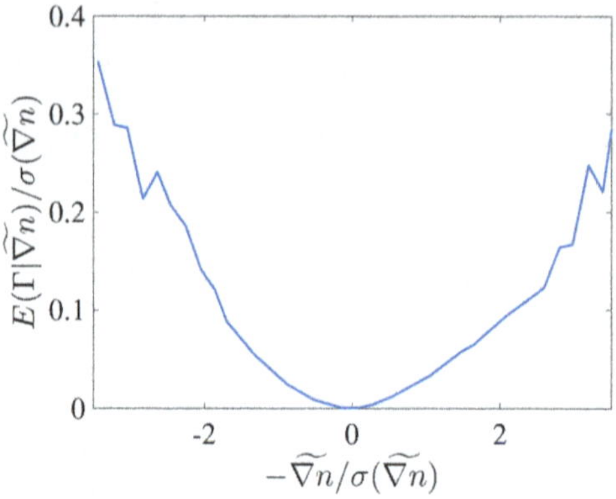

Figure 4.16. Normalized conditional expectation of the flux as a function of the normalized gradient, discharge 32805, NBI phase. (Reproduced from [6], with the permission of AIP Publishing.)

V_f, and the gradient is deduced from two radially displaced probe pins measuring the ion saturation current, I_{sat} (more details in [6]). The radial location of the probe was $\rho \simeq 0.92$. These results are typical for NBI discharges in TJ-II.

To understand this result—reproducible over a wide range of experimental conditions, as reflected in the cited references—we turn to the resistive MHD model, used frequently in this book (see appendix B). Running a simulation for profiles similar to the TJ-II experimental profiles and calculating he turbulent radial particle flux is given as $\Gamma = \tilde{v}_r\tilde{n}$, we obtain the result shown in figure 4.17 (here, $\nabla = d/dr$ and σ is the standard deviation). The expectation value of the flux, conditioned on the gradient, is similarly 'U'-shaped as the experimental curves.

As the model flux–gradient curve reproduces the experimental curve (qualitatively, at least), we can use the model to study this phenomenon a bit further. To quantify the impact of the fluctuating flux on the gradient, we computed the transfer entropy for a range of reference positions [6], although here we will only show a single result. Figure 4.18 shows the transfer entropy (with fixed $r_\Gamma = 0.594$, the position of the $\iota\!\!\!- = 8/5$ rational) versus time lag and $r = r_{\nabla n}$, in comparison with the mean poloidal flow velocity. The transfer entropy exhibits a quasi-periodic structure, associated with the MHD oscillations. Regions with different dominant frequencies can be observed: at the rational, frequencies are approximately twice as high as away from the rational. The transfer entropy is relatively low in regions with high poloidal velocity shear, where turbulent eddies will be reduced in size.

However this may be, this result shows that the fluctuating flux has an impact on the gradient at a significant distance from the flux origin (non-locality), and that this influence is not instantaneous. This delayed, non-local effect suggests that the above explanation of the left half of the 'U'-shaped flux–gradient curve is likely as follows: a large flux event originating at a position inward from the observation point may decrease the gradient at the observation point (by removing particles from a point inward from the observation point via outward particle expulsion), thus creating a correlation between large fluxes and reduced gradients. On the other hand, the

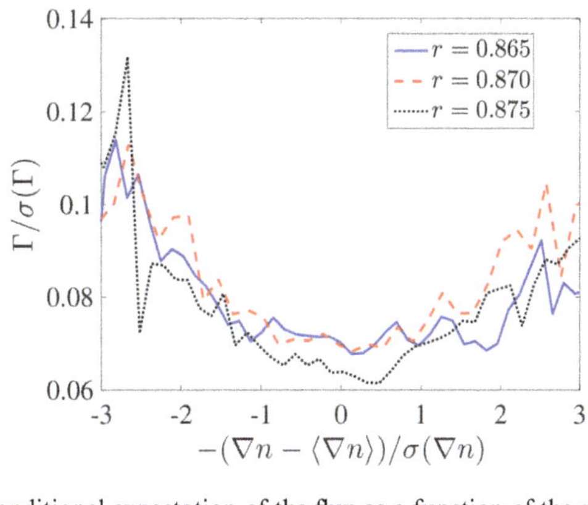

Figure 4.17. Normalized conditional expectation of the flux as a function of the normalized gradient at three radial locations for the resistive pressure-gradient-driven turbulence model. (Reproduced from [6], with the permission of AIP Publishing.)

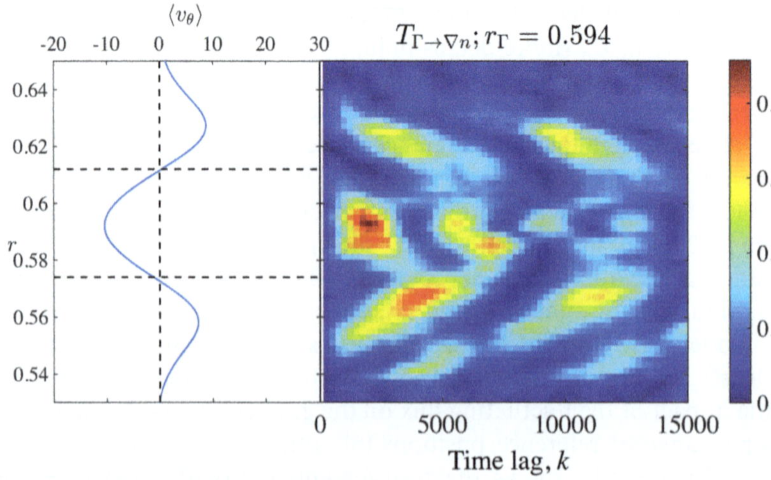

Figure 4.18. Transfer entropy $T_{\Gamma\to\nabla n}$ and poloidal flow (with $r_\Gamma = 0.594$, in normalized units, see the text). Locations corresponding to flow shear maxima, $\max_r|d\langle v_\theta\rangle/dr|$, are indicated by horizontal dashed lines. (Reproduced from [6], with the permission of AIP Publishing.)

correlation between large gradients and large fluxes (the right half of the 'U') is straightforward: gradient drives flux locally. Thus, at least part of the transport would be flux-driven (the non-local part of transport), although the 'normal' gradient-driven flux (typical of diffusive or collisional transport) would also occur. This analysis testifies to the complexity of transport.

4.6 Understanding radial transport

In chapter 1, we emphasized the importance of understanding radial transport in fusion devices. Experimentally, this question can be approached in various ways, while noting that the acquisition of data from the interior of the hot plasma, a very hostile environment, constitutes a significant difficulty.

The first and most common approach is to determine the radial profiles of the quantities appearing in the transport equations, such as $n_e(r)$ and $T_e(r)$, in a steady state situation in which the inputs (fuelling and heating) and outputs (losses through conduction, convection, and radiation) are balanced. By assuming a reasonable transport model, one can then determine (some of) the transport parameters: the local power balance between input and output allows estimating the local fluxes, and by dividing a given flux by the local gradient obtained from the corresponding conjugate profile, one obtains an estimate of the local transport coefficient. This approach implies many implicit assumptions: the statement that the flux equals the transport coefficient times the gradient ignores possible off-diagonal elements of the transport coefficient matrix (such as the impact of the density gradient on heat transport); it assumes that the flux–gradient relation is linear, which it certainly is not [38]; and that transport is strictly local, which it probably is not [39], as we will see.

Significant improvement is possible by exploring transport using perturbative methods. In this approach, a steady state plasma is perturbed (by modulating the heating or fuelling of the plasma), and the effect of the perturbation is studied. The propagation of the perturbation allows making inferences regarding the local transport coefficients. Due to the high level of turbulence, producing a sufficiently large perturbation to be able to distinguish it from the ambient fluctuations is an issue. If the perturbation is too large, the steady state situation is affected and making clean inferences is difficult. A solution to this problem is to produce regular, repeated perturbations, a technique known as 'modulation'. In this case, it is possible to extract the relevant components of the plasma response by filtering the data at the modulation frequency or its harmonics [29]. This technique has achieved a high degree of sophistication [40, 41].

Matching the results of these experimental analyses to underlying models has not been easy, as the plasma often responds in unexpected ways, as explained in section 1.7, which has sometimes been interpreted in the sense that transport must be at least partially non-linear and non-local [42, 43].

In this section, we will apply the transfer entropy to this issue. Like the modulation technique, we will find that the transfer entropy is capable of extracting relevant information from the ambient turbulent data. It achieves this feat through a statistical analysis of the impact of fluctuations at one location on the fluctuations at another location. As we will see, this allows following the propagation of small, spontaneous perturbations from one location to the next through the turbulent system. The capacity to track the propagation of perturbations implies the capacity to analyze transport. This approach potentially offers a huge advantage over the modulation approach, as it is not necessary to perturb the system externally to extract information on transport, so that the system can be examined in its 'natural' (unperturbed) state. In fact, we will see that this approach reveals new interesting aspects of transport that are generally lost in the modulation approach, as finer details tend to be overwhelmed by the relatively large modulation.

4.6.1 Heat transport in the TJ-II stellarator

In this subsection, we study heat transport in ECRH heated discharges in the TJ-II stellarator (see appendix A.2). For this purpose, we will analyze data obtained using a 12-channel electron cyclotron emission (ECE) diagnostic to measure the local electron temperature $T_e(r)$ along a midplane chord on the high magnetic field side of the plasma (at a toroidal angle of $\phi = 315°$), covering a significant part of the plasma minor radius (about 70% of the minor radius), with a radial resolution of about 1 cm [44]. We emphasize that this is one of the typical non-invasive diagnostics available at most magnetic confinement devices, capable of delivering local measurements at high temporal resolution from the hot core plasma region. Figure 4.19 shows an example of T_e measurements. The ECRH power deposited in the core of the plasma ($|\rho| \lesssim 0.2$) causes spontaneous temperature fluctuations [45, 46], possibly related to the presence of rational surfaces in the core region of the plasma where the heat is deposited [47] and the concomitant generation of fast electrons [48]. Regardless of the cause,

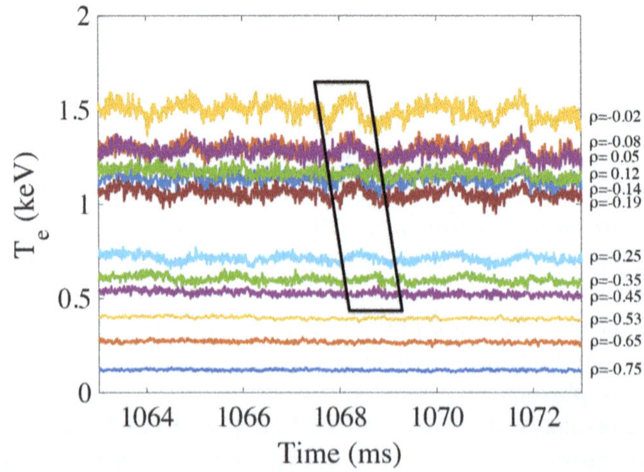

Figure 4.19. A short section of ECE data from TJ-II. (Reproduced with permission from [51]. © IOP Publishing. All rights reserved.)

these temperature fluctuations then lead to the outward propagation of small corresponding cascades or perturbations, similar to what has been reported in [49]. In this particular example, the outward propagation of a central spontaneous T_e fluctuation can be followed by the naked eye (as indicated by the parallelepiped), but in general, the analysis of the propagation of perturbations requires applying some analysis tools [50].

For this purpose, we calculate the transfer entropy between two signals, in this case between data measured by an ECE channel at a reference position ρ_{ref} (X in equation (4.6)) and data from an ECE channel at another position, ρ (Y in equation (4.6)), for a range of lags, k. We assign a time delay τ to k by multiplying it with the sampling time interval Δt: $\tau = k\Delta t$. Figure 4.20 shows an example. The ECE reference channel is taken at $\rho_{\text{ref}} \simeq -0.07$, and the other ECE channels are distributed along the minor radius $-1 \leqslant \rho \leqslant 1$. The transfer entropy between the reference ECE channel and itself is zero by definition, leading to a horizontal blue streak near $\rho = -0.07$ in the graphs, which therefore has no physical meaning. In the figure, $\rho = 0$ corresponds to the magnetic axis of the plasma, while $|\rho| = |r/a| = 1$ corresponds to the minor radius of the plasma. By convention, ECE channels with a negative ρ coordinate (the locations of which are indicated in the figure by white circles) are located on the high field side of the magnetic axis. Since these ECRH plasmas have both low normalized pressure and low net currents, the magnetic configuration is almost completely determined by the externally imposed vacuum configuration [51, 52]. This allows us to indicate the approximate location of some major rational surfaces in the figure.

The two panels in this figure ((a) and (b)) correspond to plasmas with a very different level of electron cyclotron heating power, as indicated in the caption. Comparing the low and high power cases shown in the figure, one observes a relatively smooth diagonal 'plume' of propagating perturbations in the low power case, propagating outward from $\rho = \rho_{\text{ref}}$. We note that this type of behaviour, in

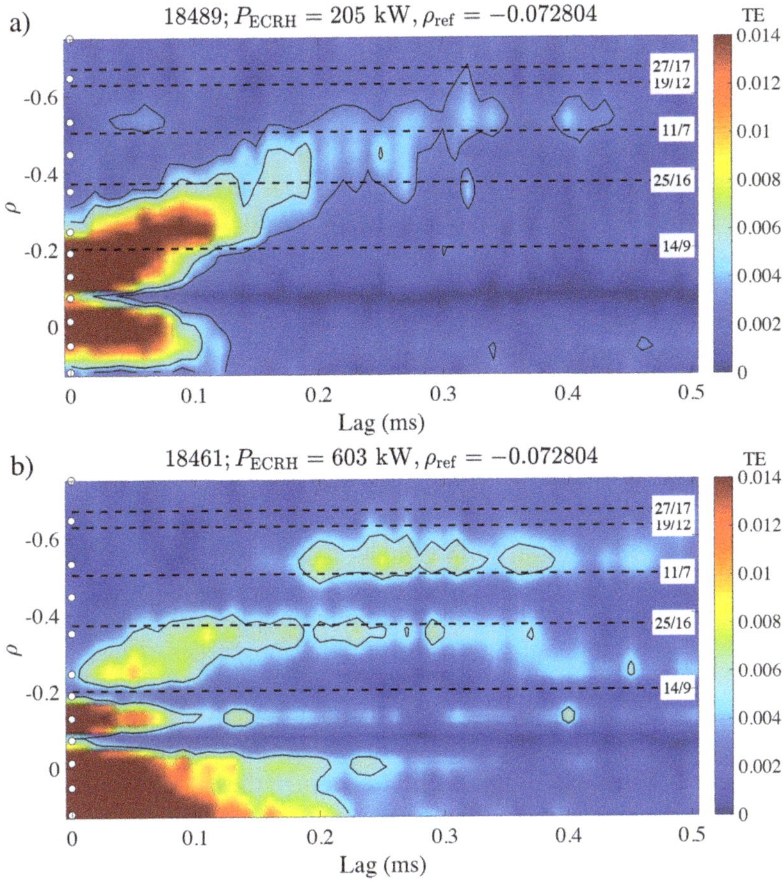

Figure 4.20. Examples of transfer entropy calculated from electron cyclotron emission (ECE) data taken at the TJ-II stellarator, using $\rho_{\mathrm{ref}} \simeq 0.07$, at (a) $P_{\mathrm{ECRH}} = 205$ kW and (b) $P_{\mathrm{ECRH}} = 603$ kW. The colour scale indicates the value of T. ECE channel positions are indicated with white circles. The approximate location of some major rational surfaces is indicated by horizontal dashed lines; the line labels specify the corresponding rotational transform of the magnetic field, *n*/*m* (toroidal per poloidal turns). (Reproduced from [53], with the permission of AIP Publishing.)

which the delay time τ between fluctuations in the reference channel at $\rho = \rho_{\mathrm{ref}}$ and other channels increases with the distance $|\rho - \rho_{\mathrm{ref}}|$ is what one would expect to see for fluctuations starting from the reference position and propagating away, a bit like ripples in a pond. The main body of the plume occurs in the range $-0.35 < \rho < -0.07$, although a rather weak continuation of the plume reaches about $\rho \simeq -0.55$, where some stagnation may be visible. This situation would be roughly consistent with 'normal' diffusive propagation.

However, in the high power case, the plume clearly stagnates at $\rho \simeq -0.35$, developing a long horizontal tail; yet, for $\tau \simeq 0.2$ ms, a second propagation branch appears at $\rho \simeq -0.55$, with an amplitude comparable to or greater than the first branch. Note that this response occurs without any detectable response at the intermediate position $\rho \simeq -0.45$, so that the perturbations seem to have 'jumped over' this intermediate position. The perturbations at $\rho \simeq -0.55$ have a stronger causal link to ρ_{ref} (i.e. a higher value of transfer entropy) than in the low power case. Apart from the radial distribution, one should also pay attention the value of the time lag at which this second branch occurs, namely $\tau > 0.2$ ms. In spite of the apparent disconnection from the first branch at $\rho \simeq -0.35$, this time delay does seem to suggest a continuation of the propagation initiated at ρ_{ref}.

The stagnation at $\rho \simeq -0.35$ and the stronger causal response at $\rho \simeq -0.55$ are quite remarkable and unexpected and, as we will see, may be related to power degradation, as perturbations seem to be better able to reach this position and influence turbulence there, possibly implying a more intense radial transport from ρ_{ref} to $\rho \simeq -0.55$ in the high power case.

One may of course wonder whether this is somehow a special case or whether this unusual behaviour is systematic. To answer this question, we have looked at a large collection of over 200 repetitive discharges at high ECRH power, in a range of magnetic configurations, as listed in table 4.1.

Figure 4.21 shows the iota profiles of the experiments listed in table 4.1. This variation in rotational transform profiles means that specific rational surfaces are located at different radial positions in different configurations, permitting the study of the impact of these rational surfaces on plasma properties.

Individual plasmas experience variations due to slight changes in ECRH heating power, gas puff, wall conditions, impurity content, radiation levels, turbulence, etc. In order to reduce the impact of such irrelevant variations and highlight the systematic aspects of transport, it is useful to average the calculated transfer entropy over similar discharges having the same magnetic configuration. Shot-to-shot variations in the transfer entropy, not systematically related to the magnetic configuration, will be suppressed by the averaging procedure. On the other hand, effects associated with the magnetic configuration will be emphasized. Prior to analysis, slow data drifts are removed by applying a high-pass filter with 100 Hz cutoff.

Table 4.1. List of experiments.

Configuration	$\iota\!\!\!-(1)$	N_{shot}
100_36	1.57	19
100_38	1.59	23
100_40	1.61	34
100_42	1.63	22
100_44	1.65	29
100_46	1.67	19
100_48	1.68	35
100_50	1.70	16
100_52	1.71	10

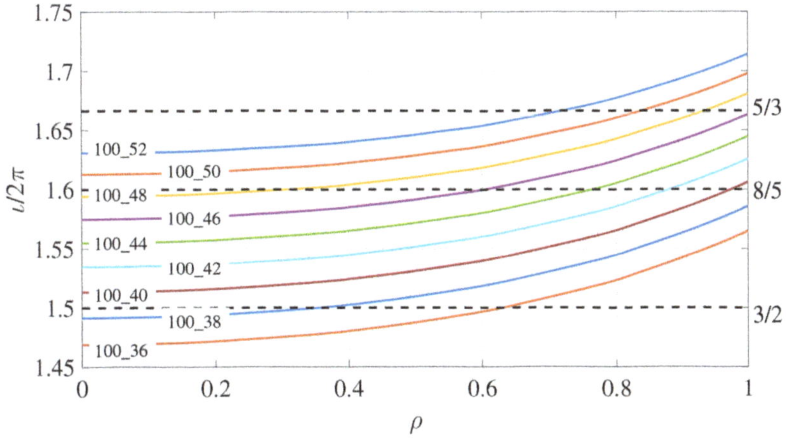

Figure 4.21. Profiles of the rotational transform, $\iota\!\!\!\!-\, = \iota/2\pi$ as a function of normalized radius, $\rho = r/a$. Major rational values are indicated by horizontal dashed lines. The line labels are identifiers of the magnetic configuration corresponding to each profile. (Reproduced with permission from [51]. © IOP Publishing. All rights reserved.)

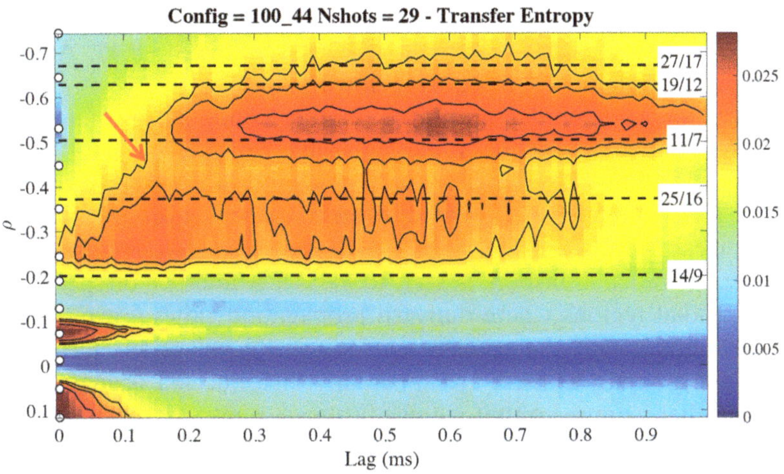

Figure 4.22. Transfer entropy T for configuration 100_44. (Reproduced with permission from [51]. © IOP Publishing. All rights reserved.)

The full set of results is presented in [51]. Here, we merely highlight a few interesting observations. Figure 4.22 shows an example for the 'standard' configuration, 100_44. Radial propagation is visible outside the ECRH power deposition zone ($|\rho| \lesssim 0.2$). As in figure 4.20, the transfer entropy graph exhibits two horizontal structures, near $\rho \simeq -0.35$ and -0.55. Apparently, the outward propagation is therefore 'held up' near the 25/16 rational surface, but succeeds nevertheless in reaching a location near 11/7 surface after making a 'jump'. Perhaps this behaviour can be better understood by focusing on the 'propagation front', i.e. the contour

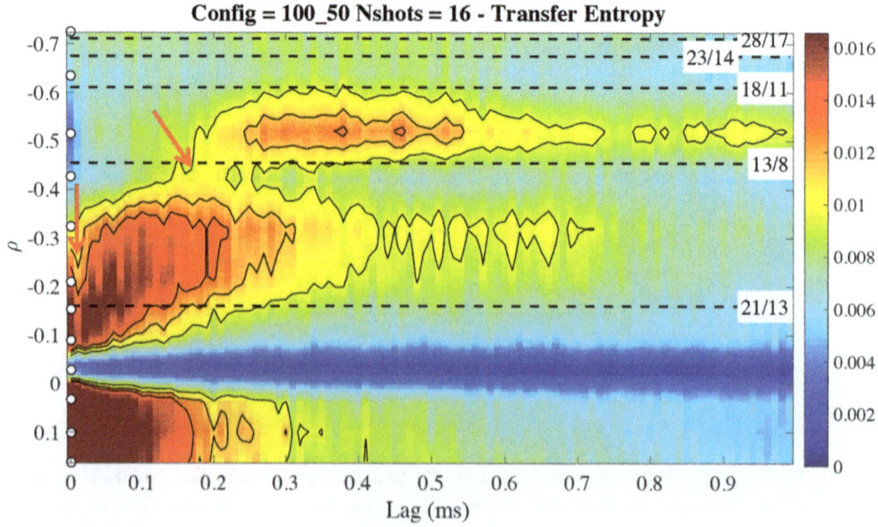

Figure 4.23. Transfer entropy T for configuration 100_50. (Reproduced with permission from [51].)

indicated by the red arrow in the figure. One observes a gradual outward propagation (a straight section) between $\rho \simeq -0.3$ and $\rho \simeq -0.45$. At that point, there is an outward 'jump' followed by a slowing down.

Figure 4.23 shows a similar graph for another configuration, namely 100_50. The outward propagation 'plume' is clearly visible between $\rho \simeq -0.1$ and -0.3, as well as the stagnation at $\rho \simeq -0.3$, and the second 'branch' at $\rho \simeq -0.52$. The propagation front (the contour indicated by the red arrow), also behaves in a similar way (with a clear 'step').

Summarizing, radial transport is seen to exhibit a number of specific features: (i) there are radial regions where outward propagation slows down, connected to (ii) radial regions where the transfer entropy graphs develop horizontal 'tails' towards long lags; and (iii) there are apparently disconnected propagation 'jumps' over certain radial regions, so that regions further out respond at similar lag values as regions further in.

The observed 'stepwise' propagation of the temperature perturbations (associated with points (i) and (ii)) can be understood from the impact of flow shear regions, which may or may not be associated with rational surfaces, on transport. Such flow shear regions constitute transport barriers for the radial propagation of turbulence. As a consequence, radial propagation may be slowed down at specific radial positions, leading to 'stick-slip' behaviour and a characteristic 'staircase' shape of the propagation front and horizontal 'tails' developing in the transfer entropy plots, corresponding to 'trapping regions'.

On the other hand, perturbations are also observed to 'jump over' certain radial regions (point (iii)), strongly suggesting a 'non-local' component of transport. Based on simulations made with a resistive MHD model, we think that this effect is related to resonances occurring between MHD modes located at different rational surfaces.

The resulting 'instantaneous, non-local' causal influence chains occurring in this model will be analyzed in more detail in section 4.7. We note that rapid transport along the boundaries of island O-point regions may also be involved [54].

These results and interpretations may be somewhat surprising and one might even be tempted to reject them on the suspicion that something must be wrong with the analysis. To address these concerns, we have analyzed a few discharges using the traditional modulation method, similar to what was reported in [55], and compared the results with the transfer entropy. In these experiments, one of the two ECRH systems was operating continuously at ∼200 kW, while the other ECRH system was modulated at low power, with a square waveform. The line average density was low, $\bar{n}_e \simeq 0.4 \times 10^{19}\ \mathrm{m}^{-3}$.

Figure 4.24 shows T_e data obtained in a discharge with configuration 100_36, having a low-order rational surface (3/2) in the edge region, and with modulated ECRH power (modulation frequency: 180 Hz, duty cycle: 30%). Only 50 ms of the available 200 ms of data are shown. The Fourier spectrum of each ECE channel is calculated, and the spectral amplitude $A(f)$ and cross phase $\phi(f)$ (between a given channel and the central channel) is obtained. Figure 4.25 and 4.26 show the amplitude and cross phase of the measurement signals at the modulation frequency and the first two harmonics. In the power deposition region, $\rho \lesssim 0.2$, the amplitude is fairly constant and the phase $\phi \simeq 0$.

The slope of the phase, $d\phi/d\rho$, is inversely proportional to the phase velocity of the outward propagating heat waves [41, 56]. Note that the slope is particularly high in the range $0.4 < |\rho| < 0.55$, suggesting that the heat waves are slower in that region than elsewhere, likely related to the presence of a barrier associated with the major rational surface (3/2) located at $|\rho| \simeq 0.63$ (see [57]). Furthermore, the slope is rather low in the range $0.55 < |\rho| < 0.8$, indicating fast propagation in the area surrounding this major rational surface, possibly indicating propagation around the O-point of a magnetic island [54].

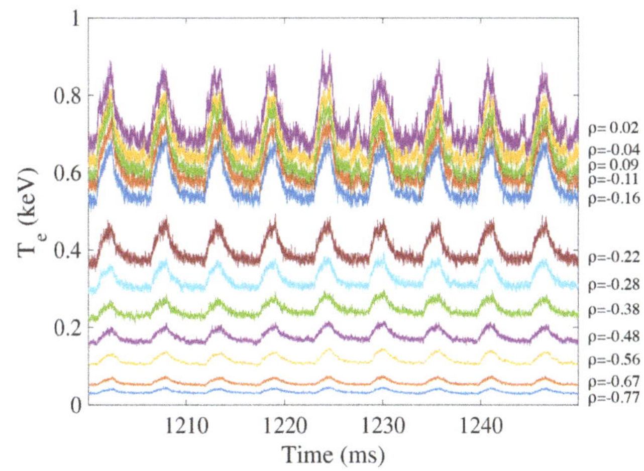

Figure 4.24. ECE data in a discharge (29 748) with modulated central ECRH power. Labels indicate the radial position ($\rho = r/a$) of the measurement. (Reproduced with permission from [51]. © IOP Publishing. All rights reserved.)

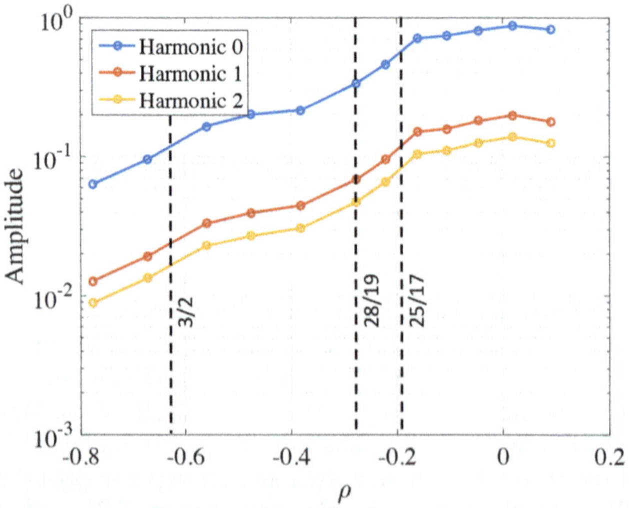

Figure 4.25. Modulation amplitude at the modulation frequency and the first two harmonics for the discharge shown in figure 4.24. The positions of a few rational surfaces in vacuum are indicated by vertical dashed lines. (Reproduced with permission from [51]. © IOP Publishing. All rights reserved.)

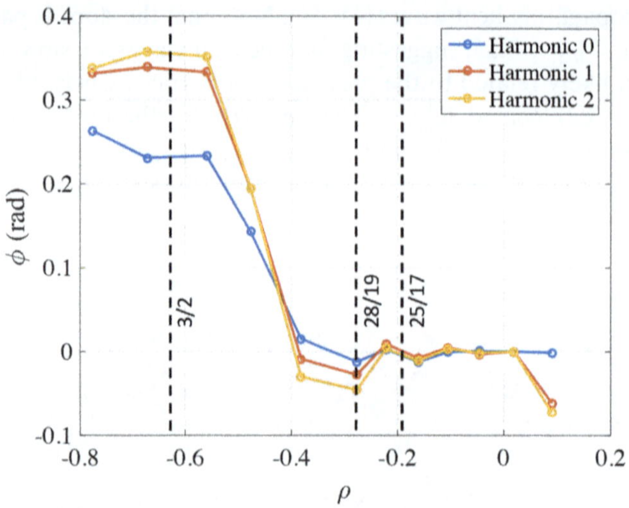

Figure 4.26. Modulation phase (relative to the central channel at $\rho \simeq 0.02$) at the modulation frequency and the first two harmonics for the discharge shown in figure 4.24 (configuration 100_36). The positions of a few rational surfaces in vacuum are indicated by vertical dashed lines. Error bars (not shown) are similar to those of figure 4.28. (Reproduced with permission from [51]. © IOP Publishing. All rights reserved.)

We then subjected these same data to the analysis technique based on the transfer entropy, in order to compare results. Figure 4.27 shows the result. Particularly interesting is the radial location $|\rho| \simeq 0.55$. Whereas in the case of spontaneously generated temperature perturbations, this location acted as a 'trapping zone' that the small, outward propagating temperature perturbations were largely unable to cross,

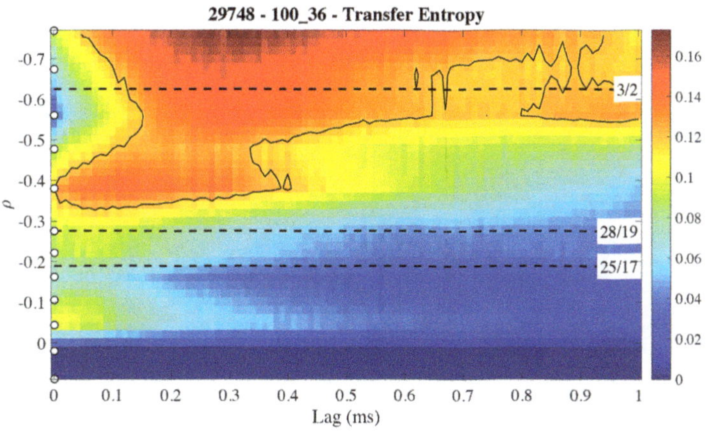

Figure 4.27. Transfer entropy T for the modulated discharge shown in figure 4.24, configuration 100_36. (Reproduced with permission from [51]. © IOP Publishing. All rights reserved.)

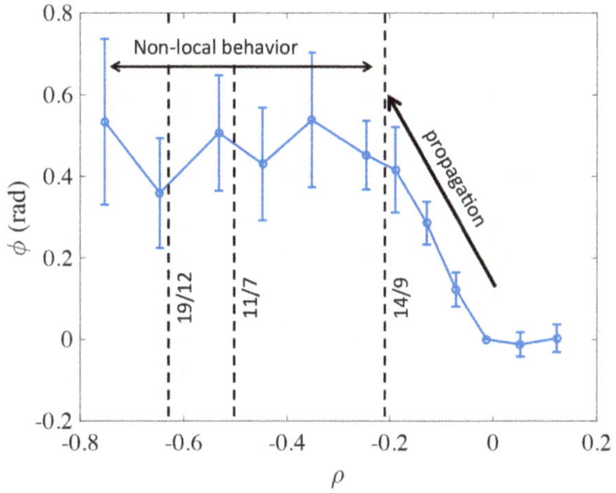

Figure 4.28. Modulation phase (relative to the central channel at $\rho \simeq -0.01$) at the second harmonic of the modulation frequency (configuration 100_44). (Reproduced with permission from [51]. © IOP Publishing. All rights reserved.)

here it constitutes only a weak barrier that is crossed easily by the much larger heat waves produced by the modulation, resulting in high values of the transfer entropy at locations outside the 3/2 rational surface ($|\rho| > 0.7$).

Still, propagation appears to slow down before reaching $|\rho| \simeq 0.5$, suggesting the existence of a minor transport barrier (possibly associated with the 3/2 rational surface) and low phase velocity (according to figure 4.26). In any case, the slowing down, seen clearly as a steep slope in the Fourier cross phase, is indeed reflected in the transfer entropy analysis.

Another ECRH modulation case is shown in figure 4.28. Here, the magnetic configuration is 100_44 and the modulation frequency is 110 Hz. The figure shows

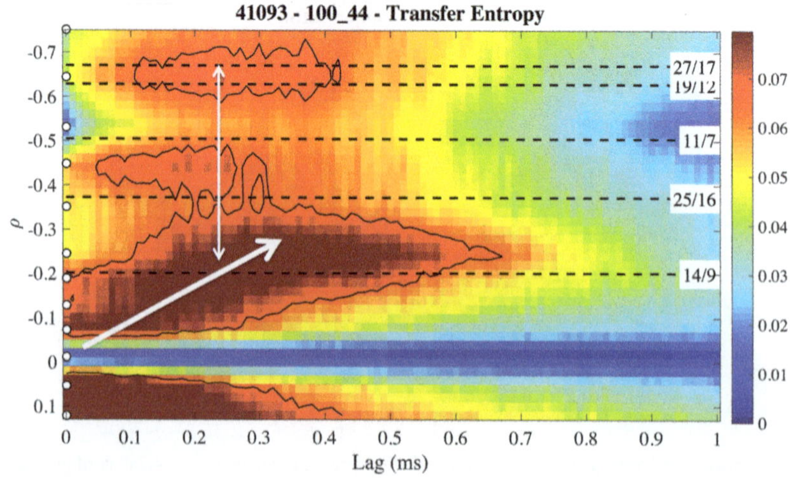

Figure 4.29. Transfer entropy T for a modulated discharge with configuration 100_44. (Reproduced with permission from [51].

the phase of the second harmonic of the modulation frequency relative to the central channel at $\rho \simeq -0.01$. There are clearly two regions: one in which the phase increases gradually with $|\rho|$, indicating propagation, and one in which the phase is approximately constant with ρ, suggesting a non-local (simultaneous) response, indicated in the figure with arrows.

The analysis of this discharge using the transfer entropy is shown in figure 4.29. This graph likewise seems to suggest the existence of a zone dominated by outward propagation ($|\rho| < 0.3$), and a zone dominated by a simultaneous response ($|\rho| > 0.3$). At $|\rho| \simeq 0.25$, a 'tail' is seen to develop in the transfer entropy, suggesting a 'trapping zone', possibly associated with the 14/9 rational surface. At the time lag the outward propagating pulses (indicated by the thick arrow) cross the 14/9 rational surface, instantaneous responses are detected further outward (specifically, along the two-headed arrow at $|\rho| \simeq 0.45$ and $|\rho| \simeq 0.65$). One could be led to think that the arrival of heat pulses at the 14/9 rational surface triggers an instability there, which couples (magnetically) to instabilities further out, thus explaining the apparent instantaneous propagation in the outward zone. Indeed, this mode coupling effect is the interpretation we currently favour (see section 4.7).

We conclude that the transfer entropy and the traditional Fourier analysis mostly agree on general lines, regarding the existence of both transport barriers and zones with rapid radial propagation. The difference resides in that the transfer entropy uses all of the available information (instead of selected information corresponding to specific frequencies), and is capable of analyzing transport *without* modulation, revealing fine details that are not accessible by the traditional perturbative method.

4.6.2 Power degradation in the TJ-II stellarator

In section 4.6.1 we saw that heat transport behaves differently at low and high heating power (see figure 4.20). This raises the question whether this difference could be related to the power degradation effect discussed in section 1.7. Figure 4.30 shows an example of electron temperature profiles, $T_e(\rho)$, at various heating levels of the ECR heating system [53]. The power degradation effect is almost visible to the naked eye: raising the power by a factor of 2.4, the electron temperature only goes up by a factor of 1.9, i.e. more than 20% less than expected for fixed transport coefficients.

The series of experiments reported here were performed on 16 April 2008. The data correspond to a set of 19 discharges at 9 different levels of heating power, P_{ECRH}. The plasma parameters (other than T_e) were kept as constant as possible throughout this set of discharges (and profiles were similar to those in [44]). Ion temperature gradients are small [58]. Figure 4.31 shows the variation of line averaged electron density, $\bar{n}_e$, from interferometry, and plasma energy content, W, obtained from a diamagnetic loop. It is seen that the line averaged electron density is fairly constant, while the variation of the plasma energy content is consistent with the expected scaling for magnetically confined plasmas, $W \propto P_{\mathrm{ECRH}}^{0.4}$, implying a scaling for the energy confinement time $\tau_E = W/(P - dW/dt) \simeq W/P \propto P_{\mathrm{ECRH}}^{-0.6}$, in accordance with standard scaling law results. Thus, this set of discharges displays the typical power degradation effect, so that we can use it to investigate this issue.

We already showed two examples of transfer entropy for these discharges in figure 4.20. The transfer entropy quantifies the propagation of 'information' rather than heat. Therefore, it is not immediately obvious how the reported findings relate to the electron heat transport coefficient, χ_e, used in traditional transport models for magnetically confined plasmas. On the other hand, diffusion is, in some sense, a 'geometric' process (the spreading in space of an initially concentrated perturbation), as reflected in the dimension of the diffusion coefficient, χ_e (m^2 s^{-1}), which involves a length and a time, but no temperature or energy. Thus, the 'spreading of

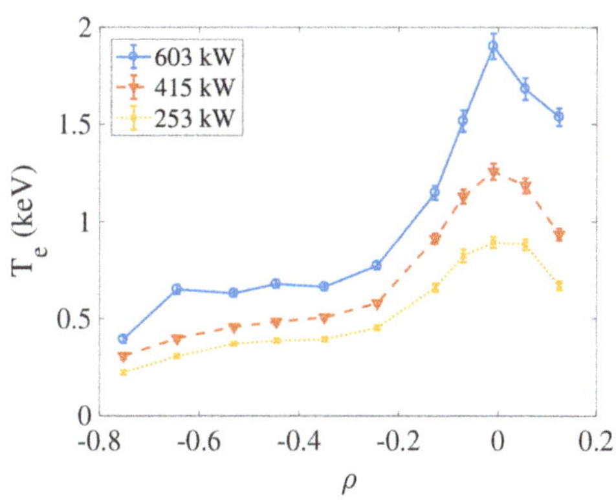

Figure 4.30. Electron temperature profiles, measured by the ECE diagnostic, at various heating power levels. (Reproduced from [53], with the permission of AIP Publishing.)

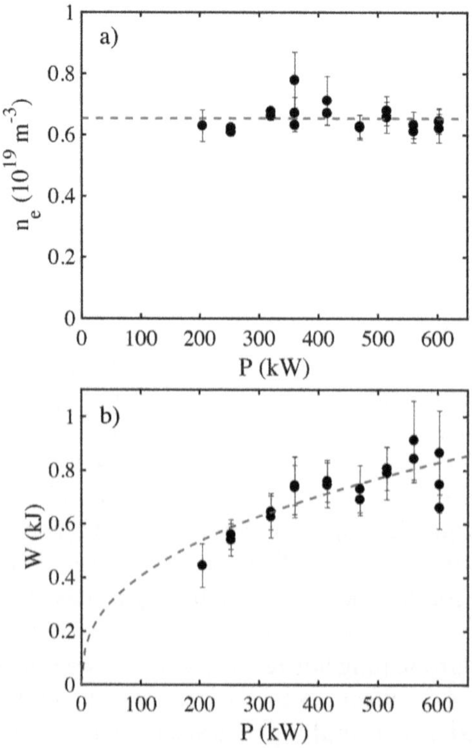

Figure 4.31. (a) Line averaged density (from interferometry) versus P_{ECRH}. (b) Plasma energy content versus P_{ECRH}. (Reproduced from [53], with the permission of AIP Publishing.)

information' measured by the transfer entropy can be related to the 'spreading of heat' measured by χ_e. From another point of view, the transport of information is mediated by the transport of (fluctuations of) heat, which is what is measured by the ECE diagnostic, hence the transport of information should be equivalent to the transport of heat in this case, or at least cannot exceed it (since the information cannot go where the heat, which 'carries' it, does not go).

To shed some light on this issue, we define an effective diffusivity, χ_{eff}, and observe how it behaves. For this purpose, we first quantify the propagation of the centre of gravity of the 'plume' of high transfer entropy versus the time lag τ by calculating the mean radius of the propagating information as follows, for fixed τ,

$$\langle \rho \rangle = \frac{\int T_{\rho_{\mathrm{rel}} \to \rho} \rho d\rho}{\int T_{\rho_{\mathrm{rel}} \to \rho} d\rho}, \tag{4.15}$$

i.e. the weighted mean of the radius, using the transfer entropy $T_{\rho_{\mathrm{rel}}}$ from the reference position to the position ρ as the weight. The integration is only performed over negative values of ρ (the high field side of the plasma). Figure 4.32 shows two examples of $\langle r \rangle = -\langle \rho \rangle a$ as a function of $\sqrt{\tau}$, based on transfer entropy graphs such

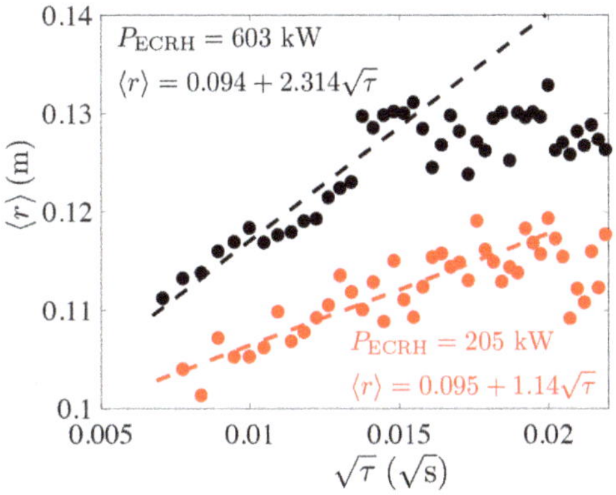

Figure 4.32. Mean radius of propagation, $\langle r \rangle$, versus the square root of the time lag, $\sqrt{\tau}$. (Reproduced from [53], with the permission of AIP Publishing.)

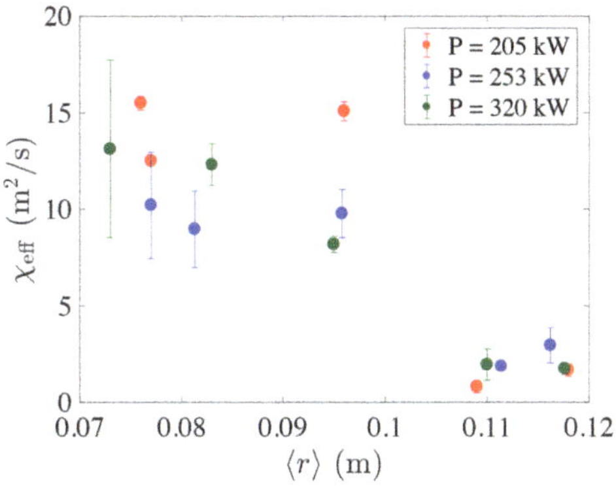

Figure 4.33. Radial dependence of the effective diffusivity for three values of P_{ECRH} in the low power range. (Reproduced from [53], with the permission of AIP Publishing.)

as those shown in figure 4.20; although here, we used the ECE channel at $\rho \simeq -0.35$ as reference in order to obtain an estimate of χ_{eff} for $|\rho| > 0.5$ ($|r| > 0.1$). A straight section of this curve allows fitting a straight line and making an estimate of the effective diffusivity (which is equal to the square of the slope of this line), and doing so we obtain values $\chi_{\mathrm{eff}} \simeq$ 1–4 $\mathrm{m}^2\ \mathrm{s}^{-1}$. We feel that this is a rather interesting result, as these values are comparable in value to the global electron heat diffusivity determined in earlier work, using modulation experiments, namely, $\chi_{\mathrm{eff}} \simeq 2\ \mathrm{m}^2\ \mathrm{s}^{-1}$ [55]. This confirms the close relationship between the electron heat diffusivity and the effective diffusivity we estimate from the transfer entropy.

We can now take this analysis one step further by estimating χ_{eff} for the whole set of discharges and using different ECE channels as a reference. This allows us to make estimates of χ_{eff} at different radial positions.

The result for the three lowest values of P_{ECRH} is shown in figure 4.33. The three cases show a similar radial dependence. The values of the effective diffusivity are high in the inner part of the plasma and drop as one moves outward (recall that the plasma minor radius is $a = 0.2$ m). We note that this is similar to what was observed in previous work, using traditional techniques (figures 13 and 14 in [55]).

Figure 4.34 shows a similar result for the three highest values of P_{ECRH}. One observes there is a clear change from figure 4.33. In the inner region, the effective diffusivity has decreased, while it has increased in the outer region. Thus, in this power range, the radial dependence of the effective diffusivity is weaker.

Let us assume, for simplicity, that the effective diffusivity is a combination of standard diffusive transport plus a variable non-local 'anomalous' component. Since ECRH power is mostly deposited in the core, the effect of the heating is to first facilitate the non-local transport in the inner part of the plasma. As the heating increases, this non-local effect expands to regions located further outward. The plasma volume increases as r^2, so even a relatively small increase in transport in the outer region will have a significant effect on global confinement.

In figure 4.35 we plotted an equivalent graph to figure 4.33 and 4.34 for intermediate values of P_{ECRH}. One observes that the transition from low to higher power behaviour possibly occurs in the range of 400 kW–500 kW.

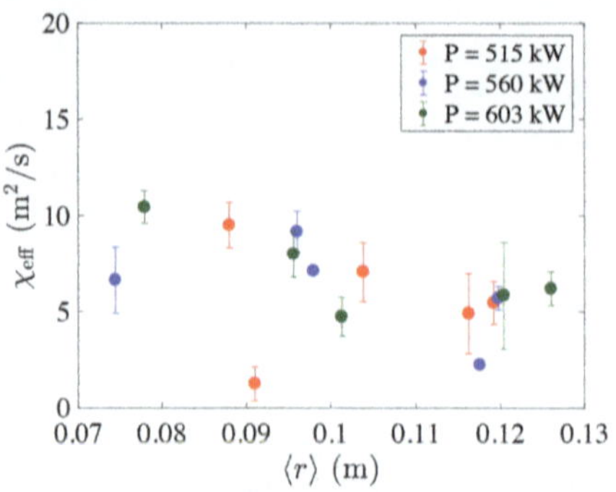

Figure 4.34. Radial dependence of the effective diffusivity for three values of P_{ECRH} in the high power range. (Reproduced from [53], with the permission of AIP Publishing.)

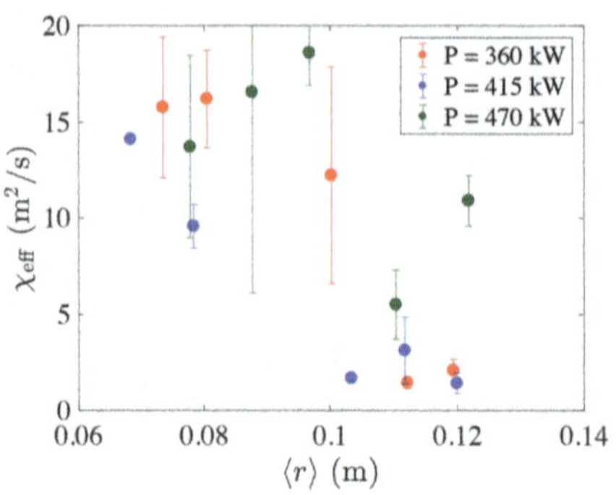

Figure 4.35. Radial dependence of the effective diffusivity for three values of P_{ECRH} in the medium power range. (Reproduced from [53], with the permission of AIP Publishing.)

This analysis of the discharges from an ECRH power scan in TJ-II suggests a mechanism to explain the deterioration of confinement with heating power. It appears that radial heat transport is not a smooth and continuous process, but involves 'stagnation zones' or 'mini-transport barriers' [59] (reminiscent of an '$E \times B$' staircase [60, 61], likely associated with localized zones with enhanced sheared flow) where transport tends to slow down. On the other hand, transport may also experience rapid radial (non-local) 'jumps', connecting apparently separate regions across the 'mini-transport barriers'.

The possible mechanism to explain the degradation of confinement with increasing heating power is as follows. When ECRH power is increased but still low, non-local transport in the inner part of the plasma is enhanced first—detected here as a high value of the effective diffusivity. As the heating increases further, this non-local effect expands towards more outlying regions of the plasma while the effective diffusivity in the core region drops. This increased importance of non-local transport in zones that are located further outward would lead to the deterioration of confinement.

We note that this rapid transport channel is consistent with the 'non-local' transport hypothesis suggested by some authors [62], while its activation with increased heating power is consistent with the 'critical gradient' hypothesis [30].

The question remains, however, to what underlying physical mechanism this 'non-local' transport component corresponds. In this respect, we suggest that this mechanism involves the radial coupling of MHD turbulence, so that the fast 'non-local' transport channel is based on the transmission of turbulent energy via the magnetic field. We note that such a mechanism would imply an important role for rational magnetic surfaces in turbulent transport, as indeed evidenced by some previous studies [63, 64].

4.6.3 Heat transport in the W7-X stellarator

We applied the same analysis technique in the W7-X stellarator (see appendix A.4 and [59]), which disposes of a similar ECE diagnostic to measure the electron temperature as TJ-II. W7-X has a larger minor radius (0.5 m versus 0.2 m) and more ECE channels (32 versus 12), thus offering better resolution than TJ-II for these studies. Qualitatively, the results discussed below mirror the results obtained for TJ-II to a large extent, suggesting that heat transport may be governed by the same principles in a small (non-optimized) device such as TJ-II and a large (optimized) device such as W7-X, even though global transport levels are very different. Both devices are low-shear stellarators, offering a low density of low-order rational surfaces which facilitates the study of the impact of such surfaces on transport. With its significantly larger minor radius and its enhanced ECE diagnostic, W7-X offers an ideal platform for these studies.

We will analyze data from a set of discharges from the first operational phase of W7-X [65]. These discharges are characterized by low electron density ($\bar{n}_e \simeq (1\text{–}2) \times 10^{19}$ m^{-3}) and are heated by electron cyclotron resonance heating (ECRH, $P_{\text{ECRH}} = (0.6\text{–}2)$ MW), resulting in so-called core electron-root

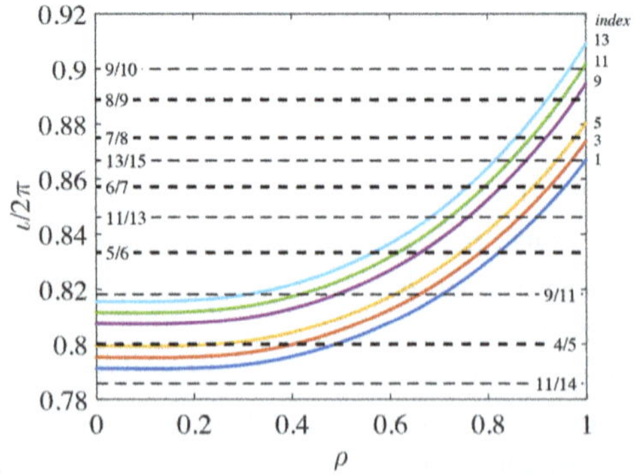

Figure 4.36. Profiles of the rotational transform, $\iota\!\!\!-$, as a function of normalized radius, $\rho = r/a$, for the set of magnetic configurations studied here. The lowest profile (index 1) corresponds to the 'standard OP1.1' configuration. Major rational values are indicated by horizontal dashed lines. (Reproduced with permission from [59]. © IOP Publishing. All rights reserved.)

confinement (CERC) plasmas [66]. The discharges correspond to a $\iota\!\!\!- = \iota/2\pi$ scan. The relevant $\iota\!\!\!-$ profiles are shown in figure 4.36. The vacuum magnetic configurations are identified by an index number, such that the six iota profiles shown correspond to index values (bottom to top): 1, 3, 5, 9, 11, and 13.

These discharges have relatively low pressure and a small amount of bootstrap current [67], which imply a shift of the flux surfaces and a slight modification of the iota profile. According to neoclassical estimates, in the radial range of interest here ($0.4 < \rho < 0.8$), the shift induced by the bootstrap current is limited to $|\Delta\rho| < 0.06$, while $\Delta\rho \simeq 0$ at $\rho \simeq 0.6$. As the features we will be discussing below have a similar radial extent, the effect of the bootstrap current and the modification of $\iota\!\!\!-$ may be considered small.

Nevertheless, the flux surfaces may shift slightly with respect to their calculated ('nominal') position obtained from an equilibrium calculation in a reference discharge, due to the cumulative effect of finite pressure (β) effects and (bootstrap) currents in the plasma. To improve the identification of the rational surfaces, we apply a simple correction based on the calculating the coherence between ECE channels. Figure 4.37 shows the mean coherence $\bar{\gamma}_{xy}$ between any two ECE channels (x and y, located at 'nominal positions' ρ and ρ', respectively), defined as

$$\bar{\gamma}_{xy} = \frac{\int S_{xy}^2(\omega)\gamma_{xy}(\omega)d\omega}{\int S_{xy}^2(\omega)d\omega}, \tag{4.16}$$

where $S_{xy}^2(\omega)$ is the spectral cross power, $S_{xy}^2(\omega) = \sqrt{|S_x^2||S_y^2|}$, and the spectral coherence is

$$\gamma_{xy}(\omega) = \frac{|\langle S_x(\omega)S_y^*(\omega)\rangle|}{\sqrt{\langle S_x^2(\omega)\rangle\langle S_y^2(\omega)\rangle}}, \tag{4.17}$$

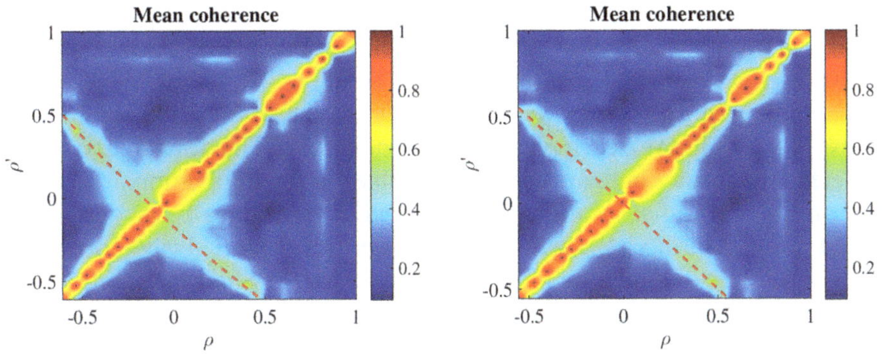

Figure 4.37. (a) Left: mean coherence between all ECE channels versus the nominal ρ values of the channels. (b) Right: the same, but after applying the ρ correction discussed in the text. Note that the dotted appearance of the auto-coherence (along the line $\rho' = \rho$) is an artefact of the interpolation method used. (Reproduced with permission from [59]. © IOP Publishing. All rights reserved.)

where rectangular brackets refer to an ensemble average over successive time intervals.

Figure 4.37(a) shows two main features: (i) a strong diagonal feature (at $\rho' = \rho$), corresponding to the auto-coherence, which is 1 by definition; and (ii) a linear feature at near right angles, close to the line $\rho' = -\rho$, corresponding to the coherence between channels on the same flux surface, but on opposite sides of the magnetic axis. This feature may be unexpected, but in fact has a simple explanation. It is due to the existence of 'filamentary' plasma structures that form as a consequence of the strong magnetic field: turbulent fluctuations tend to propagate much faster along the field lines than across them, leading to structures that are extremely elongated along the magnetic field. Thus, fluctuations on the same flux surface but on opposite sides of the magnetic axis tend to be correlated to a certain degree.

Due to the fact that the 'nominal ρ' values correspond to a discharge (number 160302.008, 3.2–3.7 s) with slightly different pressure (β) than the one studied in this case (160309.011, 0.08–0.38 s), a slight deviation from the line $\rho' = -\rho$ occurs, reflecting an effect similar to a 'Shafranov shift' of the flux surfaces. The perpendicular linear feature seen in figure 4.37(a) can be modelled as $\rho' = -\rho + b(1 - \rho^2)$, shown in this figure as a dashed line. This fit can be used to correct the 'nominal ρ' values to account for the Shafranov shift, by defining $\rho_{new} = \rho_{old} + \Delta(1 - \rho_{old}^2)$, where $\Delta = -b/2$. The result of this correction is shown in figure 4.37(b). The correction implicitly assumes that the B-field is not modified significantly with respect to the reference equilibrium (on which the 'nominal' ECE radii are based), even in the presence of the additional Shafranov shift of the flux surfaces. The figures in the remainder of this section show ρ values that are corrected for the Shafranov shift using this coherence technique. It could perhaps be considered somewhat ironic that here we are using turbulence to correct an equilibrium calculation!

As was the case in TJ-II, these ECR heated discharges have spontaneous temperature perturbations that propagate outward. To visualize this radial propagation, we again calculate the transfer entropy between a reference ECE channel and all other available ECE channels. Within each discharge, the ECRH heating power was varied. An initial 'high power' phase (ECRH power $P_{\text{ECRH}} \simeq 2.0$ MW) was followed by a low power phase ($P_{\text{ECRH}} \simeq 0.6$ MW) and a medium power phase ($P_{\text{ECRH}} \simeq 1.3$ MW), each phase lasting about 0.3 s. The discharges correspond to different magnetic configurations and profiles, such that rational surfaces are placed at different radial locations (as indicated in the figures below). The vertical axes and rational surfaces shown in the figures of this section always correspond to the nominal values of the ECE channels, corrected for the Shafranov shift according to the coherence method. The line average density was relatively low in these discharges ($\bar{n}_e \simeq 1.5 \times 10^{19}$ m^{-3}). Since we are most interested in outward propagation, we use a reference channel located close to $\rho = 0.2$, not far outside the central ECRH power deposition region.

Before delving into the physics of heat propagation, we make a short digression regarding the validity of the transfer entropy. To do so, we compare results from the transfer entropy to the standard cross correlation. This comparison is shown in figure 4.38. While the transfer entropy is a fully non-linear analysis, the correlation analysis is linear and therefore perhaps not ideally suited for the analysis of turbulence. This is reflected in the low values of the correlation, which would normally not be considered significant, except for the relatively high values occurring at time lag $\tau = 0$ (around 0.5, exceeding the maximum value of the colour scale shown). However, we note that the general pattern of the correlation for time lags $\tau > 0$ and radii $\rho > 0.2$ bears a striking similarity to the transfer entropy results, which gives confidence that the transfer entropy results are physically meaningful. The transfer entropy is much larger than its statistical significance level (around 10^{-4}). This example illustrates to what degree the non-linear transfer entropy supersedes the linear correlation when analyzing the propagation of small, irregular heat pulses.

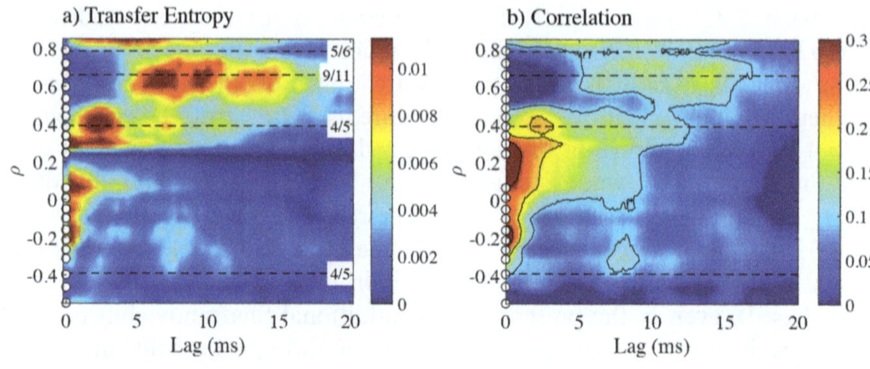

Figure 4.38. Transfer entropy and cross correlation versus ρ and time lag, τ, for discharge 20 160 309.016. Configuration index: 3 (see figure 4.36). In this and the following figures, dots on the left-hand axis indicate the location of ECE measurement channels, while horizontal dashed lines indicate the location of rational surfaces. Some contour lines have been added for clarity. (Reproduced with permission from [59]. © IOP Publishing. All rights reserved.)

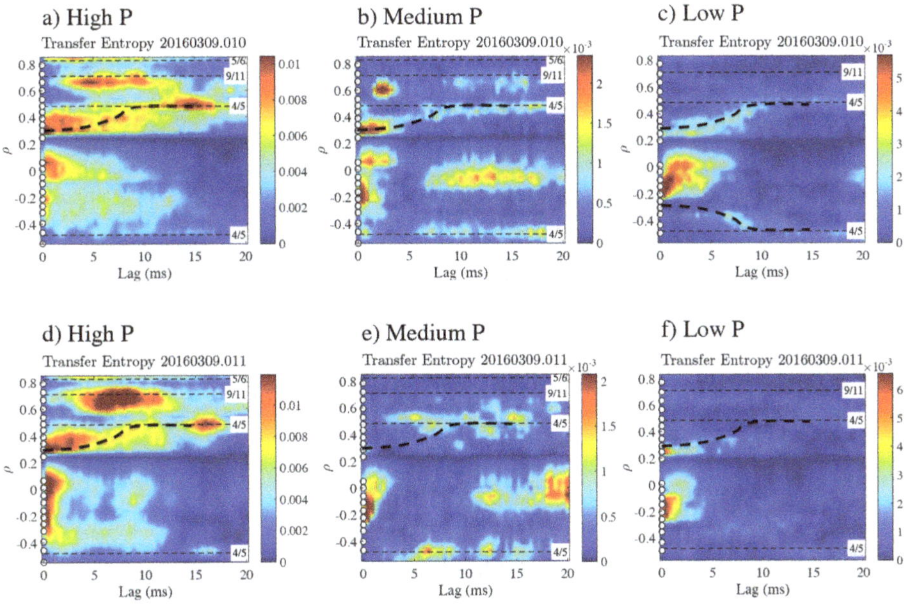

Figure 4.39. Transfer entropy versus and time lag, τ, for equivalent discharges 20160309.010 (a)–(c) and 20160309.011 (d)–(f), at three different power levels (see text). Configuration index: 1 (see figure 4.36). Radial propagation is indicated with thick dashed lines. (Reproduced with permission from [59].)

Figure 4.39 shows the result of the analysis for two discharges with the same configuration in the three power phases, which can be used to provide an indication of the reproducibility of results. Although reproducibility is not perfect, clearly some of the main features are similar among these discharges and heating levels, in particular the indicated 'plume' (thick dashed line) moving outward from $|\rho| \simeq 0.2$ to the 4/5 rational surface, very similar in shape, width, and propagation speed between the two discharges. High values of the transfer entropy at $\rho \simeq 0.85$ in the high power case should be ignored, as these can be ascribed to the low frequency tail of third harmonic emission with small optical depth, so that these measurements do not correspond to local temperatures. On the other hand, the low/high field side asymmetry in the high power phase could be caused by relatively strong emission from hot core electrons, contaminating the low field side data.

In the following, we will focus on the radial range $0 < \rho < 0.85$. We emphasize that the transfer entropy in this range tends to peak at time lag values that are consistent with typical time delays expected from the background heat transport coefficient, which gives further confidence that these results are relevant for the heat transport studies at hand.

Generally speaking, the transfer entropy reaches higher values in the high power phase, which is attributed to a larger number of perturbations originating propagation. It is noteworthy that the 4/5 rational surface seems to constitute an 'attractor' (or 'trapping zone') for this outward plume. In addition, a strong response is seen near the 9/11 rational surface (at time lags between 3 and 12 ms) in the high

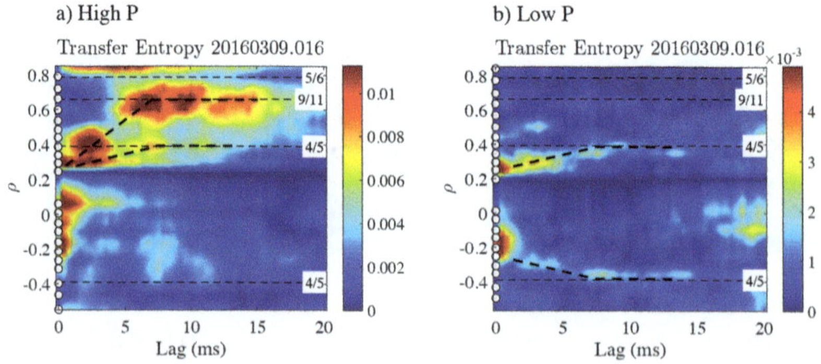

Figure 4.40. Transfer entropy versus ρ and time lag, τ, for discharge 20160309.016. Configuration index: 3 (see figure 4.36). Radial propagation is indicated with thick dashed lines. (Reproduced with permission from [59].

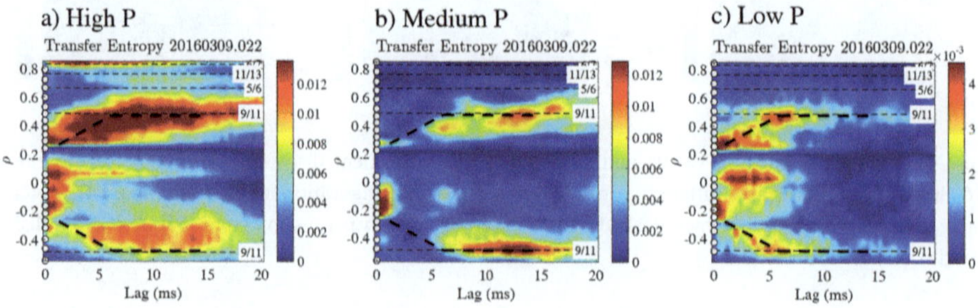

Figure 4.41. Transfer entropy versus ρ and time lag, τ, for discharge 20160309.022. Configuration index: 9 (see figure 4.36). Radial propagation is indicated with thick dashed lines. (Reproduced with permission from [59].

power cases. This suggests that the core perturbations, which may be assumed to be of larger amplitude in the high power cases, trigger a response near this 9/11 surface, located further outward. Figure 4.40 shows the results from a discharge with a different magnetic configuration (note that the rational surfaces have moved slightly with respect to the preceding figure). Comparing the low and high ECRH power phases, one observes that they have in common that some perturbations propagate outward relatively slowly to the 4/5 rational surface, which again acts as a 'trapping zone' for these perturbations. In the high power phase, there is an additional branch of radial propagation, faster and more intense (in terms of information transfer), reaching the 9/11 rational surface.

Figure 4.41 shows a similar result, again in a different magnetic configuration, for three ECRH power levels. Propagation is outward from $|\rho| \simeq 0.2$ until the position of the 9/11 rational surface is reached. The latter position acts as a 'trapping zone' for the outward propagation of information.

Summarizing the observations so far, the outward propagation of the temperature perturbations (more precisely: the causal impact of central temperature fluctuations on fluctuations further outward) is detected with clarity and shows up as a more or less 'diagonal plume' corresponding to outward propagation starting at the reference position, $|\rho| \simeq 0.2$. The typical radial propagation speed is approximately 0.5 m (the minor radius) in 20 ms, or 25 m s^{-1}. This can be compared to the result for TJ-II, where speeds in the order of 100–200 ms^{-1} were obtained [51].

The figures show that propagation is not continuous, but experiences delays ('trapping') at certain radial positions, which appear to be associated with low-order rational surfaces, within the available resolution. Also, apparent propagation 'jumps' occur, when the response at a given outward location is more intense than at some other locations further inward. This is exemplified by figure 4.40, in which two transport branches may be discerned: (a) relatively slow outward propagation up to the 4/5 rational surface, visible in both the low and high ECRH power phases, and (b) much faster radial propagation, apparently discontinuous, reaching the 9/11 rational surface, only apparent in the high power phase. This result seems to suggest that an additional transport channel is activated at increased heating power; therefore, this effect might be related to the phenomenon of power degradation.

Some figures seem to suggest that different rational surfaces may be affecting transport simultaneously. For example, in figure 4.39 and 4.40, the left graphs (high power) seem to show that both the 4/5 and 9/11 rational surfaces are associated with transport barriers or 'trapping zones'.

Now we will make another small digression. In section 4.3 we mentioned that the transfer entropy is 'directional', meaning, in this context, that it can distinguish between propagation in one direction and in the other. To check this, we simply use two different reference positions ρ_{ref} and compare the results. Figure 4.42 shows examples with reference positions at $\rho_{ref} \simeq 0.4$ and 0.8. In the first case (a), starting from position $|\rho_{ref}| \simeq 0.4$, both inward and outward propagation plumes are visible (indicated by the thick dashed lines). In the second case (b), starting from the reference position $\rho_{ref} \simeq 0.8$, a clear inward propagation plume is visible, testifying to the unique capacity of the transfer entropy to distinguish between propagation directions. Just like we saw with the outward propagating pulses, propagation appears to stagnate very close to a major rational surface (4/5). Note that the transfer entropy values are much lower in case (b), suggesting that inward

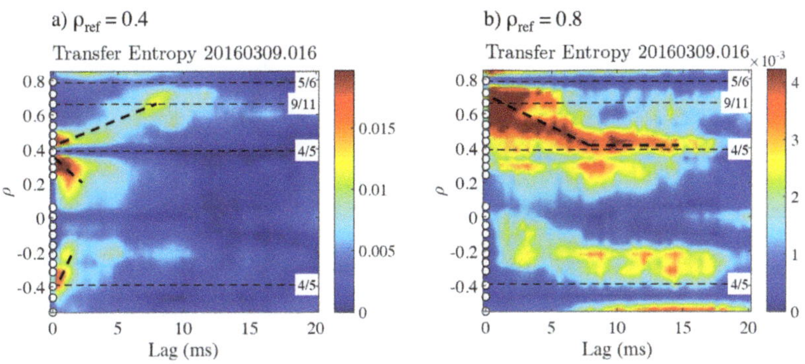

Figure 4.42. Transfer entropy versus ρ and time lag, τ, for discharge 20160309.022. Configuration index: 9 (see figure 4.36). Radial propagation is indicated with thick dashed lines. (Reproduced with permission from [59]

propagating events are scarcer—as they should be, because logically, net transport should be outward.

Next, we analyze a few discharges with modulated ECRH, similar to what we did for TJ-II. The main purpose of this exercise is to compare the transfer entropy results with the results from a traditional heat wave analysis using Fourier techniques [68], in order to understand the relevance of the transfer entropy for heat transport analysis better.

In the discharges studied here, power is modulated at $f_{\text{mod}} = 17$ Hz with a modulation depth of about 35%. Figure 4.43 shows the transfer entropy, calculated over the time window $0.35 < t < 2.0$ s, with modulated on-axis heating, alongside the results from Fourier analysis. In both types of analysis, the reference position is taken at $\rho_{\text{ref}} \simeq 0$. The Fourier results correspond to the fundamental modulation frequency, at which the amplitude (cross power) and the cross phase with respect to the reference channel ($\Delta\phi$) are determined. In order to compare with the transfer entropy results, we define a Fourier time delay on the basis of the cross phase using $\Delta t = \Delta\phi/2\pi f_{\text{mod}}$. Outward propagation of the heat waves from the core is very clear on both sides of the magnetic axis. Likewise, figure 4.44 shows the transfer entropy, calculated over the time window $0.30 < t < 1.1$ s, with modulated off-axis heating, along with the Fourier time delay and amplitude. The reference position is taken at $\rho_{\text{ref}} \simeq 0.4$. The results show heat waves propagating both inward and outward from the reference position at $\rho \simeq 0.4$ and the opposite point at $\rho \simeq -0.4$. We draw attention to the fact that the Fourier time delay values show similar radial behaviour as the local transfer entropy maxima. We believe this is an important experimental result, for it shows that the transfer entropy (an information theoretical quantity) actually reflects the physical radial heat transport of the modulated heat wave. The precision of the Fourier results (indicated by horizontal error bars) is better than the transfer entropy. This is due to the fact that the Fourier results are based on a single frequency component of the modulated $T_e(\rho, t)$ signals, namely the fundamental

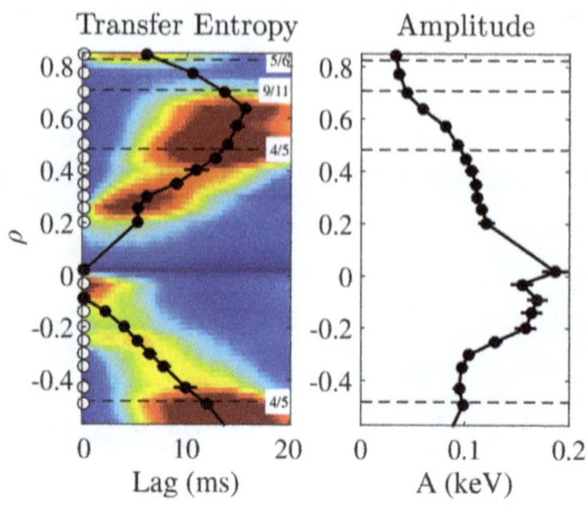

Figure 4.43. Comparison of transfer entropy (left) and Fourier analysis (time delay shown in the left panel and amplitude in the right panel as black lines with error bars) in a modulated discharge (20160302.008) with on-axis heating. (Reproduced with permission from [59]. © IOP Publishing. All rights reserved.)

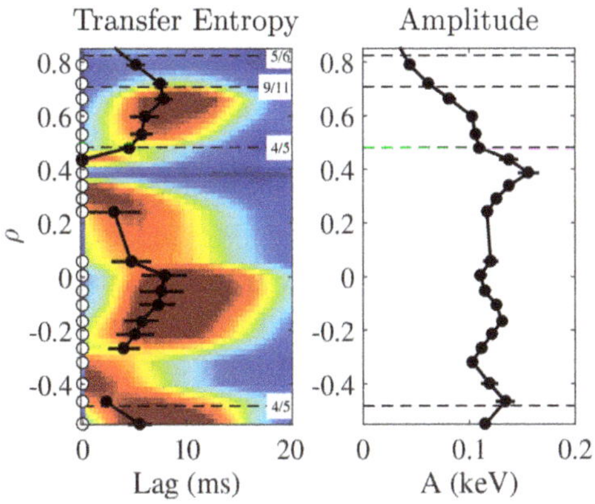

Figure 4.44. Comparison of transfer entropy (left) and Fourier analysis (time delay shown in the left panel and amplitude in the right panel as black lines with error bars) in a modulated discharge (20160308.016) with off-axis heating. (Reproduced with permission from [59]. © IOP Publishing. All rights reserved.)

frequency, whereas the transfer entropy is calculated from the unfiltered signals, and includes other frequency components as well as experimental noise. Of course, we have no intention to present the transfer entropy as an alternative to the powerful Fourier technique, but rather, we are trying to show the relevance of the transfer entropy to heat pulse propagation analysis.

The main results of this work mirror the earlier findings in TJ-II: (a) flow shear layers associated with rational surfaces have an impact on radial heat transport by constituting minor transport barriers, and (b) occasionally transport appears to jump over radial regions, resulting in non-local transport effects. Both these points imply a significant role for magnetic (MHD) turbulence in heat transport in these low density discharges. Consequently, transport is not smooth and continuous, as assumed in most (diffusive) transport models, but heat has a tendency to accumulate at trapping zones located near rational surfaces or perform rapid radial excursions. At least part of the heat transport occurs therefore in a stepwise rather than continuous fashion. The jumping effect is associated with (MHD) mode coupling, involving the exchange of energy between the plasma and the magnetic field [69, 70]. Interestingly, the jumps increase in importance as power is increased.

Spontaneous perturbations and heat waves (due to modulation) both propagate outward at a speed of around 25 m s^{-1}. By comparison, outward propagation at TJ-II occurs at a typical speed of 100–200 ms^{-1}. This difference can doubtless be associated with the better confinement properties of W7-X. A careful comparison of the analysis of ECRH modulation experiments using both a traditional Fourier technique and the transfer entropy clarifies that the time lag of the transfer entropy delivers essentially the same information as the Fourier phase delay, justifying our interpretation of the transfer entropy in terms of heat transport.

The low density ECRH heated discharges studied here (core electron-root confinement, CERC) are characterized by a positive radial electric field E_r in the core (electron root) and a negative E_r in the outer half of the plasma (ion root) [66].

The transition between these two regions constitutes a shear layer that might in principle be associated with a transport barrier. The radial position where the neoclassical radial electric field changes sign (around mid-radius) is expected to shrink by about 30% when power is reduced from its highest to its lowest value [71]. The results shown here, however, show no difference at all in the radial location of the trapping zones as heating power is changed. Hence, it appears unlikely that the neoclassical change of root is related to the trapping zones reported here. A further argument against a neoclassical explanation for the observed trapping zones is the simultaneous existence of two radial trapping zones in a single discharge, which cannot easily be explained by the single sign change of E_r predicted by neoclassical calculations. The main difference between a neoclassical flow shear layer and one associated with rational surfaces is that the latter is associated with the electromagnetic (fluctuating) component of the fields, which is thought to be essential for the effects reported here.

4.6.4 Heat transport in the JET tokamak

The preceding sections discussed heat transport in two low-shear stellarators: TJ-II and W7-X. Naturally, one may wonder whether the observed phenomena are limited to this type of device or are more general. Therefore, we now turn to a tokamak: the Joint European Torus (see appendix A.5).

JET differs from TJ-II and W7-X in several important aspects. First, since it is a tokamak, part of the magnetic configuration (the poloidal magnetic field) is generated by currents flowing in the plasma itself, implying less external control over the magnetic topology. Second, no ECR heating is available; instead, ion cyclotron resonance heating (ICRH) is used. This is important because at TJ-II and W7-X, ECR heating generated spontaneous T_e perturbations that allowed diagnosing heat transport by tracking their propagation through the plasma (statistically). Here, this method cannot be used and other perturbation sources are needed. JET discharges are usually characterized by sawtooth activity in the core region. Sawteeth are produced by rapid (timescale: $\sim$100 μs) magnetic reconnection events associated with the $q = 1$ rational surface. As a consequence of a reconnection, some heat and particles are ejected from the core, and the resulting heat pulses can be analyzed (subject to some caveats). The resulting response of $T_e(R)$ can be studied by means of an impressive, high resolution ECE radiometer with no less than 96 channels [72].

As before, we set out to study heat transport and to try to establish whether transport is smooth and continuous or not. Of course, we are not the first to do so. Beautiful work carried out at the now defunct RTP tokamak clearly and very convincingly demonstrated the existence of a multiplicity of minor transport barriers throughout the plasma [63]. The experiments consisted of a slow radial scan of the location of heat deposition, and instead of the expected continuous variation of the electron temperature profile, it was found to change in a stepwise fashion. Subsequently, a heuristic 'q-comb' model was developed to interpret the observations, based on radially localized reductions of the heat diffusion coefficient at low-order rational surfaces [73]. This simple model was capable of reproducing some of the main aspects of heat transport at RTP, but did not provide a thorough understanding of the underlying mechanics. We will see that there may be an element of truth in this model, although the appearance of the minor barriers appears to be less systematic than suggested by the q-comb model.

In this framework, we should briefly mention internal transport barriers (ITBs), which arise only transiently but are much stronger than the 'minor transport barriers' we are referring to here. In tokamaks, strong ITBs can be established by creating a core reversed magnetic shear region, so that the q-profile has a minimum, and the location of the ITB appears correlated with integral values of the safety factor, q [74, 75]. The impact of ITBs on heat transport has been studied in some detail at, e.g. Alcator C-Mod [64] and JET [76, 77], showing that the heat diffusivity drops strongly in the ITB region. ITBs have also been obtained and studied in stellarators [78], and here, too, a relationship with the magnetic configuration is suggested. The physical mechanism behind all such types of transport barrier involves the generation of zonal flows by turbulence via Reynolds stress [79], while the shear in these flows then suppresses the fluctuations. It appears that the generation of such sheared flow is facilitated near low-order rational surfaces, which explains many of these observations.

One word in this sentence may go unnoticed but is very important: the sheared zonal flow is produced 'near' (not 'at') a given rational surface. To illustrate the idea, figure 4.45 shows the radial structure of an electrostatic fluctuation potential near a rational surface, arbitrarily placed at $r/a = 0.5$, and the associated sheared flow in a very simple slab model.

This figure is no more than a cartoon, shown here to illustrate the idea of the association between fluctuations, rational surfaces, and sheared flow. If the instability eigenfunction Φ is symmetric with respect to the rational surface, the flow shear $|V'|$ will peak away from the rational surface, at a distance of the order of the width of the turbulent vortices. On the other hand, an antisymmetric eigenfunction will place the flow shear peak at the rational surface. Each type of

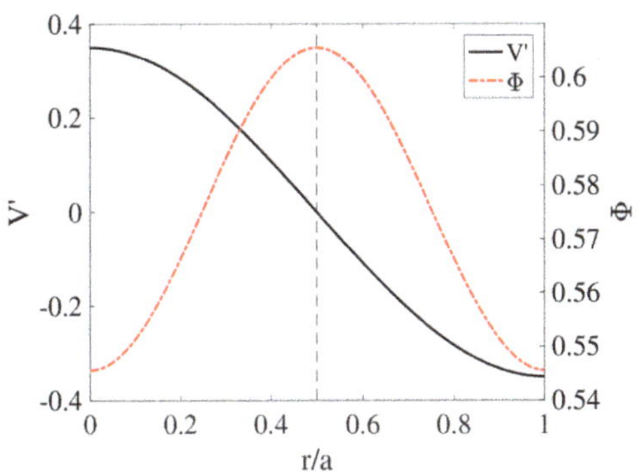

Figure 4.45. Potential fluctuation (Φ) and shear of the generated flow (V') for a simple non-linear slab model. The vertical dashed line shows the position of the singular surface. (Reproduced with permission from [80].)

instability will generate its own structure, possibly modulated by the presence of other structures nearby, and the actual situation can be much more convoluted. Nevertheless, the central idea is that the sheared flow regions are usually located near singular surfaces. Using a resistive MHD model with tracers, it has been shown [81] that such sheared flow regions act as minor transport barriers, trapping some of the tracers, while other tracers, with higher kinetic energies, perform rapid radial excursions, 'jumping over' the barriers. As the system is driven more strongly (by increasing heating power levels), on average tracers are endowed with higher energies, so that more tracers will be able to 'jump' the minor barriers. This is the mechanism we proposed to explain the deterioration of confinement in the TJ-II stellarator in section 4.6.2.

Returning to the main issue of interest, namely the analysis of the propagation of T_e perturbations, we note that JET (or any tokamak) presents specific challenges. The main challenge is the near omnipresence of sawteeth, produced in the core region due to the fact that typically, $q(0)$ is close to or even below 1. Sawteeth produce a major perturbation of the core plasma, making it impossible to interpret ECE data in the region inside the sawtooth inversion radius exclusively in terms of outward heat transport. The sawteeth-induced perturbations outside the inversion radius, however, propagate outward through a much more stable background plasma and are amenable to such analysis. The discharges we analyze below all have sawteeth and our analysis is based on the propagation of the corresponding heat pulses. In this context, we remind the reader of the caveat expressed in [82]: sawtooth perturbations may affect more than just the T_e profile and may be associated with modifications of the magnetic geometry due to the magnetic reconnection of the core region taking place during the sawtooth crash, leading to the possible misinterpretation of observed T_e perturbations. We note, however, that this misinterpretation mainly refers to the fact that the deduced *value* of the diffusion coefficient may be biased; yet the radial variation of this coefficient can still be fruitfully studied with this method, which is all we intend to do here.

Another challenge is the relatively high magnetic shear of JET, compared with the low-shear stellarator configurations studied in the cited previous work. A high magnetic shear implies that rational surfaces are packed close together, making it more difficult to distinguish the impact of each individual rational. The added difficulty of the closer spacing of rational surfaces is however partly compensated by the high radial resolution of the ECE diagnostic system at JET.

H mode plasmas with a steep edge pedestal (transport barrier) present further difficulties due to the presence of edge localized modes, among other issues, so here we will limit ourselves to L-mode plasmas.

To infer the approximate position of rational surfaces in the plasma and calculate the approximate value of the safety factor, q, at the measurement positions of the ECE channels, we will make use of the standard EFIT equilibrium reconstruction, based on magnetics only. This calculation is subject to some uncertainty [83], although in the centrally heated, steady state L-mode discharges studied here, the precision should be reasonable in the outer regions of the plasma. Due to these uncertainties, we will not attempt to identify any of the observed features with

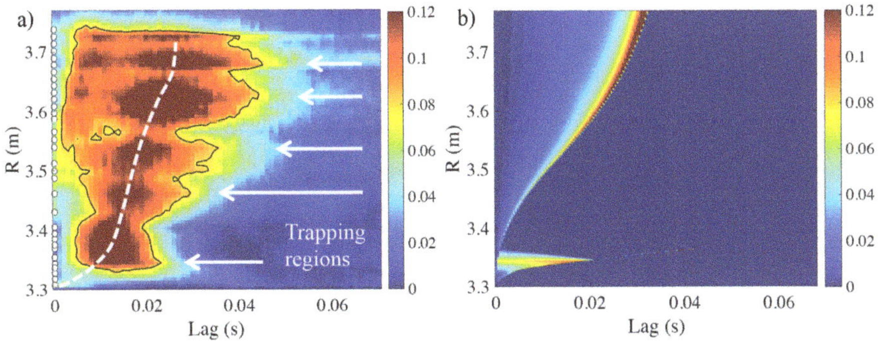

Figure 4.46. (a) Transfer entropy for JET discharge 82292. $R_{ref} = 3.30$. The colour bar indicates the value of the transfer entropy. White circles indicate the locations of ECE channels. To emphasize the shape of the high transfer entropy region, a contour at $T = 0.8$ is shown. The white dashed line indicates the overall outward propagation. White arrows indicate 'trapping regions' (see the text). (b) Example transfer entropy result for the standard diffusive spreading of a heat pulse launched at $R = 3.3$ m, assuming a constant diffusion coefficient and no convection. (Reproduced with permission from [80].)

specific rational surfaces, even though we will use the EFIT q-profile as guidance for the interpretation. Also, we will compare shots with slightly different q-profiles and compare the shift of rational surfaces in the outer regions of the plasma with the shift of the features we detect.

As before, we compute the transfer entropy between 2 signals, namely an ECE channel at a reference position R_{ref}, and an ECE channel at another position, R. In figure 4.46(a), a typical transfer entropy graph is shown for $R_{ref} = 3.30$ m, versus time lag and the R value of the other ECE channels. Typical values of the transfer entropy are well above the statistical significance level. This example graph shows that the overall transport is outward, as indicated by the white dashed line. The velocity of this propagation is consistent with the typical heat transport coefficients measured in the JET tokamak, as discussed in more detail below. Figure 4.46(b) shows the transfer entropy result for the strictly diffusive propagation of a heat pulse launched at $R = 3.3$ m in a simple one-dimensional heat transport model with a constant diffusion coefficient and no convection. The comparison highlights the contrast between the smooth and continuous propagation of information associated with heat transport in the case of pure diffusion and the complex propagation pattern in the case of the experiment.

We draw attention to the fact that the experimental propagation is modulated radially. There are radial zones where the distribution is broader horizontally than elsewhere, as indicated by white arrows in the figure. We interpret these regions as 'trapping regions', where outward transport is delayed and heat tends to accumulate. Some 'dips' are visible in the propagation plume, e.g. at $R \simeq 3.57$ and 3.66. This behaviour is very similar to what we have seen in the stellarators TJ-II and W7-X, discussed above, suggesting that this is a universal feature of heat transport in magnetic confinement devices. Figure 4.47 shows a few of the corresponding T_e time traces around the location of one these 'dips'. In this context, we remind the reader that the transfer entropy does not depend on signal amplitude (calibration).

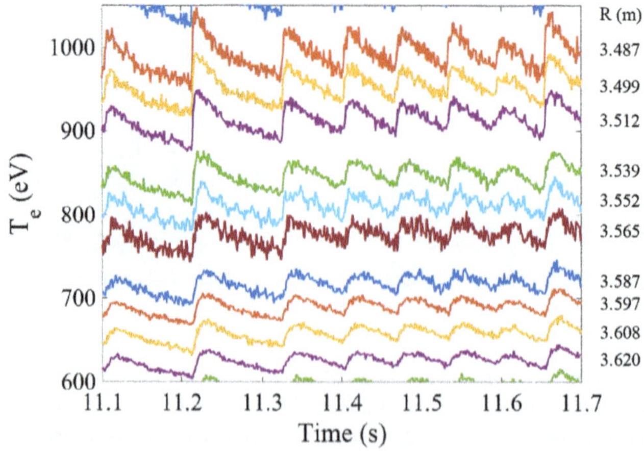

Figure 4.47. Some time traces of T_e for JET discharge 82292. The radius corresponding to each T_e time trace is indicated on the right. (Reproduced with permission from [80]. © IOP Publishing. All rights reserved.)

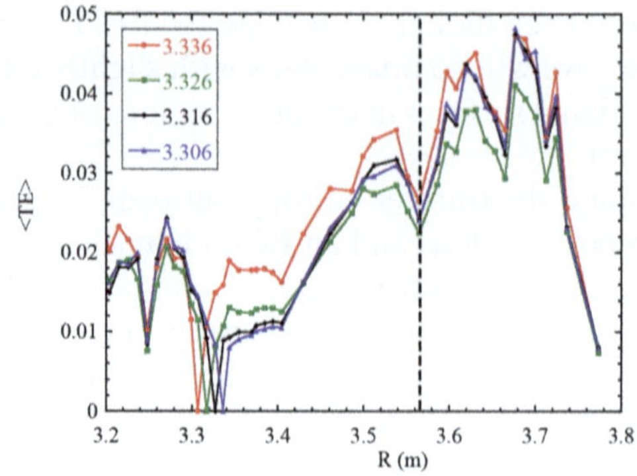

Figure 4.48. Time average of the transfer entropy over all time lags $0 \leqslant \tau \leqslant 0.2$ s for JET discharge 82292, for a few reference values $R = R_{\text{ref}}$, as indicated in the legend. (Reproduced with permission from [80]. © IOP Publishing. All rights reserved.)

In the framework of sheared flow models, 'minor transport barriers' are regions where the zonal flow shear is high, shearing apart turbulent eddies and suppressing turbulence (fully or partially); these regions would correspond to the observed 'dips'. The 'trapping regions', however, are zones in-between the minor transport barriers, where turbulence is not suppressed, but turbulent vortices exist. Propagating 'heat particles' would tend to be trapped in these vortices and 'stick' there before travelling on, leading to a broader range of time delays with high transfer entropy values. This interpretation in terms of 'minor transport barriers' and 'trapping regions' has been verified by analyzing the behaviour of tracers in a resistive MHD turbulence simulation, to be discussed below.

As in the case shown in figure 4.46(a), there may be various simultaneous barriers. To facilitate the study of the observed radial variation, we simply compute the time average of the transfer entropy versus radius, shown in figure 4.48 for a few choices of R_{ref}. One specific, deep radial minimum, corresponding to the radial 'dip' in the transfer entropy graph of figure 4.46(a), has been highlighted by means of a vertical black dashed line. It should be noted that the transfer entropy decays gradually towards this deep minimum, so that the minimum does not depend on the data of a single channel, reinforcing its robustness. In addition, one observes there are other local minima, corresponding to even softer barriers. At the reference radius, $R = R_{\text{ref}}$, the transfer entropy is 0 by definition, so the minima occurring at the various values of R_{ref} are not associated with any barrier and should be ignored.

Figure 4.48 clarifies that the locations of the transfer entropy minima do not depend on the choice of reference radius, within a reasonable range, but rather are associated with the magnetic configuration. Therefore, it is feasible to subject the location of minima in the graphs of $\langle T \rangle$ to a statistical analysis, based on the set of all available R_{ref} values for a given discharge. In many cases, we can dispose of between 35 and 65 reference radii. Thus, we can compute how often each local minimum occurs with respect to the total number of reference radii R_{ref} studied, and express it as a percentage. This number is defined as the 'persistence' of any given local minimum. We will repeat this analysis for different discharges with very similar parameters in order to verify that the locations of the significant (persistent) minima are linked to the plasma configuration.

We will estimate an effective diffusivity from the radial propagation of information [84]. For example, in figure 4.46(a), for each available ECE channel (indicated by a white dot at each corresponding value of R), we can estimate the mean time delay $\langle \tau \rangle$ from

$$\langle \tau(R) \rangle = \frac{\int \tau T(R, \tau) d\tau}{\int T(R, \tau) d\tau}. \tag{4.18}$$

Using the equilibrium from EFIT, we can convert the ECE measurement location R to a minor radius value $r = a\sqrt{\Psi_N}$, where Ψ_N is the toroidal flux, normalized to 1 at the last closed flux surface. Then, an effective diffusion coefficient can be defined by

$$\langle D \rangle = c \cdot \frac{(r - r_0)^2}{\langle \tau(r) \rangle}. \tag{4.19}$$

The coefficient c appearing in this equation is set at $c = \frac{1}{8}$, corresponding to the 'time to peak' estimate [85], although slightly different values are sometimes also used in the literature [86]. Note that this estimate of the effective diffusion coefficient is not very accurate, for two reasons. First, it is not defined for $r = r_0$ as both the enumerator and the denominator of the expression tend to zero, and the radial behaviour tends to be dominated by the enumerator $(r - r_0)^2$ for small values of $r - r_0$. Therefore, the extracted diffusion coefficient should not be taken too seriously in the region near the reference position. Second, as discussed above, it is defined exclusively on the basis of the time (or phase) delay, whereas a proper recovery of the underlying effective diffusion coefficient would require information about the perturbation amplitude as well. Nevertheless, it may serve as a means to visualize the radial variation of transport, and in this section we will use it only for this purpose.

The resulting value $\langle D \rangle$ is the *mean* diffusivity over the interval $[r_0, r]$, which can be expressed as follows:

$$\langle D \rangle_N = \frac{1}{r_N - r_0} \sum_{i=0}^{N-1} (r_{i+1} - r_i) D(r_i), \tag{4.20}$$

where D is the local diffusivity. Consequently,

$$(r_N - r_0)\langle D \rangle_N - (r_{N-1} - r_0)\langle D \rangle_{N-1} = (r_N - r_{N-1}) D(r_{N-1}), \tag{4.21}$$

from which the local effective diffusivity, $D(r_{N-1})$ follows. Of course, when $\langle D \rangle$ does not depend strongly on r, the mean diffusivity and the local diffusivity are nearly the same.

Having established these methods, we now study the L-mode discharges listed in table 4.2 [87]. These very similar discharges are heated with neutral beam injection in addition to the standard Ohmic heating of around 1 MW and have $B_T \simeq 2.5$ T, $I_p \simeq 2.5$ MA, resulting in $q_{95} \simeq 3.2$. Their q-profiles are practically identical (figure 4.49). The stored energy is about 0.8 MJ and the line averaged electron density $2.7 - 2.9 \times 10^{19}$ m^{-3}.

Table 4.2. JET L-mode discharges analyzed.

Discharge	Time range (s)	$P_{\rm NBI}$ (MW)	$\bar{n}_e$ (10^{19} m^{-3})
82291	13–16	1	2.7
82292	9–12	1	2.9
82293	9–12	1	2.8
82294	9–12	1	2.8

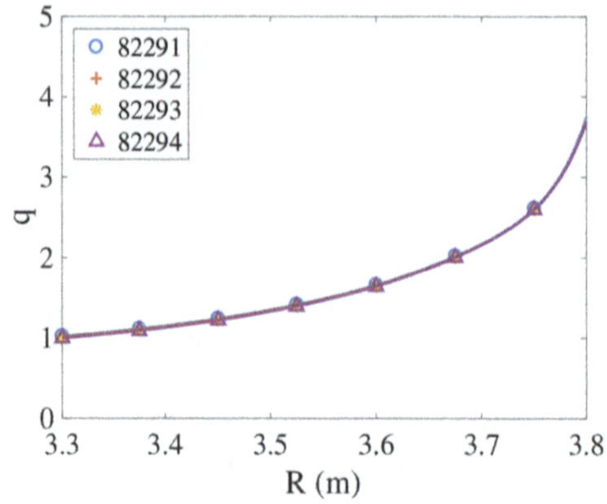

Figure 4.49. q-profiles for four similar L-mode discharges. Here and in subsequent graphs, the R axis indicates the radial position along the ECE line of sight. (Reproduced with permission from [80].

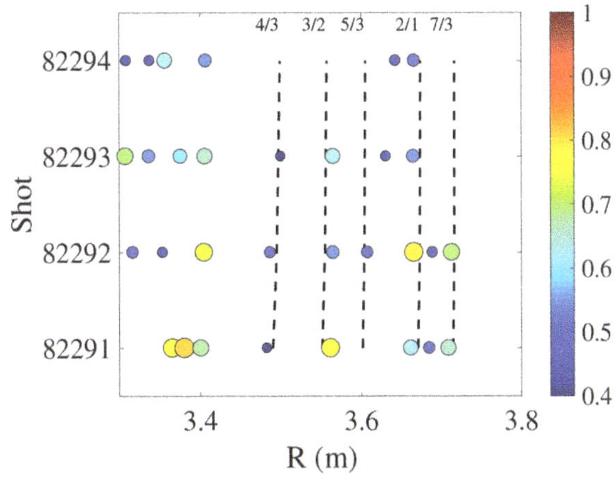

Figure 4.50. Comparison of the distribution of minima between the four L-mode discharges with a similar q-profile and two discharges with different q. The diameter of the circles and their colour are proportional to the barrier persistence (minimum value shown: 40%). Dashed lines indicate the approximate location of a few low-order rational surfaces (according to EFIT, with limited precision), as indicated by the labels at the top of the figure. (Reproduced with permission from [80]. © IOP Publishing. All rights reserved.)

For each of the discharges in table 4.2, we calculated the transfer entropy using all possible reference positions R_{ref}. Then we identified the locations of the minima from the time average of the transfer entropy, $\langle T \rangle$, and quantified their persistence, as discussed above. The result is shown graphically in figure 4.50. It is clear that positions of many minima are quite persistent within a given shot (reflected by the size and colour of the symbols). In addition, the minima are often reproducible from shot to shot. An example of this is the dashed line labelled 2/1, which is associated with a minor barrier in all discharges. The apparent near-coincidence of minima and selected rational surfaces should not be over-interpreted; on the one hand, the precision of the EFIT estimate of q is limited, and on the other, minor barriers do not need to coincide with the locations of rational surfaces.

Figure 4.51 shows the mean effective diffusion coefficient for the first four discharges of table 4.2 with a similar q-profile, calculated according to equation (4.21). The deduced effective diffusivities vary only little between these similar discharges, as indicated by the error bars, suggesting that the method is rather robust. Figure 4.52 shows a comparison between the effective diffusivities of two of the shots used to elaborate figure 4.51 and the corresponding barrier persistence. Although both quantities are computed from the transfer entropy, the persistence depends on its amplitude and the effective diffusivity on its temporal distribution, hence these two estimates are independent. In a number of cases, the radial structures in D_{eff} are correlated with the positions of the minor barriers, quantified using the persistence statistic: in the top figure, at $r \simeq 0.625$ and 0.725 ($R \simeq 3.56$ and 3.66 m, respectively) and in the bottom figure, at $r \simeq 0.73$ ($R \simeq 3.67$ m).

Next, we study some L-mode D plasmas with vacuum toroidal magnetic field $B_T \simeq 3.3$ T, plasma current $I_p \simeq 2$ MA and $q_{95} \simeq 5$ (see [88] for more information).

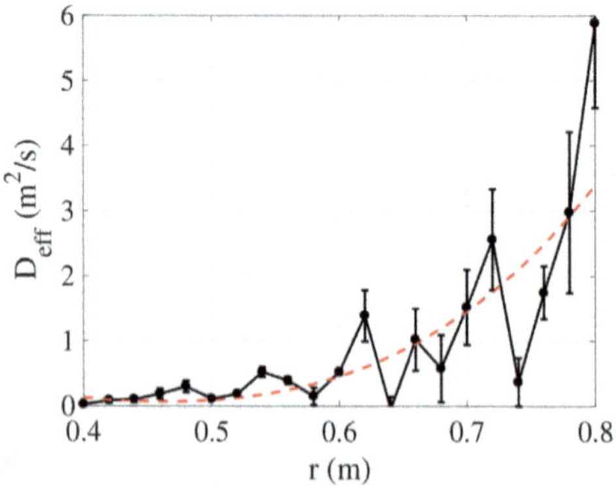

Figure 4.51. Mean effective diffusion coefficient for the discharges of table 4.2 (8229x). Error bars indicate the standard deviation. The red dashed line is a low-order polynomial fit, shown to guide the eye. (Reproduced with permission from [80]. © IOP Publishing. All rights reserved.)

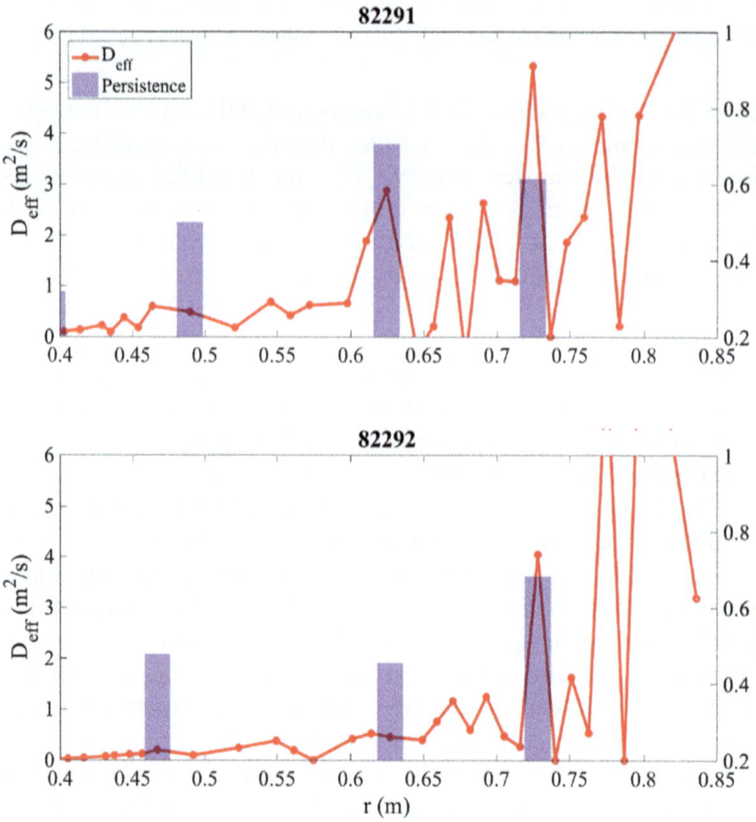

Figure 4.52. Effective diffusion coefficient and barrier persistence for two similar discharges. (Reproduced with permission from [80]. © IOP Publishing. All rights reserved.)

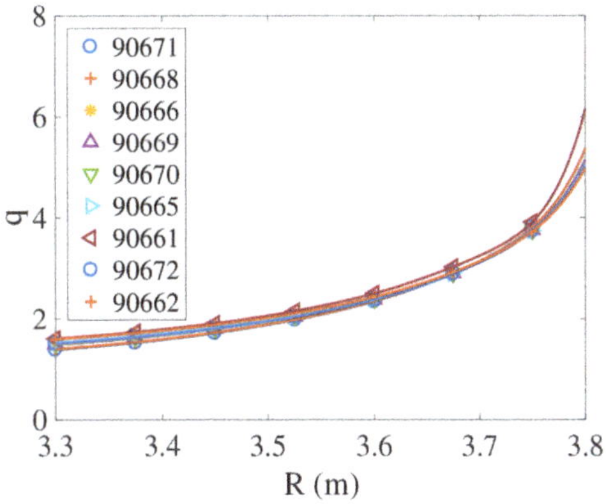

Figure 4.53. q-profiles of the L-mode power scan. (Reproduced with permission from [80].)

Table 4.3. JET L-mode power scan discharges.

Discharge	Power (MW)	$\bar{n}_e$ (10^{19} m^{-3})
90671	4.3	2.7
90668	4.4	2.7
90666	6.8	2.6
90669	7.0	2.6
90670	7.6	2.6
90665	7.7	2.5
90661	7.7	2.5
90672	8.0	2.5
90662	8.9	2.6

The corresponding q-profiles are shown in figure 4.53; they are very similar. The plasmas were heated with ICRH at varying power levels (see table 4.3). We analyze the time interval $9 \leqslant t \leqslant 11$ s, in which the plasmas were roughly in steady state.

An example T_e profile is shown in figure 4.54. The ECE measurements are only reliable for $R < 3.8$; at higher radii, the signal obtained no longer represents a local temperature measurement (due to small optical depth), and the corresponding channels are not considered in the analysis here. The sawtooth region extends up to about $R \simeq 3.22$, while our analyses focus on the heat pulse propagation region $R > 3.3$.

Figure 4.55 shows a comparison of the transfer entropy obtained for two discharges from the power scan of table 4.3. There are some interesting similarities, such as the 'dip' (barrier) occurring near $R \simeq 3.73$ m. Be that as it may, the difference that we consider most significant is that the propagation in the high power case seems to be faster (time lags are shorter) and able to 'jump over' the barrier to a

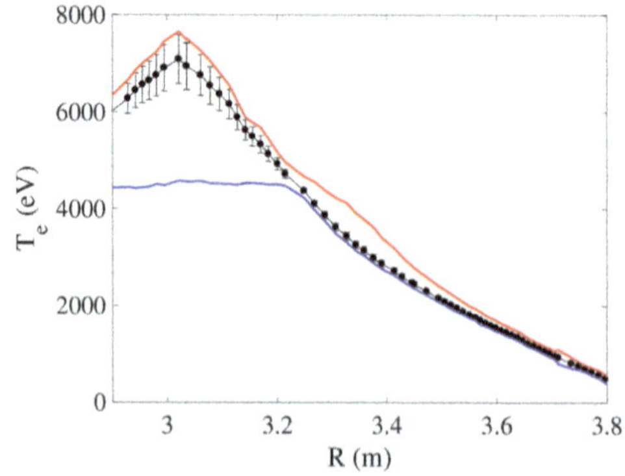

Figure 4.54. Black dots: ECE T_e-profile of a discharge of the power scan (90672), averaged over the time window of interest. The bars shown indicate the standard deviation of T_e in this time window, not its error. Upper continuous red line: maximum T_e; bottom continuous blue line: minimum T_e over the time window. (Reproduced with permission from [80].)

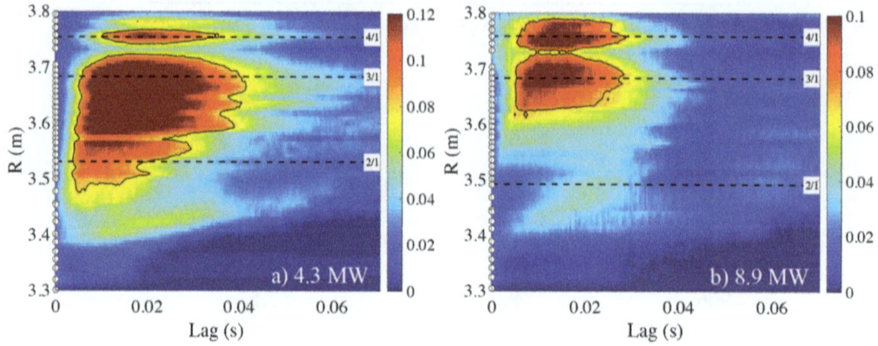

Figure 4.55. Comparison of transfer entropy between two discharges from the power scan (table 4.3), calculated with $R_{\rm ref} = 3.3$ m and a very different power level (as indicated). (Reproduced with permission from [80].)

larger degree, penetrating further into the edge plasma region. This behaviour is consistent with observations made on TJ-II.

Next, we calculate the local effective diffusivity using equation (4.19) and equation (4.21) for a few of the discharges. The diffusivity increases with increasing heating power, see figure 4.56. This is consistent with the expected behaviour in the framework of the phenomenon of power degradation, and provides confidence that the method for estimating the heat diffusivity from the transfer entropy 'works', at least qualitatively.

Applying the algorithm to determine the persistence of the local minima of $\langle T \rangle$, we obtain figure 4.57. The location of some rational surfaces according to the EFIT reconstruction are indicated; again, due to their limited precision, one should not over-interpret the indicated rational values, although we do consider the shot-to-shot variation significant. The dashed line labelled 3/1 is accompanied by barriers for

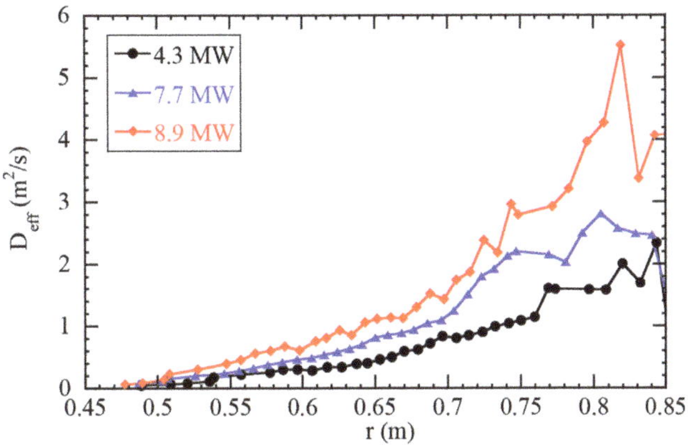

Figure 4.56. Dependence of effective diffusion on heating power. (Reproduced with permission from [80].

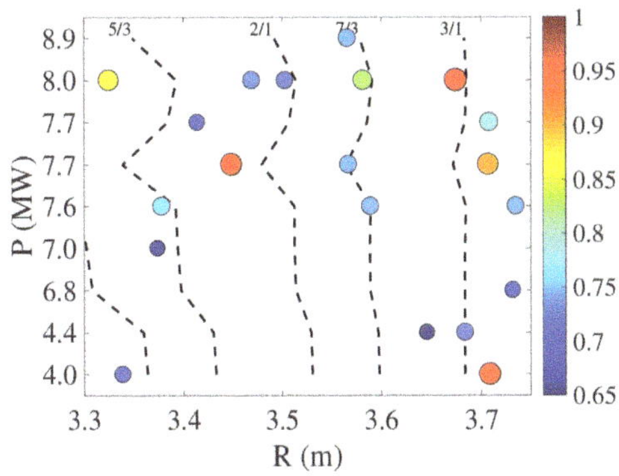

Figure 4.57. The locations of persistent minor transport barriers. The left axis indicates the power of each discharge listed in table 4.3. The diameter of the circles and their colour are proportional to the barrier frequency (minimum value shown: 65%). Dashed lines indicate the approximate location of a few low-order rational surfaces (according to EFIT, with a limited precision), as indicated by the labels at the top of the figure. (Reproduced with permission from [80].

nearly all power levels. The radial positions of the minor barriers near the dashed line labelled 7/3 vary by a small amount between discharges (thus excluding instrumental effects as the cause of the barriers) and match the radial variation of the approximate location of the rational surface for the higher power levels. In the interior of the plasma ($R < 3.6$), one observes that the density of barriers has a tendency to increase as the power level is increased.

Figure 4.58 shows the effective diffusion coefficient and the persistence for three discharges at different heating levels. Again, the radial structures seen in the effective diffusion coefficient sometimes appear related to the location of persistent minor transport barriers: namely, at $r \simeq 0.76, 0.83$ (top figure), at $r \simeq 0.7, 0.77, 0.84$ (middle figure), and at $r \simeq 0.745, 0.83$ (bottom figure). The radial resolution and

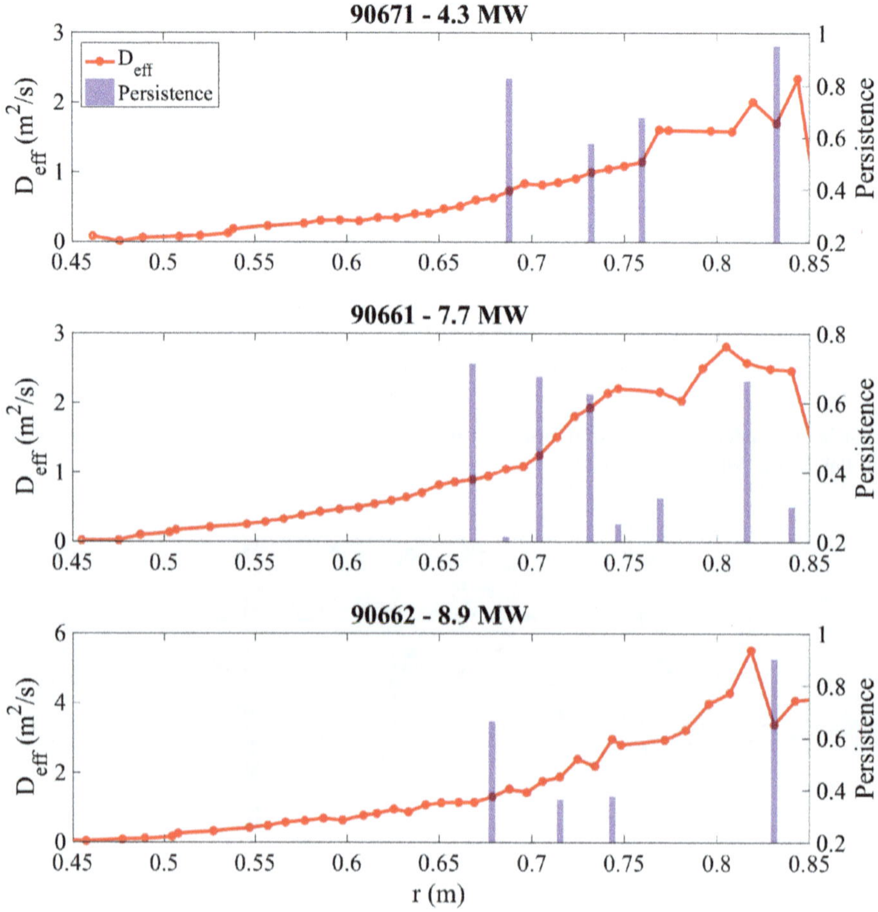

Figure 4.58. Minor transport barrier frequency and effective diffusion coefficient. (Reproduced with permission from [80].)

the available statistics are insufficient to determine whether a high persistence systematically corresponds to a low effective diffusivity, as one might expect.

To summarize, we have used the high resolution ECE diagnostic and the transfer entropy to detect minor transport barriers in JET L-mode discharges with good radial resolution [80]. We have shown that the transport of heat is not smooth and continuous; instead, there are 'trapping regions' separated by 'minor transport barriers'. Heat may 'jump over' these barriers, and when heating power is increased, this 'jumping' behaviour becomes more prominent.

By analyzing a set of similar discharges, we have shown that the method of detecting minor transport barriers is effective and robust. The minor barriers appear to be associated with low-order rational surfaces and change position as the q-profile changes.

We have also been able to extract an effective diffusion coefficient from the transfer entropy results. The effective diffusion coefficient values and their overall radial variation are consistent with values reported in literature for JET. The detailed radial variation displays structures, associated with local variations of heat transport, that are linked to the detected minor transport barriers.

In a series of L-mode discharges with varying heating power levels, the effective diffusivity changes consistently with the typical power degradation effect observed in magnetic confinement devices. The mechanism underlying this deterioration of confinement may be associated with the increased intensity of the 'jumping' effect observed. This may be an interesting observation in the framework of the effort to understand the mechanisms underlying power degradation in fusion plasmas.

4.7 Modelling

In order to better understand the experimental results presented in the preceding sections, we will make use of a resistive MHD model. The equations of the model are given in appendix B. This model is a two-fluid resistive MHD turbulence model, based on the reduced MHD equations [89]. The dominant instability corresponds to pressure-gradient-driven modes. The geometry of the system is a periodic cylinder, so the geometry is simplified with respect to the experimental devices studied in the preceding sections. Boundary conditions are also simplified. Because of these simplifications (geometry, equations, and boundary conditions), one should not expect a detailed reproduction of the experimental turbulence. However, we will see that the model provides some tantalizing clues that help us to understand and interpret the experimental results.

In this section, we will focus on the evolution of the averaged electron temperature which is described by (also see appendix B)

$$n_{\mathrm{eq}}\frac{\partial\langle T_{\mathrm{e}}\rangle}{\partial t} = -\frac{1}{r}\frac{\partial}{\partial r}\left(r\Gamma_Q\right) + \chi_{0\perp}n_{\mathrm{eq}}\frac{1}{r}\frac{\partial}{\partial r}\left(r\frac{\partial\langle T_{\mathrm{e}}\rangle}{\partial r}\right) + S_Q, \tag{4.22}$$

where the heat flux Γ_Q is

$$\Gamma_Q = n_{\mathrm{eq}}\langle\tilde{v}_{\mathrm{r}}\tilde{T}_{\mathrm{e}}\rangle - \frac{S}{\bar{\omega}_{ci}}T_{\mathrm{eq}}\langle\tilde{B}_{\mathrm{r}}\tilde{J}_z\rangle - n_{\mathrm{eq}}\left\langle\tilde{B}_{\mathrm{r}}\chi_{\parallel}\nabla_{\parallel}T_{\mathrm{e}}\right\rangle. \tag{4.23}$$

The angular brackets indicate an average over poloidal and toroidal angles. v_{r} and B_{r} are the radial velocity and magnetic field, respectively. The source S_Q is chosen so as to keep the integrated electron temperature constant. To study turbulence under conditions similar to TJ-II, we impose the rotational transform of configuration 100_44, see figure 4.21. Figure 4.59 displays the initial density and electron temperature profiles used in the simulations. They are obtained by self-consistently evolving the equations until a steady state is obtained. These profiles are characterized by flat regions around low-order rational surfaces, which is the result of plasma self-organization, as explained elsewhere [31].

To study electron temperature transport in the framework of this model, an instantaneous electron temperature perturbation is imposed at a specific radial location. The perturbation has a Gaussian shape with a maximum amplitude of $\Delta T_{\mathrm{e}} = 1.5$ and width given by $\sigma^2 = 0.005$. Then, the temporal evolution of the system

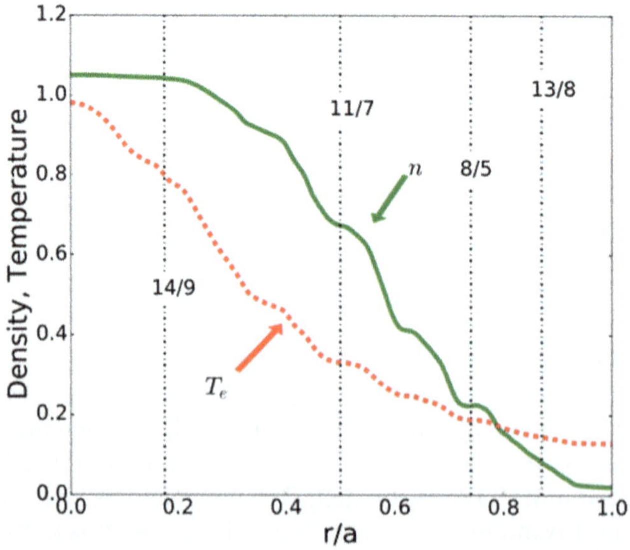

Figure 4.59. Density (n) and electron temperature (T_e) profiles used in the simulations. Vertical dashed lines represent main rational surfaces. (Reproduced from [84], with the permission of AIP Publishing.)

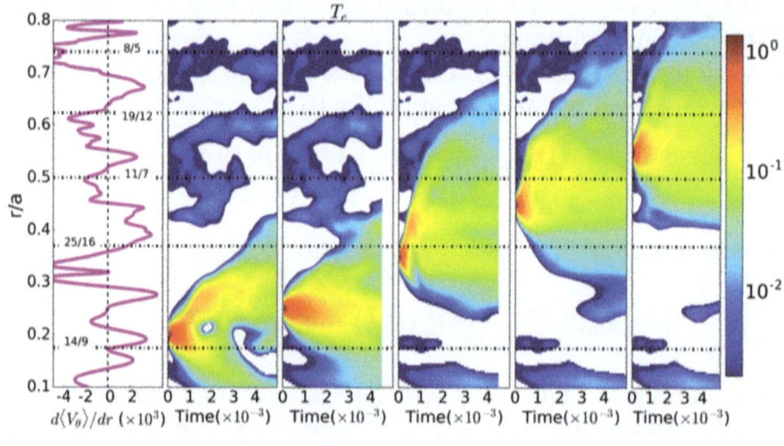

Figure 4.60. The left panel displays the flow shear ($d\langle V_\theta\rangle/dr$) of the flux-averaged poloidal velocity. Horizontal dashed lines indicate the location of main rationals. The five panels on the right show the time evolution of a heat perturbation launched at different radial locations. (Reproduced from [84], with the permission of AIP Publishing.)

is studied. When the perturbation spreads out from its initial location, its amplitude decays. So, in order to study the propagation of the perturbation at different radial positions, it is useful to initialize perturbations at different radial locations. Each perturbation is introduced in the same initial steady state.

Figure 4.60 illustrates the evolution of perturbations. In the left panel, the flow shear $d\langle v_\theta\rangle/dr$ of the flux-averaged poloidal velocity $\langle v_\theta\rangle$ immediately before the pulses is shown. The average poloidal velocity changes only slightly during the time intervals studied. The horizontal dashed lines indicate the position of the main rational surfaces. The other five panels represent the temporal evolution of the perturbations. For a better visualization of the pulses, the initial temperature profile (immediately before the perturbation) has been subtracted. The initial location of the pulses is, from left to right, $r/a = 0.2$, 0.25, 0.35, 0.45 and 0.55.

Although the perturbations are the same and all of them propagate radially outward, they do so at different rates. Clearly, the pulses in the outermost locations seem to propagate faster, which is consistent with experiments. The first pulse, launched from $r/a = 0.2$, initially propagates radially but then seems to slow down around $r/a \simeq 0.28$, where a strong shear in the poloidal flow exists. This point can therefore be considered a minor transport barrier. The second pulse, launched from $r/a = 0.25$, exhibits a similar slowing down at the same position, so that most of the pulse is trapped. The third pulse, launched from $r/a = 0.35$, shows a faster radial propagation than the preceding ones, while we observe a small trapping region around $r/a \simeq 0.40$, where a strong flow shear is present. The pulse propagates until $r/a \simeq 0.58$, where it is slowed down, probably because of the flow shear at that radial location. The fourth perturbation, launched from $r/a = 0.45$, experiences the same slowing down at $r/a \simeq 0.58$. Therefore, it is likely that this position also corresponds to a minor transport barrier. The final perturbation, launched from $r/a = 0.55$, seems to be partially trapped from the start, probably due to the barrier detected by the previous two perturbations. However, it continues to propagate radially until $r/a \simeq 0.75$, where it slows down due the strong flow shear in that region.

The right panels of figure 4.60 show that the overall spreading of the initial perturbations spread is similar to what would occur in a diffusion-type system. However, more complex behavior is also visible, involving localized 'slowing down', thus creating 'steps' in the pulse front. These 'steps' roughly coincide with maxima of the flow shear.

Now we apply the transfer entropy method to these numerical data. For this purpose, the transfer entropy is computed between the time trace of a radial reference point, taken equal to the point from which the perturbation is launched, and the time trace of any other radial point. The result is shown in figure 4.61. In the left panel, the radial shear of the temporally averaged poloidal velocity $\langle v_\theta \rangle_t$ is represented. The angular average of the poloidal velocity has been averaged in time over 0.1 τ_R (a time much longer than the duration of a pulse). Note that the time-averaged profile is similar to the instantaneous profile shown in figure 4.60, as the system is in steady state. The other five panels show the transfer entropy results obtained by calculating the TE between a reference point (white arrow) and all the other discrete radial locations in the system. For better visualization, the values of transfer entropy in each panel are normalized to their maximum values.

The TE is high in the red areas. Note that these high entropy regions are not continuous, but they do reflect an overall outward propagation in the radial direction. The slope of connected regions with high TE is indicative of the propagation speed of the perturbation and one observes that propagation is faster in the outer regions. Interestingly, there are also high TE areas that are mainly

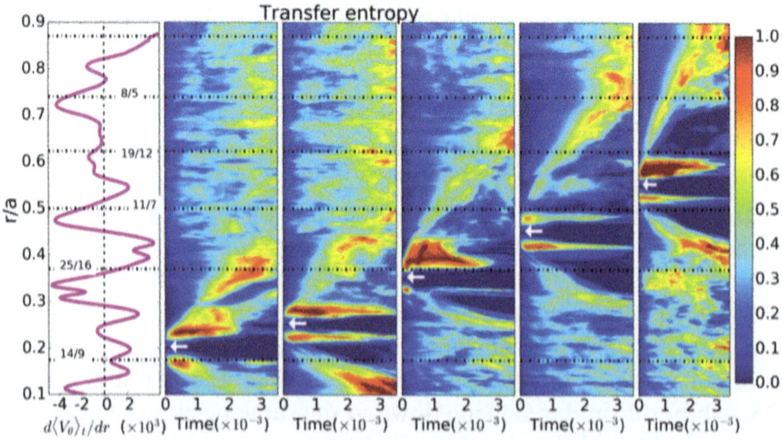

Figure 4.61. The left panel represents the radial shear of the temporal averaged poloidal velocity $\langle v_\theta \rangle_t$. Horizontal dashed lines indicate the location of main rationals. The five panels on the right show the transfer entropy for each pulse of figure 4.60. (Reproduced from [84], with the permission of AIP Publishing.)

oriented horizontally, very similar to what one observes in the experiment (see preceding sections). These horizontal areas clearly suggest a local slowing down of the propagation (or 'trapping'). These features lead us to conclude that trapping regions (mini transport barriers) exist from $r/a \simeq 0.25$ to $r/a \simeq 0.30$ (based on the first and second perturbation). Likewise, there is another trapping region around $r/a \simeq 0.4$ (observed in the first, second, and third pulse) and $r/a \simeq 0.6$ (visible in the fifth perturbation). In all cases, the transfer entropy also appears to jump between (skip over) some radial regions.

In conclusion, electron temperature perturbations propagate radially outwards, similar to what is observed in the experiments. In addition, one observes that the propagating perturbations are slowed down in 'trapping' regions with increased flow shear, associated with low-order rational surfaces and magnetic islands. The propagation front is characterized by a similar 'staircase' as is observed in the experiment. The transfer entropy appears to be especially responsive to these 'trapping regions', which show up as horizontal areas of high transfer entropy in the spatiotemporal maps. Furthermore, the transfer entropy sometimes appears to indicate a radial influence that appears to 'jump' or 'skip' radially over certain regions (i.e., displaying large values at the same time lag but at very different radial positions), again similar to the experimental observations. This 'stepwise', 'jumping' behavior is not immediately evident from the evolution of the temperature perturbation, which is largely continuous; however, this non-continuous behaviour is highlighted by the analysis based on the transfer entropy. Therefore, this technique reveals underlying behavior that is not easily accessible using standard analysis techniques.

A fully analogous study of the one presented in this section was also made for W7-X [59]—with similar results.

4.8 Summary

This chapter is almost exclusively dedicated to applications of the transfer entropy. We have exploited this technique for two main purposes: one, to study the interaction between the various (fluctuating) fields in plasma turbulence, and two, to study the propagation of perturbations, and hence transport.

The transfer entropy originates in the field of information theory. We have chosen this technique simply because it served our purposes, but we note that a range of similar, alternative techniques is available [4].

By applying this technique to Langmuir probe data from the TJ-K stellarator (section 4.5.1), we were able to detect the interactions between fields relevant in the scope of zonal flow theory: the zonal potential, the global turbulent particle flux, and the turbulence amplitude. The technique reveals the direction of these interactions (causality), and we found that this directionality depends on the timescale. The existence of several timescales is consistent with the predator–prey behaviour expected and observed in this type of systems.

Next, by applying the technique to magnetic field fluctuations detected by a Mirnov coil and measurements from a Doppler reflectometer at L–H confinement transitions in the TJ-II stellarator (section 4.5.2), we were able to show that magnetic fluctuations likely play an important role in generating or facilitating the development of the plasma flows triggering the confinement transition, which subsequently suppress magnetic activity.

Motivated by the experimental observation of a U-shaped flux–gradient relation in the TJ-II stellarator, in section 4.5.3, we applied this technique to resistive MHD simulations in which similar behaviour was observed. This study showed that the fluctuating flux impacts the fluctuating gradient at a significant distance from the flux event origin and that it is likely this long range effect that can explain the observed U-shaped flux–gradient curves. In other words, profile regulation through flux events is intrinsically connected with long range effects.

We also point the reader forward to section 5.7, where causal effects between various fields are studied during confinement transitions, with spatiotemporal resolution.

The use of the transfer entropy to study the propagation of perturbations, and hence transport, turned out to be very fruitful. The technique is extremely sensitive, which allowed tracking the propagation of rather small perturbations, facilitating the study of radial heat transport based on ECE electron temperature measurements without the need of external power modulation, at least in low density, ECR heated plasmas. In section 4.6, we applied this technique to ECE measurements in three devices: the TJ-II and W7-X stellarators, and the JET tokamak.

In the case of the two stellarators, the propagation of perturbations could be visualized by taking a relatively central ECE channel as reference and computing the transfer entropy between the reference and other ECE channels. The resulting 'propagation plume' revealed rather complex behaviour. At specific radial locations, the 'plume' appeared to stagnate, leading to long horizontal structures in the two-dimensional (position versus lag) transfer entropy graphs—which are interpreted as minor transport barriers. On the other hand, propagation often appeared to continue after 'jumping' over intermediate transfer entropy valleys.

In both devices, the observed behaviour was compared to more traditional analyses based on ECRH modulation and Fourier analysis, and it was shown that the transfer entropy results are consistent with the traditional results. Further work at the TJ-II stellarator revealed that the minor transport barriers and the rapid radial transport effect (seen as the 'jumping' effect in the two-dimensional graphs) might provide an explanation for power degradation: namely, as power increases, the rapid radial transport component was observed to increase in importance.

Whereas the minor transport barriers are likely linked to sheared flows forming near relatively low-order rational surfaces, the nature of the rapid transport effect is not entirely clear. It is hypothesized that this effect could be linked to the coupling of MHD modes associated with different rational surfaces, which would allow the fast radial transmission of fluctuation energy. This is also suggested by simulations performed using a resistive MHD model. The model also showed that temperature perturbations will behave more smoothly than the transfer entropy, suggesting it will be more difficult to reveal this complex behaviour by analyzing the temperature perturbations directly.

At JET, a different approach was followed: sawtooth heat pulses were used to analyze heat transport outside the inversion radius. Here also, radial structure was observed in the propagation plume visualized in the two-dimensional transfer entropy graph. Further analysis revealed that the corresponding minor transport barriers could be linked to the q-profile. An effective diffusion coefficient could be derived from the transfer entropy, and its value was consistent with traditional estimates and the power degradation effect.

References

[1] Burrell K H 1999 Tests of causality: experimental evidence that sheared $E \times B$ flow alters turbulence and transport in tokamaks *Phys. Plasmas* **6** 4418

[2] Wiener N 1956 The theory of prediction *Modern Mathematics for Engineers* (New York: McGraw-Hill)

[3] Granger C W J 1980 Testing for causality: a personal viewpoint *J. Econ. Dyn. Control* **2** 329

[4] Hlaváčková-Schindler K, Paluš M, Vejmelka M and Bhattacharya J 2007 Causality detection based on information-theoretic approaches in time series analysis *Phys. Reports* **441** 1

[5] Schreiber T 2000 Measuring information transfer *Phys. Rev. Lett.* **85** 461

[6] van Milligen B P, Carreras B A, García L, Martin de Aguilera A, Hidalgo C, Nicolau J H and the TJ-II Team 2016 The causal relation between turbulent particle flux and density gradient *Phys. Plasmas* **23** 072307

[7] Pastor I, Pérez-García V M, Encinas F and Guerra J M 1993 Ordered and chaotic behavior of two coupled van der Pol oscillators *Phys. Rev.* E **48** 171

[8] van Milligen B P, Birkenmeier G, Ramisch M, Estrada T, Hidalgo C and Alonso A 2014 The causal relation between turbulent particle flux and density gradient *Nucl. Fusion* **54** 023011

[9] Pilarczyk P, García L, Carreras B A and Llerena I 2012 A dynamical model for plasma confinement transitions *J. Phys. A: Math. Theor.* **45** 125502
[10] Birkenmeier G and Ramisch B 2013 Experimental evidence of turbulent transport regulation by zonal flows *Phys. Rev. Lett.* **110** 145004
[11] Wagner F *et al* 1982 Regime of improved confinement and high beta in neutral-beam-heated divertor discharges of the ASDEX tokamak *Phys. Rev. Lett.* **49** 1408
[12] Wobig H and Kisslinger J 2000 Viscous damping of rotation in Wendelstein 7-AS *Plasma Phys. Control. Fusion* **42** 823
[13] Scott B D 2005 Energetics of the interaction between electromagnetic $E \times B$ turbulence and zonal flows *New J. Phys.* **7** 92
[14] Diamond P H, Itoh S-I, Itoh K and Hahm T S 2005 Zonal flows in plasma—a review *Plasma Phys. Control. Fusion* **47** R35
[15] Fujisawa A 2009 A review of zonal flow experiments *Nucl. Fusion* **49** 013001
[16] Wagner F *et al* 2005 W7-AS: one step of the Wendelstein stellarator line *Phys. Plasmas* **12** 072509
[17] Estrada T *et al* 2010 L–H transition experiments in TJ-II *Contrib. Plasma Phys.* **50** 501
[18] van Milligen B P, Estrada T, Carreras B A, Ascasíbar E, Hidalgo C, Pastor I, Fontdecaba J M, Balbín R and the TJ-II Team 2016 The causal impact of magnetic fluctuations in slow and fast L–H transitions at TJ-II *Phys. Plasmas* **23** 072305
[19] Estrada T *et al* 2009 Sheared flows and transition to improved confinement regime in the TJ-II stellarator *Plasma Phys. Control. Fusion* **51** 124015
[20] Estrada T, Hidalgo C, Happel T and Diamond P H 2011 Spatiotemporal structure of the interaction between turbulence and flows at the L–H transition in a toroidal plasma *Phys. Rev. Lett.* **107** 245004
[21] Estrada T, Ascasíbar E, Blanco E, Cappa A, Castejón F and Hidalgo C 2015 Limit cycle oscillations at the L–I–H transition in TJ-II plasmas: triggering, temporal ordering and radial propagation *Nucl. Fusion* **55** 063005
[22] Colchin R J *et al* 2002 Physics of slow L–H transitions in the DIII-D tokamak *Nucl. Fusion* **42** 1134
[23] Hidalgo C *et al* 2009 Multi-scale physics mechanisms and spontaneous edge transport bifurcations in fusion plasmas *Europhys. Lett.* **87** 55002
[24] Happel T, Estrada T, Blanco E, Tribaldos V, Cappa A and Bustos A 2009 Doppler reflectometer system in the stellarator TJ-II *Rev. Sci. Instrum.* **80** 073502
[25] Happel T, Estrada T, Blanco E, Hidalgo C, Conway G D and Stroth U 2011 Scale-selective turbulence reduction in H-mode plasmas in the TJ-II stellarator *Phys. Plasmas* **18** 102302
[26] Estrada T *et al* 2012 Spatial, temporal and spectral structure of the turbulence-flow interaction at the L–H transition *Plasma Phys. Control. Fusion* **54** 124024
[27] van Milligen B P *et al* 2016 The role of magnetic islands in modifying long range temporal correlations of density fluctuations and local heat transport *Nucl. Fusion* **56** 016013
[28] Hidalgo C, Silva C, Carreras B A, van Milligen B, Figueiredo H, García L, Pedrosa M A, Gonçalves B and Alonso A 2012 Dynamical coupling between gradients and transport in fusion plasmas *Phys. Rev. Lett.* **108** 065001
[29] Ryter F *et al* 2001 Experimental studies of electron transport *Plasma Phys. Control. Fusion* **43** A323
[30] Garbet X *et al* 2004 Profile stiffness and global confinement *Plasma Phys. Control. Fusion* **46** 1351

[31] Ichiguchi K and Carreras B A 2011 Multi-scale MHD analysis incorporating pressure transport equation for beta-increasing LHD plasma *Nucl. Fusion* **51** 053021
[32] ITER Physics Expert Groups on Confinement and Transport *et al* 1999 Ch 2: Plasma confinement and transport *Nucl. Fusion* **39** 2175
[33] Hoang G T, Bourdelle C, Garbet X, Giruzzi G, Aniel T, Ottaviani M, Horton W, Zhu P and Budny R V 2001 Experimental determination of critical threshold in electron transport on Tore Supra *Phys. Rev. Lett.* **87** 125001
[34] Gentle K W, Austin M E and Phillips P E 2004 Direct measurement of the electron energy flux versus electron temperature gradient in tokamak discharges *Plasma Sci. Technol.* **6** 2130
[35] DeBoo J C *et al* 2012 Electron profile stiffness and critical gradient studies *Phys. Plasmas* **19** 082518
[36] Carreras B A and Lynch V E 2006 On the use of critical gradient models in fusion plasma transport studies *Phys. Plasmas* **13** 062301
[37] Kasuya N, Sugita S, Inagaki S, Itoh K, Yagi M and Itoh S-I 2015 Simulation study of hysteresis in the gradient-flux relation in toroidal plasma turbulence *Plasma Phys. Control. Fusion* **57** 044010
[38] Lopes Cardozo N J 1995 Perturbative transport studies in fusion plasmas *Plasma Phys. Control. Fusion* **37** 799
[39] Gentle K W *et al* 1995 An experimental counter-example to the local transport paradigm *Phys. Plasmas* **2** 2292
[40] Sattin F, Escande D F, Auriemma F, Urso G and Terranova D 2014 Difficulties and solutions for estimating transport by perturbative experiments *Plasma Phys. Control. Fusion* **56** 114008
[41] van Berkel M, Zwart H J, Tamura N, Hogeweij G M D, Inagaki S, de Baar M R and Ida K 2014 Explicit approximations to estimate the perturbative diffusivity in the presence of convectivity and damping. I. Semi-infinite slab approximations *Phys. Plasmas* **21** 112507
[42] Mantica P, Galli P, Gorini G, Hogeweij G M D and de Kloe J Jun 1999 Nonlocal transient transport and thermal barriers in Rijnhuizen Tokamak Project plasmas *Phys. Rev. Lett.* **82** 5048–51
[43] Garbet X *et al* 2004 Physics of transport in tokamaks *Plasma Phys. Control. Fusion* **46** B557
[44] de la Luna E, Sánchez J, Tribaldos V and Estrada T 2001 Multichannel electron cyclotron emission radiometry in TJ-II stellarator *Rev. Sci. Instrum.* **72** 379
[45] Happel T *et al* 2015 Core turbulence behavior moving from ion-temperature-gradient regime towards trapped-electron-mode regime in the ASDEX Upgrade tokamak and comparison with gyrokinetic simulation *Phys. Plasmas* **22** 032503
[46] Hidalgo C *et al* 2016 On the influence of ECRH on neoclassical and anomalous mechanisms using a dual heavy ion beam probe diagnostic in the TJ-II stellarator *Proc. 26th IAEA Fusion Energy Conf.* EX/P7-44
[47] Estrada T, de la Luna E, Ascasíbar E, Jiménez J A, Castejón F, García-Cortés I, López-Fraguas A, Sánchez J and Tribaldos V 2002 Transient behaviour in the plasma core of TJ-II stellarator and its relation with rational surfaces *Plasma Phys. Control. Fusion* **44** 1615
[48] García L, Ochando M A, Carreras B A, Carralero D and Hidalgo C 2016 Effect of fast electrons on the stability of resistive interchange modes in the TJ-II stellarator *Phys. Plasmas* **23** 062319

[49] Politzer P A 2000 Observation of avalanchelike phenomena in a magnetically confined plasma *Phys. Rev. Lett.* **84** 1192

[50] van Milligen B P *et al* 2002 Ballistic transport phenomena in TJ-II *Nucl. Fusion* **42** 787

[51] van Milligen B P, Nicolau J H, García L, Carreras B A, Hidalgo C and the TJ-II Team 2017 The impact of rational surfaces on radial heat transport in TJ-II *Nucl. Fusion* **57** 056028

[52] van Milligen B P *et al* 2011 Integrated data analysis at TJ-II: the density profile *Rev. Sci. Instrum.* **82** 073503

[53] van Milligen B P, Carreras B A, Hidalgo C, Cappa Á and the TJ-II Team 2018 A possible mechanism for confinement power degradation in the TJ-II stellarator *Phys. Plasmas* **25** 062503

[54] Spakman G W *et al* 2008 Heat pulse propagation studies around magnetic islands induced by the dynamic ergodic divertor in TEXTOR *Nucl. Fusion* **48** 115005

[55] Eguilior S, Castejón F, de la Luna E, Cappa A, Likin K and Fernández A 2003 Heat wave experiments on TJ-II flexible heliac *Plasma Phys. Control. Fusion* **45** 105

[56] Jacchia A, Mantica P, De Luca F and Gorini G 1991 Determination of diffusive and nondiffusive transport in modulation experiments in plasmas *Phys. Fluids* B **3** 3033

[57] Ryter F, Camenen Y, DeBoo J C, Imbeaux F, Mantica P, Regnoli G, Sozzi C and Stroth U 2006 Electron heat transport studies *Plasma Phys. Control. Fusion* **48** B453

[58] Fontdecaba J M *et al* 2004 Energy-resolved neutral particle fluxes in TJ-II ECRH plasmas *Fusion Sci. Technol.* **46** 271

[59] van Milligen J M, Hoefel F, Nicolau R, Hirsch R, García R, Carreras R, Hidalgo C and the W7-X Team 2018 Study of radial heat transport in W7-X using the transfer entropy *Nucl. Fusion* **58** 076002

[60] Dif-Pradalier G *et al* 2015 Finding the elusive $E \times B$ staircase in magnetized plasmas *Phys. Rev. Lett.* **114** 085004

[61] Choi M J *et al* 2019 Experimental observation of the non-diffusive avalanche-like electron heat transport events and their dynamical interaction with the shear flow structure *Nucl. Fusion* **59** 086027

[62] Ida K *et al* 2015 Towards an emerging understanding of non-locality phenomena and non-local transport *Nucl. Fusion* **55** 013022

[63] Lopes Cardozo N J *et al* 1997 Electron thermal transport in RTP: filaments, barriers and bifurcations *Plasma Phys. Control. Fusion* **39** B303

[64] Wukitch S J *et al* 2002 Double transport barrier experiments on Alcator C-Mod *Phys. Plasmas* **9** 2149

[65] Wolf R C *et al* 2017 Major results from the first plasma campaign of the Wendelstein 7-X stellarator *Nucl. Fusion* **57** 102020

[66] Dinklage A *et al* 2016 Core confinement in Wendelstein 7-X limiter plasmas *European Physical Society Conf. on Plasma Physics, ECA* **40A** O2.107

[67] Pedersen T S *et al* 2017 Key results from the first plasma operation phase and outlook for future performance in Wendelstein 7-X *Phys. Plasmas* **24** 055503

[68] Höfel U *et al* 2016 First measurement on electron heat transport by heatwaves in the core plasma of Wendelstein 7-X *European Physical Society Conf. on Plasma Physics, ECA* **40A** P4.008

[69] Callen J D and Kissick M W 1997 Evidence and concepts for non-local transport *Plasma Phys. Control. Fusion* **39** B173

[70] Pustovitov V D 2012 Nonlocal effects in energy balance in an equilibrium plasma during its fast heating/cooling in tokamaks and stellarators *Plasma Phys. Control. Fusion* **54** 124036

[71] Pablant N A *et al* 2018 Core radial electric field and transport in Wendelstein 7-X plasmas *Phys. Plasmas* **25** 022508
[72] de la Luna E *et al* 2004 Electron cyclotron emission radiometer upgrade on the Joint European Torus (JET) tokamak *Rev. Sci. Instrum.* **75** 3831
[73] Schilham A M R, Hogeweij G M D and Lopes Cardozo N J 2001 Electron thermal transport barriers in RTP: experiment and modelling *Plasma Phys. Control. Fusion* **43** 1699
[74] Joffrin E *et al* 2003 Internal transport barrier triggering by rational magnetic flux surfaces in tokamaks *Nucl. Fusion* **43** 1167
[75] Militello F, Romanelli M, Connor J W and Hastie R J 2011 Generation of zonal perturbations and transport barriers in plasmas *Nucl. Fusion* **51** 033006
[76] Mantica P *et al* 2002 Perturbative transport experiments in JET low or reverse magnetic shear plasmas *Plasma Phys. Control. Fusion* **44** 2185
[77] Marinoni A *et al* 2006 Analysis and modelling of power modulation experiments in JET plasmas with internal transport barriers *Plasma Phys. Control. Fusion* **48** 1469
[78] Fujisawa A 2002 Transport barriers and bifurcation characteristics in stellarators *Plasma Phys. Control. Fusion* **44** A1
[79] Carreras B A, Lynch V E and García L 1991 Electron diamagnetic effects on the resistive pressure-gradient-driven turbulence and poloidal flow generation *Phys. Plasmas* **3** 1438
[80] van Milligen B P, Carreras B A, de la Luna E and Solano E R 2019 Radial variation of heat transport in L-mode JET discharges *Nucl. Fusion* **59** 056006
[81] García L, Carreras B A and Llerena L 2017 Relation of plasma flow structures to passive particle tracer orbits *Nucl. Fusion* **57** 116013
[82] Mantica P and Ryter F 2006 Perturbative studies of turbulent transport in fusion plasmas *Compt. Rend. Phys.* **7** 634
[83] Brix M, Hawkes N C, Boboc A, Drozdov V and Sharapov S E 2008 Accuracy of EFIT equilibrium reconstruction with internal diagnostic information at JET *Rev. Sci. Instrum.* **79** 10F325
[84] Nicolau J H, García L and Carreras B A 2018 Applicability of transfer entropy for the calculation of effective diffusivity in heat transport *Phys. Plasmas* **25** 102304
[85] Soler M and Callen J D 1979 On measuring the electron heat diffusion coefficient in a tokamak from sawtooth oscillation observations *Nucl. Fusion* **19** 703
[86] Fredrickson E D *et al* 1986 Heat pulse propagation studies in TFTR *Nucl. Fusion* **26** 849
[87] Ivanova-Stanik I, Aho-Mantila L, Wischmeier M and Zagórski R 2016 COREDIV and SOLPS numerical simulations of the Nitrogen seeded JET ILW L-mode discharges *Contrib. Plasma Phys.* **56** 760
[88] Bonanomi N, Mantica P, Giroud C, Angioni C, Manas P and Menmuir S 2018 Light impurity transport in JET ILW L-mode plasmas *Nucl. Fusion* **58** 036009
[89] Strauss H R 1976 Nonlinear, three-dimensional magnetohydrodynamics of noncircular tokamaks *Phys. Fluids* **19** 134

IOP Publishing

Analysis of Turbulence in Fusion Plasmas

Boudewijn Ph van Milligen and Raul Sanchez

Chapter 5

Intermittence

In chapter 2 we noted that the fluctuating particle flux, measured by Langmuir probes in the plasma edge, is often 'bursty'. The presence of occasional large bursts leads to a probability distribution function with 'long tails', and this phenomenon also appears to be associated with values of the Hurst coefficient exceeding 0.5, reflecting the temporal distribution of these bursts. Even so, saying a signal is 'bursty' is a bit vague. Is it possible to be more precise about this 'bursty' behaviour? To make progress, we again need a suitable analysis tool: in this case, a quantifier of intermittence. A signal is said to be intermittent when it displays 'activity' only during a fraction of the time, such that the 'active' time windows are distributed (randomly) over time. Here, the 'activity' typically would correspond to a temporal increase of the RMS variation of the signal.

To clarify the idea, let us have a look at an example. The Ikeda map is a simple dynamical system that experiences a transition to intermittent behaviour as its parameters are varied. This map generates a sequence of points z_n, $n = 1, \dots, N$ via the equation

$$z_{n+1} = a + bz_n \exp\left(i\kappa - \frac{i\eta}{1 + |z_n|^2}\right), \tag{5.1}$$

where z_n is a complex number. We will analyse the sequence Re (z_n). We select the parameters $a = 0.85$, $b = 0.9$, and $\kappa = 0.4$ and vary η over a range where intermittence appears (due to a 'crisis' occurring at $\eta_c = 7.268\,848\,94\ldots$) [1]. Figure 5.1 shows an example of such data for a few η values. When η is small ($\eta = 6.70$), the motion is regular and not intermittent. Increasing η to 7.26, the fluctuations become irregular, with short interruptions so that the signal might be considered 'weakly intermittent', but as long as $\eta < \eta_c$, the fluctuations vary within a clearly established range (roughly between 0 and 1). However, when $\eta > \eta_c$, occasional short-lived bursts appear that exceed this range. These bursts occur when the trajectory switches to a different 'branch' or 'chaotic attractor' of the system. This 'bursty' behaviour,

doi:10.1088/978-0-7503-4856-0ch5

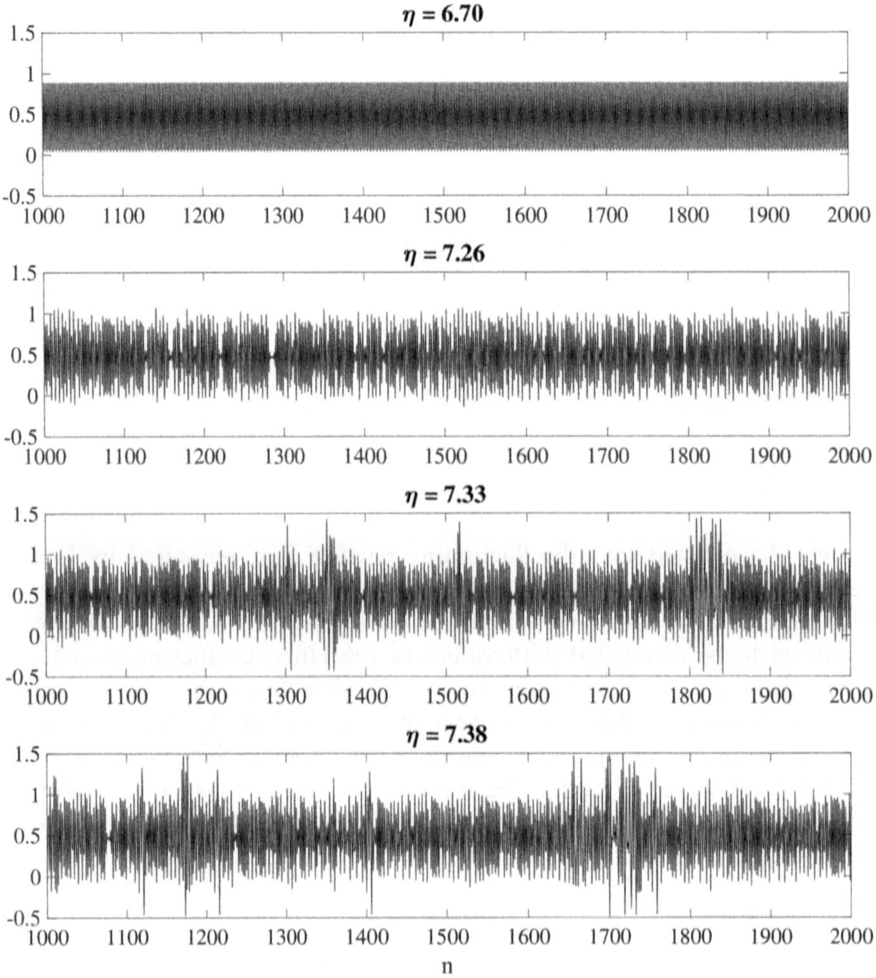

Figure 5.1. Data $\mathrm{Re}(z_n)$ from the Ikeda map, equation (5.1), with $a = 0.85$, $b = 0.9$, $\kappa = 0.4$ and values of η as indicated above each plot.

interrupting the 'steady' background fluctuations, is what is typically meant by 'intermittence'.

As a side remark, note the similarity of this behaviour with the experimental observations shown in figure 2.2 in chapter 2, where we observed that the 'burstiness' of the ion saturation current measurements varied with the magnetic well in TJ-II.

5.1 Antecedents

Early on in fluid turbulence studies, well before the appearance of the field of 'chaos' as an established scientific discipline, it was noted that fluid turbulence often exhibits this type of 'bursty' behaviour. In an attempt to quantify this phenomenon, Batchelor proposed a simplified model for velocity fluctuations in a turbulent fluid, in which periods of quiescence alternated with bursts of fluctuations having a

Gaussian distribution ([2], p 183). Assuming the duration of the bursty behaviour occupied a fraction γ of time, one obtains $K = 3/\gamma$, where K is the kurtosis (or flatness, see section 2.1). Since $\gamma \leqslant 1$ by definition, one would expect $K > 3$ for intermittent signals. In any case, this idea suggested that the determination of the kurtosis would provide information about the burstiness of the signal. A number of studies were based on this idea [3]. However, we note that this proposed quantifier of intermittence relies on very specific and somewhat unrealistic assumptions, namely quiescence interrupted by fluctuation bursts, while these fluctuations are modelled by a Gaussian random variable. More generally, the kurtosis is simply an indicator of the shape of the probability distribution function, which does not reflect temporal intermittence at all. This produces doubts about the general applicability of this approach, as many known intermittent systems do not possess a Gaussian distribution.

In view of the fact that actual experimental signals from turbulence are rarely truly 'quiescent', Frisch described a method to distinguish the bursty part of the signal from the non-bursty part ([4], p 122). By applying a high-pass frequency filter to the data, temporal sections of the signal without much high frequency content would then become essentially 'quiescent'. The inverse of K calculated from this filtered signal would thus provide a measure of the time the high frequency activity is 'on' (the above γ). In this approach, the cutoff frequency f_c of the applied high-pass filter enters as an analysis parameter; 'intermittence' would correspond to the situation where K varies with f_c, whereas a constant K as a function of f_c would signal the absence of intermittence. More recently, this method was taken one step further by calculating K as a function of wavelet scale (i.e. frequency) [5]. Although these ideas improve on the original quantifier, K, they do not result in a clearly defined parameter to indicate intermittence. The next section will present an alternative approach that focuses on the time-varying nature of the signal, rather than the statistical distribution.

5.2 The intermittence parameter

Chaos theory provides more profound insight into the properties of intermittence. Turbulence is often assumed to be governed by multiplicative processes, in which larger 'eddies' or fluid elements transform or break down into smaller pieces in a hierarchical manner (constituting a 'cascade'). Particles moving in such a system would occasionally be trapped in large eddies, and occasionally in small eddies. Considering each eddy an 'attractor', the resulting complex motion is 'multifractal', as the orbit moves from one attractor to the next. In a multifractal system, particularly the small scales will behave intermittently [6, 7]. Broadly speaking, the degree of multifractality is related to the fractal dimension of the underlying dynamics, the configuration of turbulent cascades and the resulting spectra.

From these considerations, it appears that the degree of multifractality is a prime quantifier of turbulence. It is therefore somewhat surprising that very little has been done to determine this quantifier for turbulence in fusion plasmas.

From an algorithmic point of view, we will see that the quantification of the degree of multifractality is based on one of the hallmark methods of chaos theory:

namely, a ‘box counting’ method is used, in which the time series of length N is subdivided into smaller sections (‘boxes’) of length $n < N$, after which the scaling of a specific quantifier with n is studied. We mention this point because this type of algorithm is not easily fooled by trivial temporal distributions, as might be the case with the methods based on the kurtosis mentioned in the preceding section. As will be clarified below, the degree of multifractality is expressed in terms of the so-called intermittence parameter $C(1)$, which can attain values in the range from 0 to 1.

The method for calculating the intermittence parameter is described in [6–8], which we summarize here for convenience. Given a (stationary[1]) time series $X = \{x_i, i = 1, \dots, N\}$ that has been sampled at a constant sampling rate, we calculate the measure

$$\epsilon(1, i) = \frac{(x_i - \langle x_i \rangle)^2}{\langle (x_i - \langle x_i \rangle)^2 \rangle}, \qquad i = 1, \dots, N, \tag{5.2}$$

where $\langle x_i \rangle = (\sum_{i=1}^{N} x_i)/N$. This quantity can be understood to represent the ‘turbulent energy’ of the fluctuations, normalized to its mean. This measure is averaged over subblocks of data of length $n < N$, as follows:

$$\epsilon(n, i) = \frac{1}{n}\sum_{j=0}^{n-1} \epsilon(1, i + j). \tag{5.3}$$

It can be understood as the n-point variance of the signal, normalized by the N-point variance. One can see how this quantity measures the ‘burstiness’ of a signal at a scale n. We then calculate the q-moments, $\langle \epsilon(n, i)^q \rangle$. We note that this step is somewhat reminiscent of the statistical moments (mean, standard deviation, skewness, kurtosis) and therefore of the kurtosis method mentioned earlier, but here the moments are calculated for the normalized n-point variance of the signal, rather than the signal itself. In a given range of n-values, these moments are expected to scale like [9]

$$\langle \epsilon(n, i)^q \rangle \propto n^{-K(q)}, \tag{5.4}$$

where $K(1) \equiv 0$. If the time series X is monofractal, the function $K(q)$ is asymptotically linear in q, otherwise the series is multifractal[2].

Since we are interested in the turbulent (high frequency) characteristics of the fluctuations, we determine $K(q)$ from a linear fit of a set of values of $\log\langle \epsilon^q \rangle$ versus $\log n$, for a range of values of $n = 1, \dots, n_1$, where n_1 is small, e.g. $n_1 = 6$, and for a few values of q (see below). The parameter $K(q)$ is robust in the sense that the choice of n_1 is not critical, although ocular inspection of the $\log\langle \epsilon^q \rangle$ versus $\log n$ curves is recommended to check that the curves behave as a power law for small n.

The parameter $C(q)$ is defined as [9]

$$C(q) = \frac{K(q)}{q - 1}. \tag{5.5}$$

[1] Simple non-stationarity such as linear signal drift should be removed prior to analysis.

[2] Here, K and q should not be confused with the kurtosis and the safety factor, respectively.

Of special interest is the so-called 'intermittence parameter' $C(1)$, which must be calculated using L'Hôpital's rule

$$C(1) = \left. \frac{dK(q)}{dq} \right|_{q=1} \tag{5.6}$$

due to the singularity of equation (5.5) at $q = 1$. Its value ranges from 0, for a monofractal time series, to 1.

To find the intermittence parameter $C(1)$, equation (5.6), one needs to calculate $K(q)$ values for various values of $q \neq 1$ and use those to estimate $C(1)$. It is not possible to obtain $K(q)$ for $q < 1$ due to the absence of a clear linear scaling region according to equation (5.4). Using finite $q > 1$ values, one can use the following three approximate expressions to estimate the slope at $q = 1$ obtained by Taylor expansion in q with the condition $K(1) = 0$:

$$\begin{aligned} C_0(1) &= \left. \frac{dK}{dq} \right|_{q=1} = 4K(1.5) - K(2) \\ C_1(1) &= \left. \frac{dK}{dq} \right|_{q=1} = 2K(2) - \frac{1}{2}K(3) \\ C_2(1) &= \left. \frac{dK}{dq} \right|_{q=1} = \frac{8}{3}K(1.5) - \frac{1}{6}K(3). \end{aligned} \tag{5.7}$$

These three estimates of the derivative yield similar results for $C(1)$ [10].

Returning to our simple dynamical system, the Ikeda map, equation (5.1), we estimate the value of two intermittence quantifiers, the kurtosis and $C(1)$. Figure 5.2 shows these two quantifiers for a range of values of η. The black lines with circles show K (top) and $C(1)$ (bottom) for $\mathrm{Re}(z_n)$. One observes that both quantifiers evolve similarly with η. It is known that intermittence is increasing towards the right (see figure 5.1), which agrees with the upward trend of the curves. For $\eta \lesssim 6.95$, the behaviour is not intermittent at all, but quasi-periodic. This is correctly reflected in the $C(1)$ parameter, which is zero in that region; also we obtain $K = 1$, which corresponds to the lower bound of this parameter. Based solely on this comparison, one might conjecture that K and $C(1)$ deliver equivalent information.

This is, however, not the case. To show that the two parameters are not equivalent, we sort the data array $\mathrm{Re}(z_n)$ in ascending order, yielding y_n such that $y_{n+1} \geqslant y_n$. Next, we define a new array such that $y_n' = y_n$ for n even, and $y_n' = y_{N'-n}$ for n odd. Here $N' = 2\lfloor (N+1)/2 \rfloor$. In other words, even elements of y' are sorted in ascending order, and odd elements in descending order. This systematically reordered data series y_n' is thus stationary and not intermittent by construction, even though the set of data values it contains is the same as the original series, Re (z_n). We then apply the analysis to the reordered series y_n', shown as red lines with crosses in the figure. It is seen that K is exactly the same as before, which is logical as the reordering of the data has not modified the probability distribution of the data.

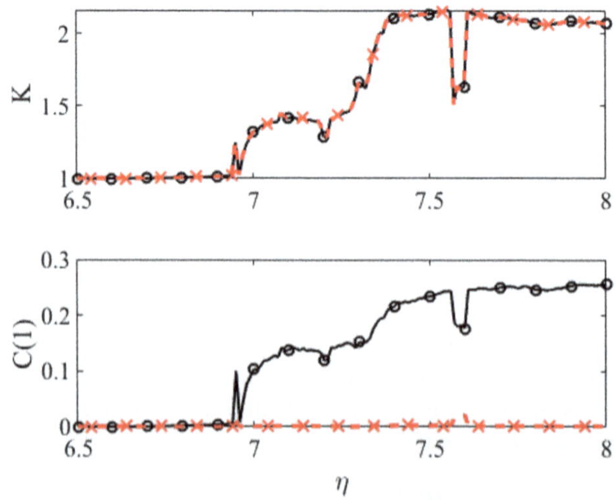

Figure 5.2. Intermittency quantifiers applied to the Ikeda map with $a = 0.85$, $b = 0.9$, $\kappa = 0.4$ and values of η as indicated on the horizontal axis. Black circles: applied to original data $\mathrm{Re}(z_n)$. Red crosses: applied to reordered data y_n'.

But $C(1)$ has dropped to zero, as the reordered array is not intermittent. In this artificial example $C(1)$ clearly outperforms K as a quantifier of intermittence in the sense that it does not produce false positives. It is therefore potentially more sensitive to actual intermittent behaviour, while K might produce misleading results.

To emphasize this point through a practical example, we now make a comparison of $C(1)$ and K for floating potential measurements from a Langmuir probe at TJ-II, inserted in the edge region of a plasma heated by ECRH. The floating potential and the edge radial electric field E_r were modulated by means of a biasing probe, using a square wave modulation waveform. The resulting modulation of the poloidal plasma velocity v_θ affects turbulence, as reported in more detail elsewhere [11]. For now, we merely use these data to compare the methods. Figure 5.3 shows a time trace of the measured floating potential, clearly reflecting the applied modulation (top), as well as the intermittence parameter $C(1)$ and kurtosis K calculated from these data. At the times when the modulated voltage changes from 'low' to 'high', the signal is of course not stationary and results corresponding to those times should be ignored. At other times, however, the floating potential signal is stationary and allows analysis. One observes that $C(1)$ responds clearly to the modulation. This can be explained from the fact that the Langmuir probe measures a mixture of the temporal and spatial structure of the turbulence, which depends on the poloidal plasma velocity v_θ. The next section will discuss this effect in more detail. The important point to note here is that K does not follow the modulation clearly.

More generally, the results for intermittence from turbulence simulations and measurements presented in the following sections could not be reproduced clearly if K was used instead of $C(1)$. Therefore, in spite of the potential usefulness of the kurtosis as a quantifier for intermittence in simple dynamical systems, we prefer using $C(1)$ for the study of actual turbulence, as it appears to be more robust and reliable.

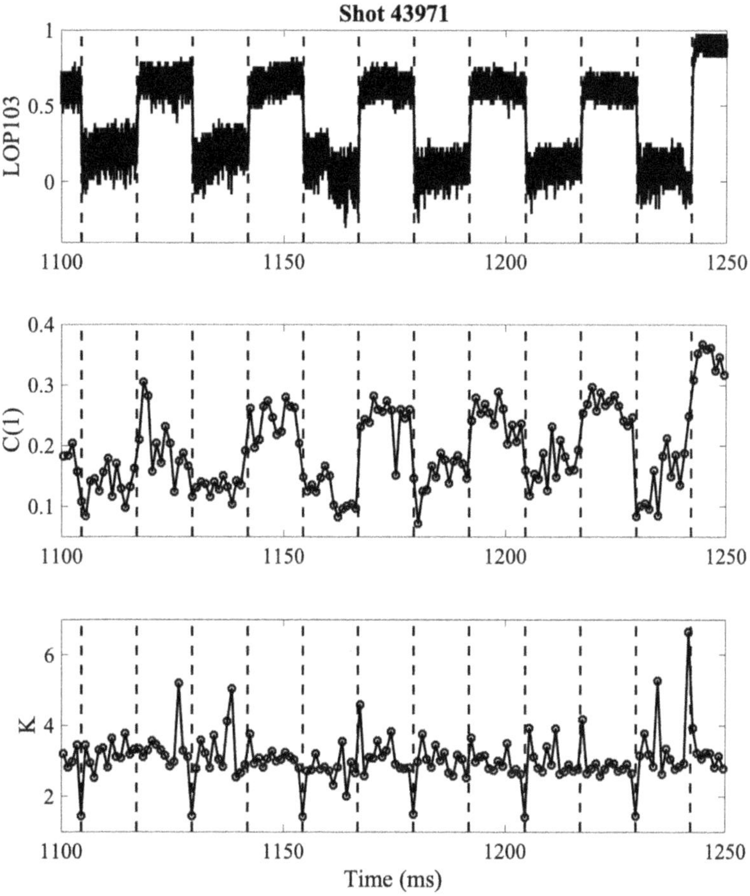

Figure 5.3. Modulated floating potential (top) and calculated intermittence parameter, $C(1)$ (middle), and kurtosis, K (bottom). Vertical dashed lined indicate the flanks of the square wave modulation.

In fact, the calculation of the intermittence parameter $C(1)$ is rather insensitive to random noise. To test this statement, we have taken an experimental time series with varying values of $C(1)$, added random Gaussian noise to it, and recalculated $C(1)$. Figure 5.4 shows the standard deviation between the recalculated value of $C(1)$ and the value of $C(1)$ from the original data. The amplitude of the additive noise is expressed as a percentage of the RMS value of the data. Even for noise levels as high as 100%, $C(1)$ varies by less than 10% from its value without noise, attesting to the robustness of the calculation.

5.3 Numerical study of intermittence in a turbulence model

To improve our understanding of the behaviour of the intermittence in a turbulent context, similar to what is encountered in fusion plasmas, we will make use of a two-fluid resistive magneto-hydrodynamic turbulence model, details of which are given in appendix B. Below, we will compare turbulence runs performed using a modelling

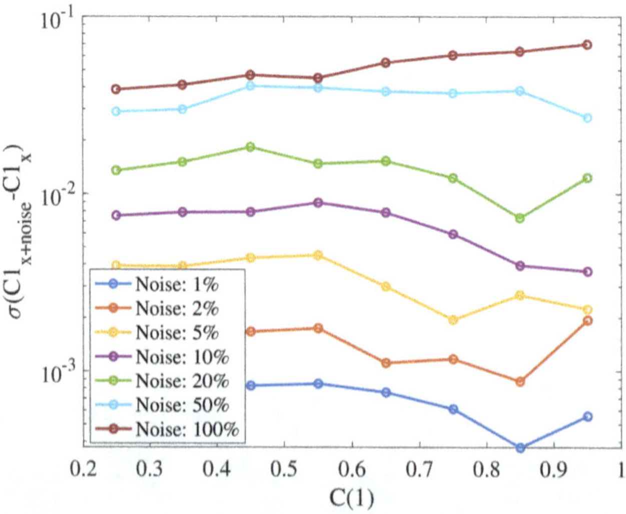

Figure 5.4. Standard deviation between $C(1)$ calculated from data $x(t)$ with added random Gaussian noise, and $C(1)$ calculated from data $x(t)$ without added noise. The horizontal axis specifies the value $C(1)$ of the data without noise. The noise amplitude is expressed as a percentage of the RMS value of the data $x(t)$. (Reproduced with permission from [12]. © IOP Publishing. All rights reserved.)

set-up that reflects some of the relevant profiles of two stellarators: the flexible heliac TJ-II (appendix A.2) and the optimized stellarator W7-X (appendix A.4). For TJ-II, we have used the configuration labelled 100_46, with an important low-order rational surface, $\iota\!\!\!-= \iota/2\pi = 8/5$, located at $\rho \simeq 0.6$. For W7-X, we have used the standard OP1.1 configuration [13], with an important low-order rational surface, $\iota\!\!\!-= \iota/2\pi = 4/5$, located at $\rho \simeq 0.48$. For both devices, we define the normalized minor radius as $\rho = \sqrt{\psi/\psi_{\mathrm{LCFS}}}$, where ψ is the toroidal magnetic flux and ψ_{LCFS} its value at the plasma boundary. The corresponding rotational transform profiles (in a vacuum) are shown in figure 5.5. The dominant helicities for these two configurations are 8/5 and 4/5, respectively.

Figure 5.6 shows the normalized electron temperature fluctuations $\tilde{T}_e/\langle T_e(\rho = 0)\rangle$ at a given time in the evolution as a function of ρ at zero poloidal and toroidal angles. In the case of TJ-II (a), the temperature oscillation is seen to be predominantly symmetrical around the rational surface (even parity). However, in the case of W7-X (b), the oscillation appears to be a mixture of even and odd parity modes, with predominance of the odd mode.

The single helicity calculations yields time traces of the temperature $T(\rho, t)$ for 1000 radial positions ρ. The dynamical evolution in a single helicity calculation is dominated by a filamentary vortex along the magnetic lines at the resonant surface. All the components of this vortex have mode numbers n/m that match the local value of the rotational transform, $\iota\!\!\!-$, and the time evolution of this structure, located at a fixed position, has a monofractal nature. However, due to Reynolds stress an averaged poloidal flow, $\langle V_\theta \rangle$, i.e. a zonal flow, is also generated. If the basic modes have an even parity with respect to the rational surface, as is the case of the TJ-II

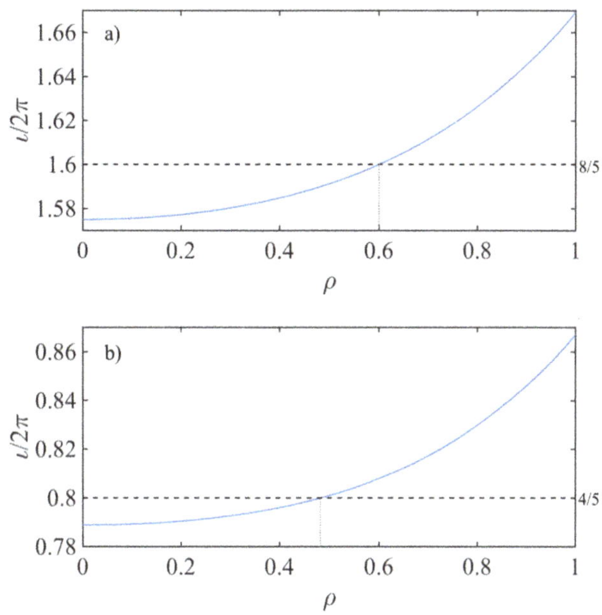

Figure 5.5. (a) Rotational transform profile (TJ-II, configuration 100_46). (b) Rotational transform profile (W7-X, standard OP1.1 configuration). The rational surfaces corresponding to the single helicity calculations are indicated by dashed lines; the corresponding rational values are indicated on the right. (Reproduced with permission from [10].)

calculation (figure 5.6), the zonal flow amplitude is practically zero at the centre of the vortex and grows as one moves away from it. The effect of the poloidal rotation induced by this zonal flow is to mix the local time evolution of the turbulence with the poloidal spatial variation. As a consequence, one obtains a mixing of two fractal structures, leading to an increase of the intermittence parameter as the measured signal becomes more multifractal. This effect is clearly observed in the TJ-II case shown in figure 5.7, where two peaks are seen at $\rho \simeq 0.53$ and 0.67, on both sides of the rational surface.

When the modes have an odd parity, as in the case of the calculation for W7X, the vortex oscillates radially around the location of the rational surface and a small amount of zonal flow is generated near that location as a consequence. This effect is understood to be the cause of the small central peak, at $\rho \simeq 0.46$, of the intermittence parameter between the two major peaks at $\rho \simeq 0.4$ and 0.51, observed in the W7-X case shown in figure 5.7. At a larger distance from the rational surface, the increased averaged poloidal flow leads to a similar increase of the multifractal behaviour and hence the intermittence parameter as in the TJ-II case.

Zonal flows may also affect the multifractality of the measured fluctuations through the low frequency components of the zonal flow, although this effect is generally weak compared with the poloidal mixing.

Thus, there exists a strong relation between the averaged poloidal flow and the intermittence parameter due to the mixing of temporal and poloidal fractal properties induced by the flow. This is exemplified in figure 5.8. Nevertheless, there is no

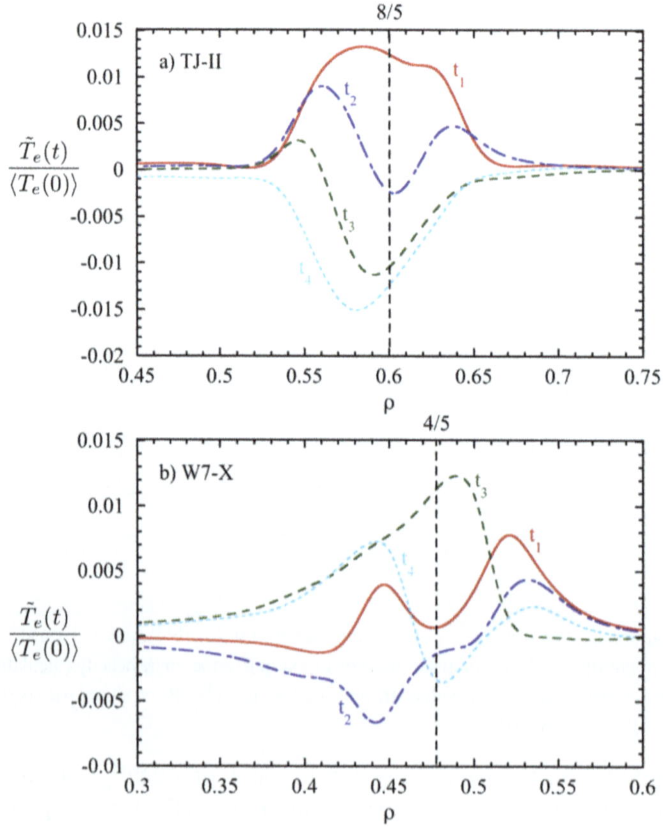

Figure 5.6. Dominant single helicity temperature mode for (a) TJ-II configuration 100_46 and (b) the standard W7-X configuration. The rational surfaces corresponding to the single helicity mode are indicated by vertical lines; the corresponding rational values are indicated at the top. The time steps between successive profiles are (a) 0.002 5 and (b) 0.001 5, in units of τ_R. (Reproduced with permission from [10]. © IOP Publishing. All rights reserved.)

one-to-one relationship between the poloidal flow and the intermittence parameter as the intermittence also depends on the fractal properties of the poloidal structures, which may vary with radial position.

The relationship between intermittence and mean poloidal flow velocity is further tested by a separate single helicity calculation for TJ-II, in which the generation of poloidal flow is not allowed, by imposing poloidal/toroidal symmetry on the fluctuations. The result is shown in figure 5.9. Without poloidal flow, the intermittence level is very low, i.e. at noise level. From all of the above, we conclude that the detected relatively high values of $C(1)$ must result from the mixing of temporal and poloidal fractal properties caused by the poloidal flow. This is a very important observation, as it implies that fluctuations produced by a single helicity without flow would lead to values of $C(1)$ close to zero, a fact that will be referred to repeatedly in the following.

The intermittence parameter also bears some relationship with the mean transfer entropy, T, calculated using a reference position of $\rho_{\text{ref}} = 0.2$, averaged over the

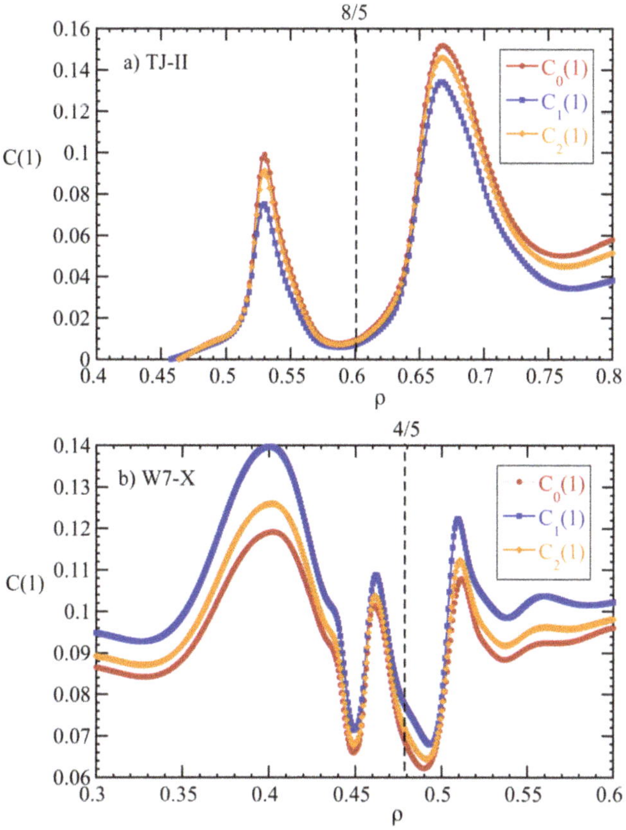

Figure 5.7. Three estimates of the intermittence parameter, equation (5.7), for the single helicity calculations: (a) TJ-II configuration 100_46 and (b) the standard W7-X configuration. The vertical lines indicate the location of the corresponding rational surfaces. (Reproduced with permission from [10]. © IOP Publishing. All rights reserved.)

available time lags. Figure 5.10 shows graphs of the intermittence parameter $C(1)$ and the mean transfer entropy T (with reversed vertical axis in order to facilitate comparison). In the case of TJ-II, there exists a clear anticorrelation between the two quantities over the radial range $0.53 < \rho < 0.72$, roughly centred on the 8/5 rational surface located at $\rho \simeq 0.6$. In the case of W7-X, a similar anticorrelation is observed in the range $0.38 < \rho < 0.45$, but on the outside of the rational surface, near $\rho \simeq 0.48$, the relation between the two curves indicates a correlation rather than an anticorrelation. However, ignoring fine radial details, in both cases the maximum of T is centred on the rational surface ($0.55 < \rho < 0.64$ for TJ-II and $0.43 < \rho < 0.52$ for W7-X).

Recall that we interpreted a low value of T as a minor transport barrier [14]. A transport barrier is a region where the zonal flow shear is sufficiently strong to reduce turbulence, so that the interaction between zonal flow and turbulence will be intense [15] and intermittence is large. The maxima of T occur near the singular surfaces, where the turbulence filaments are centred and which constitute 'trapping regions'.

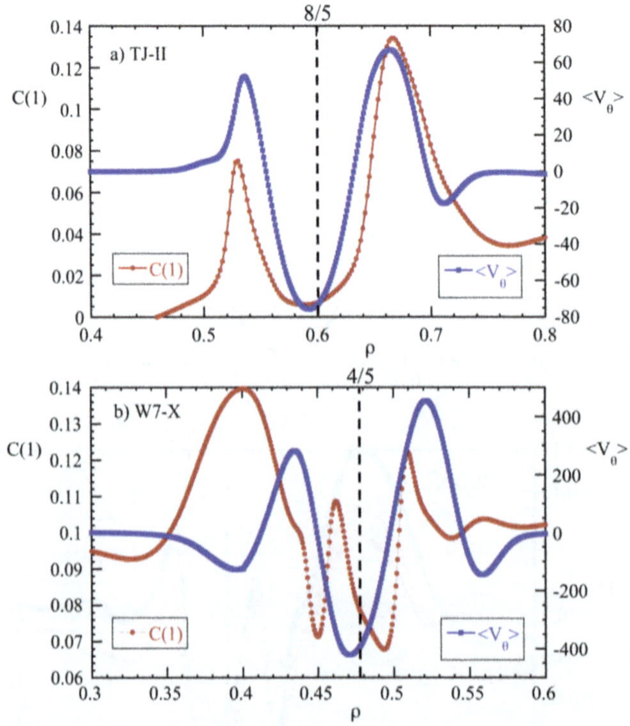

Figure 5.8. Radial dependence of the intermittence parameter and the averaged poloidal velocity $\langle V_\theta \rangle$, in the single helicity calculations for (a) TJ-II configuration 100_46 and (b) the standard W7-X configuration. (Reproduced with permission from [10]. © IOP Publishing. All rights reserved.)

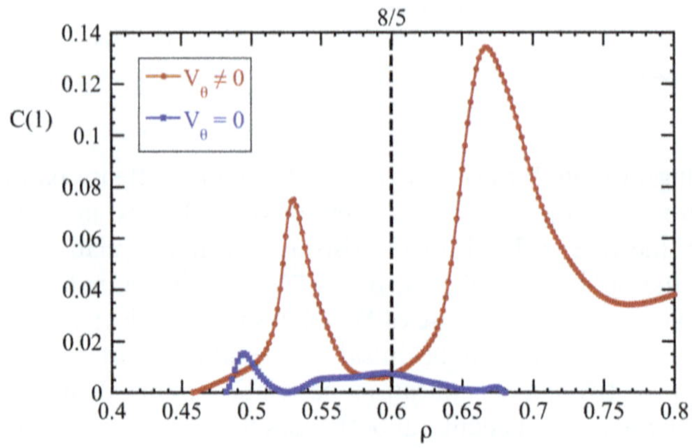

Figure 5.9. Radial dependence of the intermittence parameter in the single helicity calculations for TJ-II configuration 100_46, with and without poloidal velocity. $\langle V_\theta \rangle$ (Reproduced with permission from [10]. © IOP Publishing. All rights reserved.)

To enhance the realism of the modelling effort, we next proceed to perform multiple helicity calculations for two TJ-II configurations: 100_44 and 100_46. Figure 5.11 shows the rotational transform profiles for these configurations.

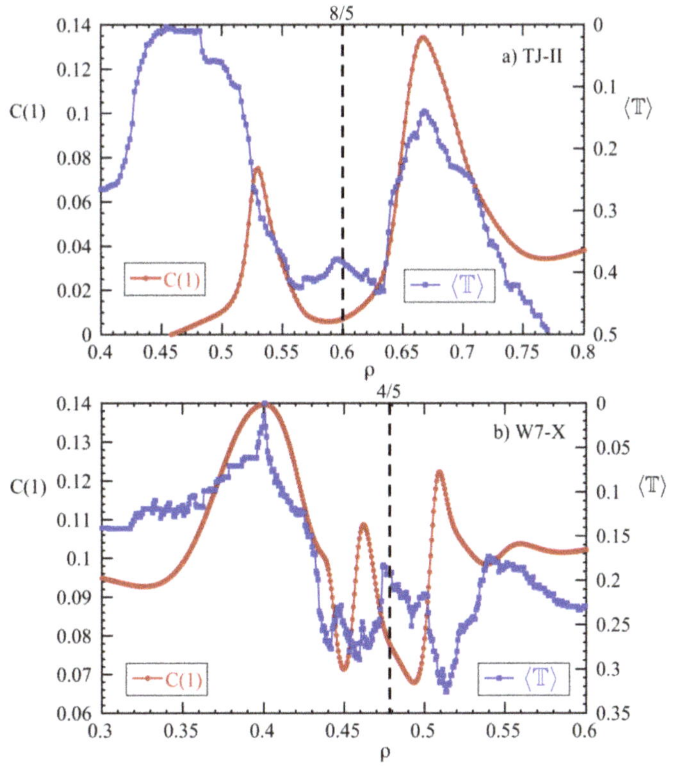

Figure 5.10. Radial dependence of the intermittence parameter in the single helicity calculations, compared to the mean transfer entropy T (note that the right vertical axis is reversed), (a) for TJ-II configuration 100_46 and (b) the standard W7-X configuration. (Reproduced with permission from [10]. © IOP Publishing. All rights reserved.)

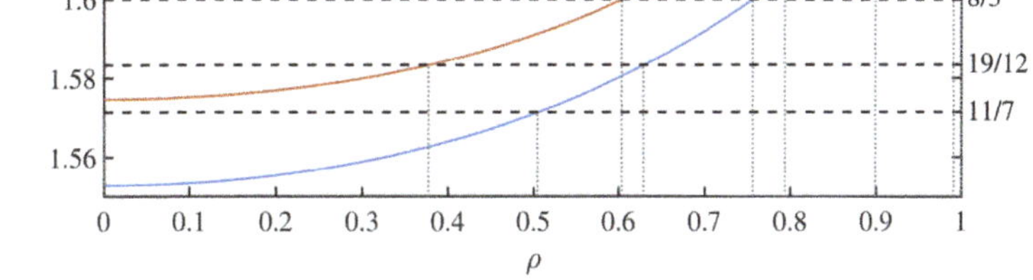

Figure 5.11. Rotational transform profiles for TJ-II configurations labelled 100_44 and 100_46. Several rational surfaces are indicated. (Reproduced with permission from [10]. © IOP Publishing. All rights reserved.)

The location of some low-order rational surfaces is indicated. The data sets produced by the calculations cover 450 radial positions at 1000 time points each.

Figure 5.12 shows *C*(1) versus normalized radius for two magnetic configurations of TJ-II. The position of some low-order rational surfaces is indicated, showing that minima of *C*(1) tend to be centred at the rational surfaces. However, in some cases the rational surface coincides with a narrow local maximum of *C*(1), which can perhaps be ascribed to the parity of the local mode.

Figure 5.13 shows the relatively clear anticorrelation between the intermittence parameter *C*(1) and the transfer entropy *T*. Note that the vertical axis of *T* is reversed to facilitate comparison. Recall that in the single helicity case, figure 5.10, the anticorrelation held mainly near the corresponding single rational surface. Here, the presence of many rational surfaces leads to the persistence of the anticorrelation over most of the radial range. Intermittence maxima occur at the position of the minor transport barriers (minima of *T*), while minima correspond to the 'trapping zones'.

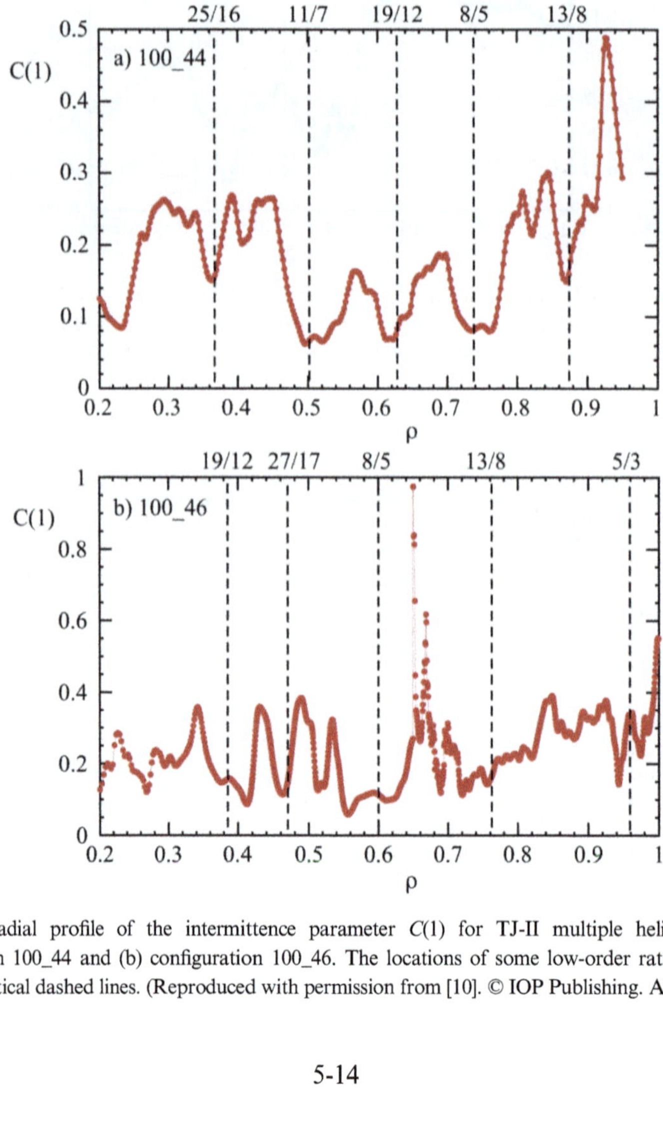

Figure 5.12. Radial profile of the intermittence parameter *C*(1) for TJ-II multiple helicity calculations: (a) configuration 100_44 and (b) configuration 100_46. The locations of some low-order rational surfaces are indicated by vertical dashed lines. (Reproduced with permission from [10]. © IOP Publishing. All rights reserved.)

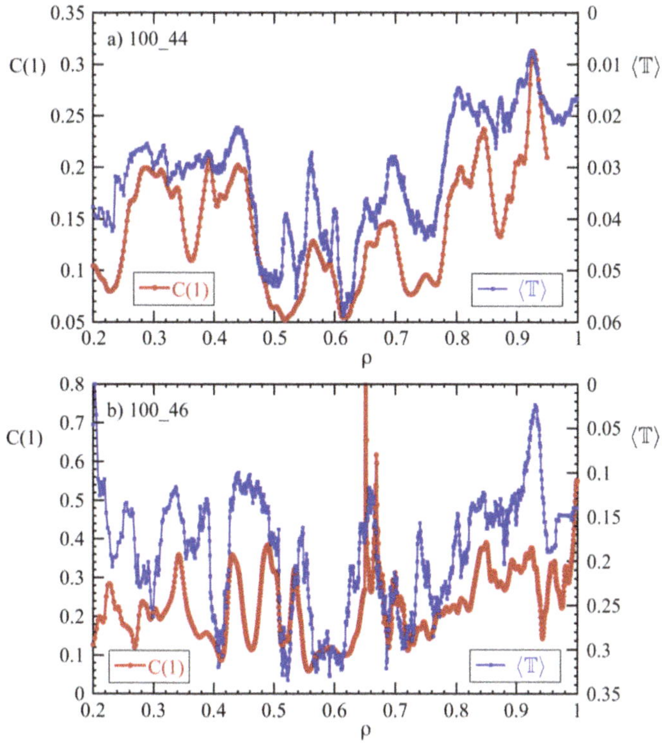

Figure 5.13. TJ-II multiple helicity calculation for (a) configuration 100_44 and (b) configuration 100_46, comparison of intermittence $C(1)$ and transfer entropy (note that the right vertical axis is reversed). (Reproduced with permission from [10].)

In summary, both the intermittence parameter $C(1)$ and the mean transfer entropy T respond to the magnetic structure underlying the turbulence in these calculations, in particular to the location of the low-order rational surfaces. There exists a strong anticorrelation between $C(1)$ and T, presumably caused by sheared flow as explained above. The maxima of $C(1)$ appear to be correlated to the location of the transport barriers.

To shed more light on this issue, we repeated the calculation of this same case without poloidal flow. In this case, intermittence is modified but not reduced to zero, see figure 5.14. It is interesting to observe that intermittence is high near the system boundary, in accordance with [16]. This is reasonable, as this is the region with relatively high magnetic shear where the low-order magnetic surfaces are closer together, so that the helicities interact more strongly. In the remainder of the radial range, switching off the poloidal flow reduces but does not lead to the disappearance of intermittence, suggesting that the mixing of different helicities is an alternative source of intermittence. Thus, we conclude that there exist at least two sources of intermittence, namely the fluctuations of the local averaged flow and the interaction between different helicities.

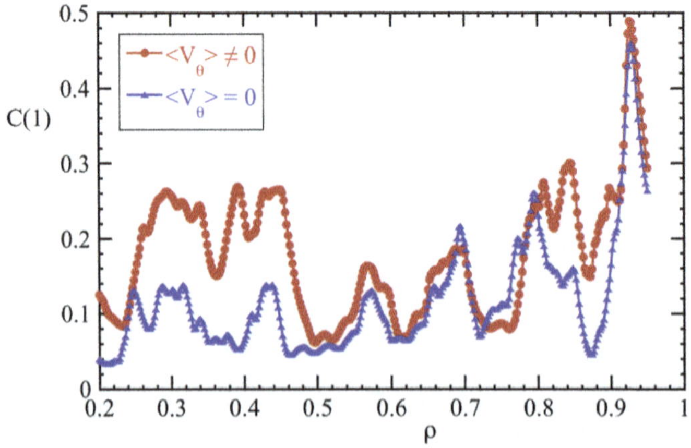

Figure 5.14. TJ-II multiple helicity (configuration 100_44), *C*(1), with and without mean poloidal flow. (Reproduced with permission from [10]. © IOP Publishing. All rights reserved.)

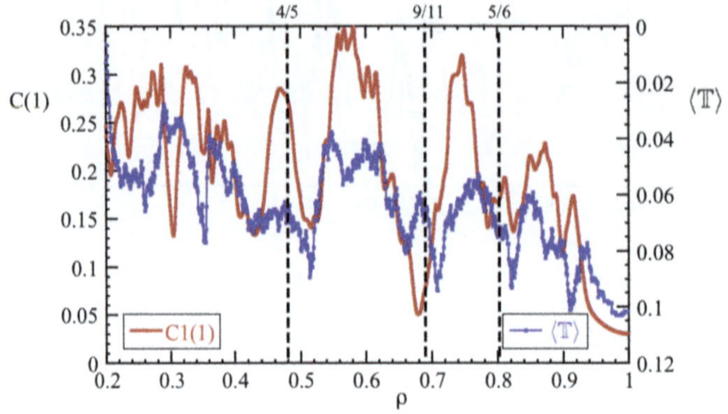

Figure 5.15. Results from calculations for W7-X showing the intermittence parameter, *C*(1), and the mean transfer entropy, *T* (note that the right vertical axis is reversed). (Reproduced with permission from [10]. © IOP Publishing. All rights reserved.)

We have performed a similar analysis for the standard OP1.1 magnetic configuration of W7-X, mentioned above. Figure 5.15 shows that the minima of *C*(1) are again correlated with the singular surfaces, although sometimes the rational surface coincides with a minimum, and sometimes with a narrow maximum within a broad minimum, indicating that the corresponding modes may have a different parity. Again, *C*(1) and *T* show a fairly clear anticorrelation. The intermittence maxima are close to the position of the minor transport barriers, while the minima correspond to the 'trapping zones'.

Figure 5.16 zooms in on the region around the 4/5 rational surface. The 4/5 trapping region extends from $\rho = 0.35$ to $\rho = 0.55$. There is a broad peak of the mean transfer entropy, *T*, in this region. In this case, the intermittence parameter *C*(1) has a broad minimum but with a peak in the middle, similar to the single helicity case.

Between $\rho = 0.37$ and 0.55, *C*(1) exhibits a broad valley with a peak near the central position at $\rho \simeq 0.475$, where the main rational surface is located. In this same

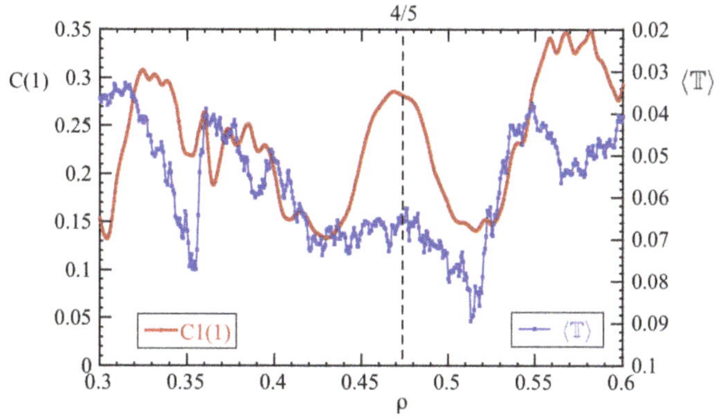

Figure 5.16. Results from calculations for W7-X showing the intermittence parameter, $C(1)$, and the mean transfer entropy, T (note that the right vertical axis is reversed). (Reproduced with permission from [10].

region, T displays a broad peak at $0.42 < \rho < 0.53$. The local maxima of T at $\rho \simeq 0.42$ and 0.53 coincide with local minima of $C(1)$. We note that this behaviour is similar to the single helicity case, see figure 5.10.

Summarizing these results, we performed simplified *nonlinear single helicity calculations*, in which we found that the intermittence parameter $C(1)$ has a minimum in the region of the turbulent vortices located at low-order rational surfaces if the mode has even parity. If the mode associated with the vortex has odd parity relative to the rational surface, the vortex moves from one side to the other of the rational surface and as a consequence narrow local maximum may appear inside the broad minimum of $C(1)$.

The intermittence parameter $C(1)$ is clearly affected by the mean poloidal flow velocity $\langle V_\theta \rangle$. $C(1)$ increases as one moves away from the rational surface, associated with an increase of $\langle V_\theta \rangle$. This relationship was also tested by running the calculations without poloidal flow, in which case the intermittence parameter dropped to very small values. We therefore conclude that in single helicity calculations, the intermittence (multifractal nature) is generated by mixing between the fractality of the time evolution of the turbulence and its fractal poloidal structure. Consequently, the intermittence measure allows separating the region of the basic vortex structure from the associated zonal flows.

The intermittence parameter is also related to the mean transfer entropy, T. In previous work, we found that maxima of T tend to correspond to 'trapping regions' for radial transport (where turbulence filaments are strong), whereas minima of T tend to correspond to 'minor transport barriers' (zonal flow regions) [17, 18]. Indeed, in the single helicity calculations, maxima of T were found to correspond to minima of $C(1)$ and vice versa. From this, we conclude that maxima of $C(1)$ tend to be correlated to the location of minor transport barriers, whereas minima of $C(1)$ typically occur at the 'trapping regions'. The latter implies that the 'trapping zones' are closer to monofractality than the surrounding plasma.

In *multiple helicity turbulence calculations*, the additional interaction between helical modes was found to also contribute to intermittence, thus leading to a more complex picture in which intermittence was generated both by the mixing between poloidal flow fluctuations and turbulence and the mutual interaction between helicities. Preliminary calculations in toroidal geometry have shown that the correspondence of minima of *C*(1) with low-order rational surfaces is maintained even in the presence of toroidal coupling.

In general, for a clear response of the intermittence parameter near rational surfaces, it is expected that several conditions must be fulfilled. First, there must be active instabilities (i.e. the turbulence drive must be active, generally meaning that the relevant thermodynamic gradients must be steep enough). Second, the main low-order rational surfaces must be spaced sufficiently far apart, meaning that the magnetic shear must be low enough. This condition is probably easier to fulfil in low-shear stellarators such as those studied here than in typical tokamaks. Third, flow shear due to physical effects unrelated to magnetic surfaces (such as neoclassical sheared flows) may distort or perturb the sheared flow layers associated with rational surfaces.

5.4 Intermittence in the W7-X stellarator

The numerical results from the preceding section provide us with the necessary basic understanding to interpret experimental intermittence results. The next step is to see what results are obtained when one calculates the intermittence for experimental data.

We first analyze data from the electron cyclotron emission (ECE) temperature diagnostics obtained during the initial operational phase of W7-X (OP1.1), in which a rotational transform scan was performed in centrally heated ECRH discharges. This experiment, performed on 9 March 2016, is described in some detail in [13, 18–21]. Figure 5.17 shows the rotational transform profiles used in the configuration scan, labelled by an 'index number'.

Here, we examine the set of high power phases (ECRH power $P_{\mathrm{ECRH}} \simeq 2.0$ MW), computing the intermittence parameter *C*(1) for each of the available ECE channels.

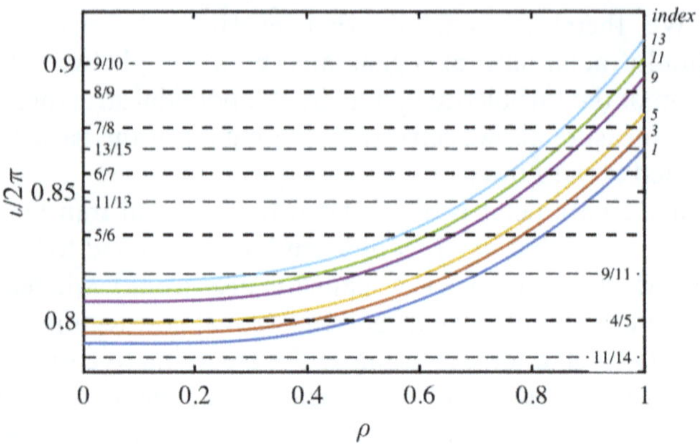

Figure 5.17. Rotational transform profiles of the W7-X configuration scan studied. The index number identifying the magnetic configurations is shown on the right. Some low-order rational surfaces are shown by means of horizontal dashed lines, labelled with the rotational transform value. (Reproduced with permission from [10]. © IOP Publishing. All rights reserved.)

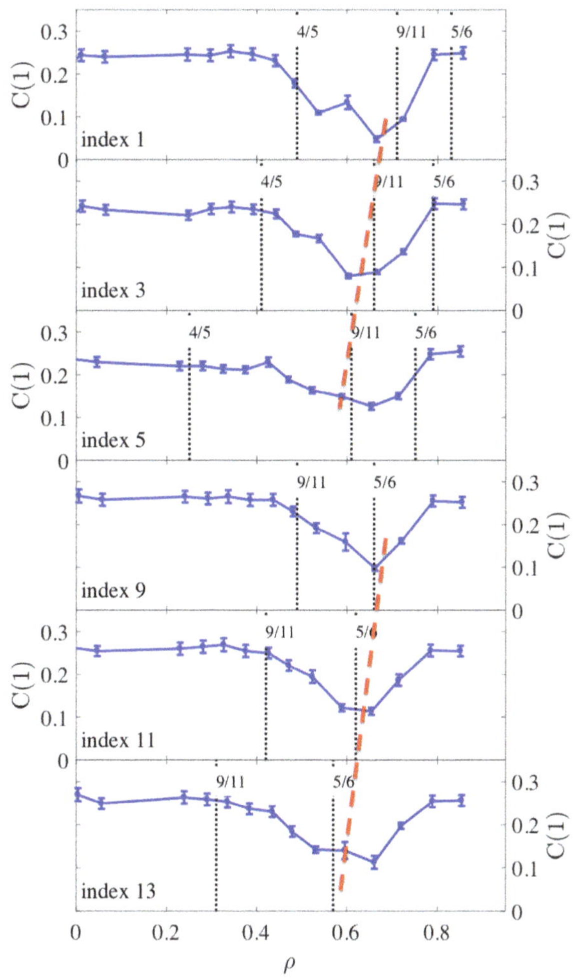

Figure 5.18. Intermittence parameter $C(1)$, calculated from ECE data in the high power phase, for discharges in the W7-X configuration scan. The approximate location of rational surfaces is indicated by vertical dashed lines, labelled with the rational value. Red dashed lines indicate the radial movement of the minima of $C(1)$. (Reproduced with permission from [10].)

$C(1)$ has clear radial minima, as shown in figure 5.18. Similar minima can also be seen in the medium and low power phases, although they are less pronounced. One observes that $C(1)$ has a systematic tendency to show a minimum in the region $0.5 < \rho < 0.7$, the gradient region. Interestingly, the minima of $C(1)$ shift radially from shot to shot in the configuration scans (as indicated by the red dashed lines) and appear to follow the location of the low-order rational surfaces, which therefore *also* affect the minima of $C(1)$. This effect is not easily explained from neoclassical theory, but matches our findings above: the presence of the rational surfaces affects the characteristics of turbulence.

The anticorrelation between the intermittence parameter and the entropy transfer that we observed in the numerical calculations is also present in these experimental results. Figure 5.19 compares the intermittence parameter, $C(1)$, of the low time scales with the mean transfer entropy, T, for two power levels (low and high) in discharge 20160319.016 (note that the right vertical axes are reversed). More information about the calculation of the transfer entropy in this discharge can be found in [18] and chapter 4. The mean transfer entropy was calculated using reference channels $0 < \rho_{\text{ref}} < 0.8$ and over time lags $0 < \tau \leqslant 0.02$ s. The anticorrelation between $C(1)$ and T is clear.

The analysis of these experimental ECE data from the W7-X stellarator thus shows that the observations made in the numerical study above also appear to hold to some degree in the experiment: in regions where instabilities are present, the parameter $C(1)$ computed from the ECE electron temperature measurements shows a minimum that appears to 'follow' low-order rational surfaces as they shift through the plasma (on a shot-by-shot basis). Likewise, the profile of the intermittence parameter $C(1)$ and the transfer entropy T show a similar anticorrelation as the calculations. We emphasize that this is the first time that relevant results on intermittence are reported for the hot interior region of a plasma.

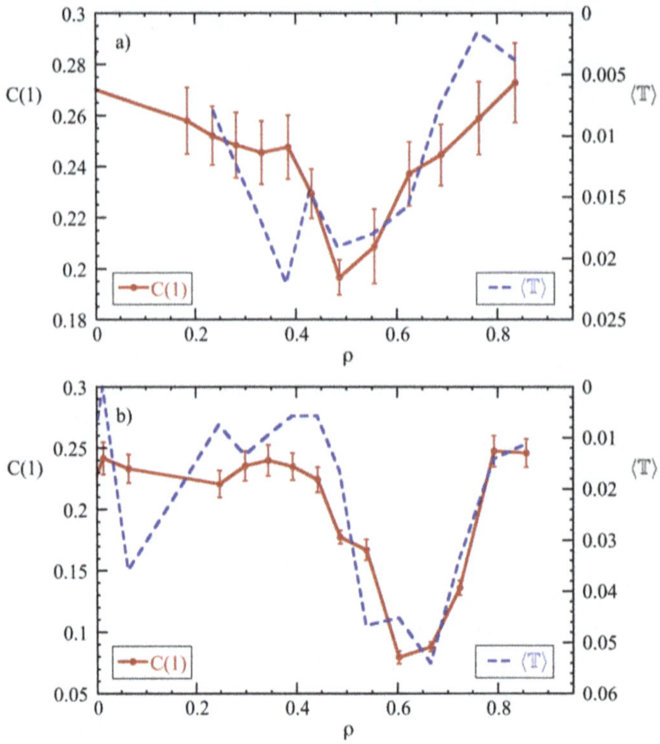

Figure 5.19. Intermittence parameter $C(1)$ and mean transfer entropy T, calculated from ECE data in the (a) low and (b) high power phases, for a discharge in the W7-X configuration scan (configuration index: 3). Note that the axis of T is reversed. (Reproduced with permission from [10]. © IOP Publishing. All rights reserved.)

5.5 Intermittence in the TJ-II plasma edge

While the experimental study at the W7-X stellarator of the preceding section was promising, the radial resolution of the ECE diagnostic did not permit detecting much detail in the radial structure. For this purpose, we returned to our familiar Langmuir probe at TJ-II, this time in experiments carefully designed to study the effect of rational surfaces by performing slow scans of the rotational transform [22].

In these experiments, the edge value of the rotational transform profile (and hence the full profile) was scanned dynamically. This extraordinary capacity is a characteristic feature of this 'flexible heliac' device [23]. The scans were performed in the course of a time interval of 150 ms during a discharge, while internal plasma currents were kept small. Using Langmuir probes placed at a fixed position in the plasma edge region, we could thus detect the impact of the varying rotational transform on the intermittence parameter calculated from potential fluctuations, by causing low-order rational surfaces to move across the probe position. By analyzing the intermittence in time intervals of 10 ms, the corresponding effective radial resolution achieved was about 3 mm, this certainly not being the lower limit of the possibilities offered by this technique.

We performed a series of dynamic configuration scans in low density ECR heated plasmas (nominal power: 2 × 250 kW) by varying the rotational transform at the edge, $\iota\!\!\!-(1) = \iota(1)/2\pi$, slowly and linearly in time (by modulating the external coil currents), while using plasma current control to keep currents inside the plasma small ($|I_p| < 0.5$ kA). Due to the low current and low plasma pressure ($\beta < 0.1\%$), Shafranov shifts of the flux surfaces are insignificant [24]. Thus, the magnetic configuration is very tightly controlled, particularly in the external part of the plasma, $\rho = r/a > 0.8$, where the influence of plasma currents on the magnetic configuration is weak [25]. Table 5.1 shows the five scans that were performed, while figure 5.20 shows the corresponding $\iota\!\!\!-$ profiles. The shape of the $\iota\!\!\!-$ profiles is very similar except for an offset.

The first four scans were made at low line average electron density ($\bar{n}_e < 0.6 \times 10^{19}$ m^{-3}), so that the plasma is in the neoclassical electron root, in which the radial electric field E_r is predominantly positive throughout the plasma [26]. The fifth scan consists of a number of discharges in which $\bar{n}_e$ was gradually raised from about 0.4 to about

Table 5.1. Main characteristics of the configuration scans discussed here. The linear scans occur between $t = 1100$ ms and $t = 1250$ ms.

Scan number	Start configuration	End configuration	Start $\iota(1)/2\pi$	End $\iota(1)/2\pi$	Nr. discharges
1	100_46_65	101_42_64	1.672	1.630	10
2	101_42_64	100_46_65	1.630	1.672	5
3	101_42_64	101_38_62	1.630	1.593	10
4	101_38_62	101_42_64	1.593	1.630	10
5	101_38_62	101_42_64	1.593	1.630	21

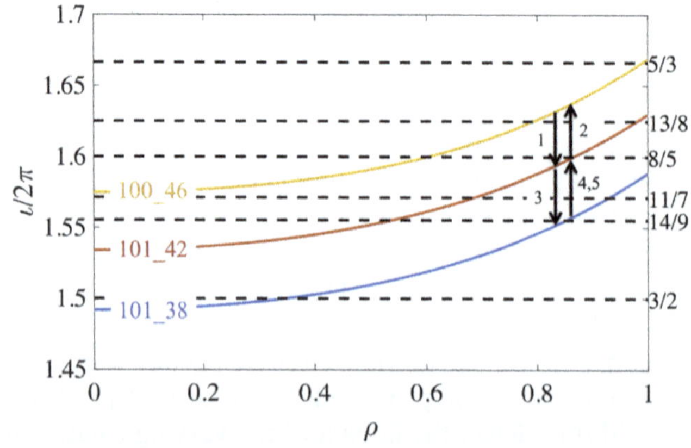

Figure 5.20. The rotational transform, $\iota/2\pi$, for various relevant configurations (labelled by the first two numbers of the configuration labels of table 5.1). Some rational values of $\iota/2\pi$ are indicated by means of horizontal dashed lines. The 5 scans are indicated symbolically with black arrows. (Reproduced with permission from [22]. © IOP Publishing. All rights reserved.)

$0.75 \times 10^{19}\ \mathrm{m}^{-3}$, crossing the critical density at around $0.6 \times 10^{19}\ \mathrm{m}^{-3}$ and passing from electron to ion root. This change in confinement state is characterized by a change of sign of the radial electric field, E_r, specifically in the edge region of the plasma.

The interpretation of the experimental results is facilitated by comparing them with calculations made using a resistive MHD turbulence model (see appendix B). The model performs turbulence simulations in a simplified cylindrical geometry, but with profiles of density, temperature, and rotational transform similar to the experimental situation. Once the numerical calculation reached steady state, we then extracted temporal data of the fluctuating plasma potential at specific spatial points, which can be thought of as 'synthetic Langmuir probes'.

Performing a rotational transform scan similar to that of the experiment is very expensive, as in that case a new turbulence simulation run must be performed for each choice of rotational transform. So, instead, we took a single numerical calculation and moved the synthetic probes radially though the plasma. One should bear in mind that this implies that the gradients of the density and temperature profiles are not constant at the probe locations, but vary according to the radial positions of the synthetic probes, which is an important difference with the experimental situation. Correspondingly, this approach prevents us from precisely reproducing the experimental values of the intermittence parameter, but helps us understand the observed radial structure of maxima and minima.

The numerical calculations were performed for configuration 100_44_64 (halfway between the endpoints of the scans of table 5.1). We ran the simulation until steady state was obtained and then continued until we obtained sufficient data in the steady state phase to apply the multifractal analysis to the various fluctuating quantities. Figure 5.21 shows the intermittence parameter for the potential fluctuations, $C(1)$, as a function of radius. The intermittence parameter shows minima at various low-order rational surfaces, consistent with results obtained in previous work [10] for

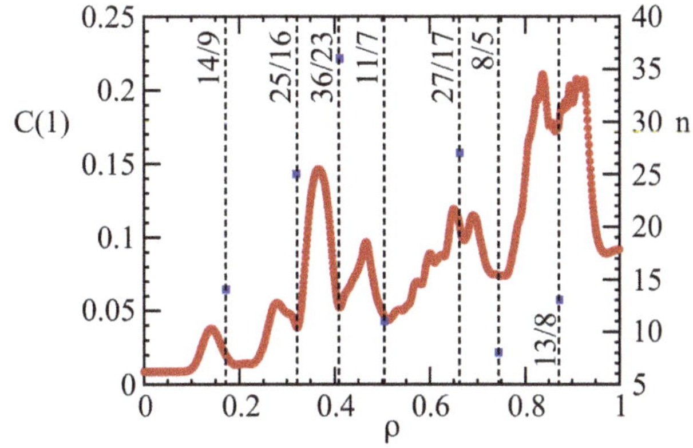

Figure 5.21. Intermittence parameter, *C*(1), calculated from turbulence simulation using the resistive MHD model. Some low-order rationals are indicated (the *n* value of the rationals *n*/*m* is indicated with blue squares, corresponding to the right axis). (Reproduced with permission from [22]. © IOP Publishing. All rights reserved.)

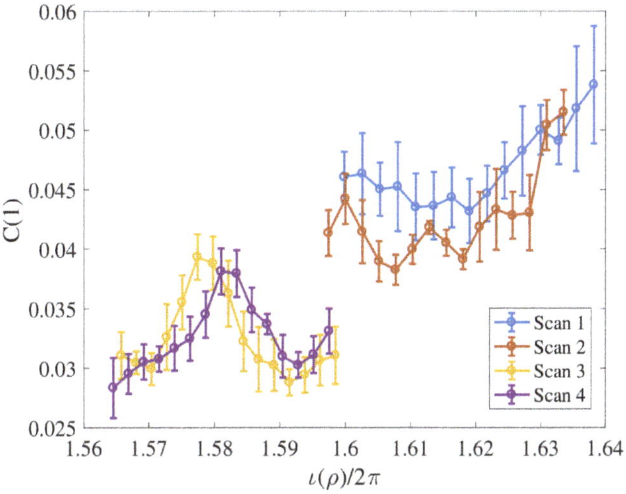

Figure 5.22. Intermittence parameter, *C*(1), calculated from a Langmuir probe signal in the experiments. The abscissa shows the estimated rotational transform at the nominal probe positions (uncorrected, see text). (Reproduced with permission from [22]. © IOP Publishing. All rights reserved.)

electron temperature fluctuations. This simulation result provides a guide for the analysis of experimental results in the following.

From the floating potential measured with one of the probes (probe D) in TJ-II, we compute the intermittence parameter *C*(1). In the following, we assume that the effect of electron temperature fluctuations is small, so that the results for the experimental floating potential may be compared to the results for the plasma potential from the simulation. Figure 5.22 shows the value of *C*(1) versus ι at the nominal position of the probe, ρ. To estimate ι at the nominal position of probe D, we use the expression $\iota(\rho) = 0.165 + 0.879\iota(1)$, deduced from figure 2 of [25].

As discussed above, we expect that the main low-order rational produces a clear minimum in the intermittence parameter $C(1)$. Considering that the uncertainty in the probe position ($\leqslant$1 cm) corresponds to an error in ι of up to about 0.01 at the probe location, it seems likely that the minimum observed to occur at $\iota = 1.592$ corresponds, in fact, to $\iota = 1.6$, implying that the nominal ι should be corrected by adding 0.008.

After applying this correction, we compare the experimental result for the intermittence parameter to the result obtained from the model, as shown in figure 5.23. This figure also indicates the location of a few important low-order rational surfaces. Although the ordinates corresponding to the two curves are different, nevertheless the agreement of the general behaviour is rather striking. The experimental and model curves for $C(1)$ do not only match regarding the main minimum at $\iota = 1.6$, but also with respect to the general trend (between $\iota = 11/7$ and 13/8) and the secondary minimum near $\iota = 21/13$. The drop of the $C(1)$ curve of the model for $\iota > 1.63$ is a boundary effect of the simulation. The difference in amplitude of the two curves can be explained from the fact that the numerical calculation does not model the full process of the changing external currents (i.e. the magnetic configuration), nor the fact that the experimental electron density is roughly constant at the fixed measurement points. In the numerical calculation, the whole plasma is kept in a fixed state, with radially varying electron density. Furthermore, the experimental data may of course be affected by noise and other instrumental effects.

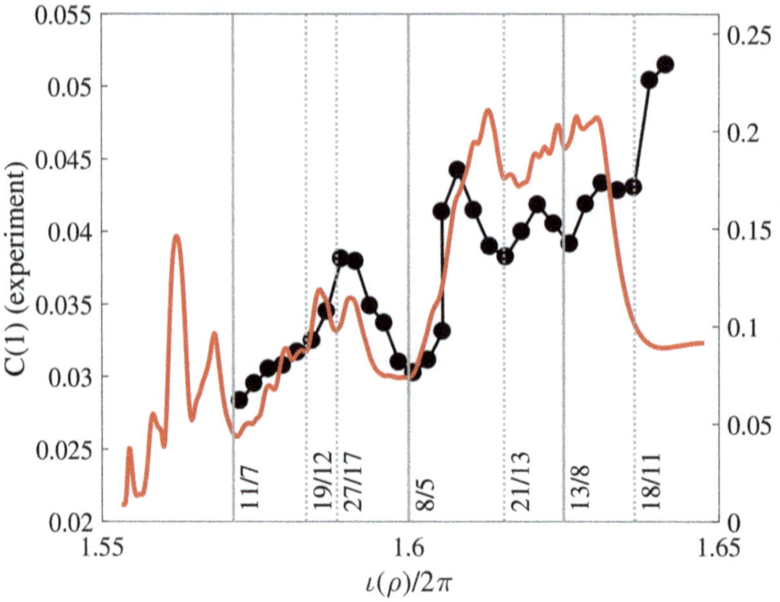

Figure 5.23. Intermittence parameter, $C(1)$, calculated from a Langmuir probe signal in the electron root rotational transform scan experiments, compared to the modelling result. The experimental data are identical to those of scans 2 and 4 shown in figure 5.22, but their rotational transform values have been corrected by the ad hoc shift of +0.008, as motivated in the text. (Reproduced with permission from [22]. © IOP Publishing. All rights reserved.)

We also applied the analyses of the preceding sections to the fifth scan of table 5.1. This scan is similar to scan 4, except that the line averaged electron density, $\bar{n}_e$, was slowly raised on a shot-by-shot basis, while kept approximately constant in each discharge. Figure 5.24 shows the variation of the line averaged density $\bar{n}_e$ and the radial electric field, E_r, in this series of shots. Here, E_r was estimated as the negative radial gradient of the floating potential at the probe, $-\nabla_r\phi_f$. The error bars indicate the standard deviation of the corresponding quantity, calculated from the variation in the relevant time window ($1100 \leqslant t \leqslant 1250$ ms). These graphs highlight the systematic density variation that was achieved in this series of shots, and the resulting gradual change in the mean radial electric field. Below and above the critical density at $\bar{n}_e \simeq 6.5 \times 10^{19}$ m^{-3}, the density was kept rather constant in time, leading to small values of the standard deviation, but near the critical density, spontaneous density variations were produced as a consequence of the fact that the

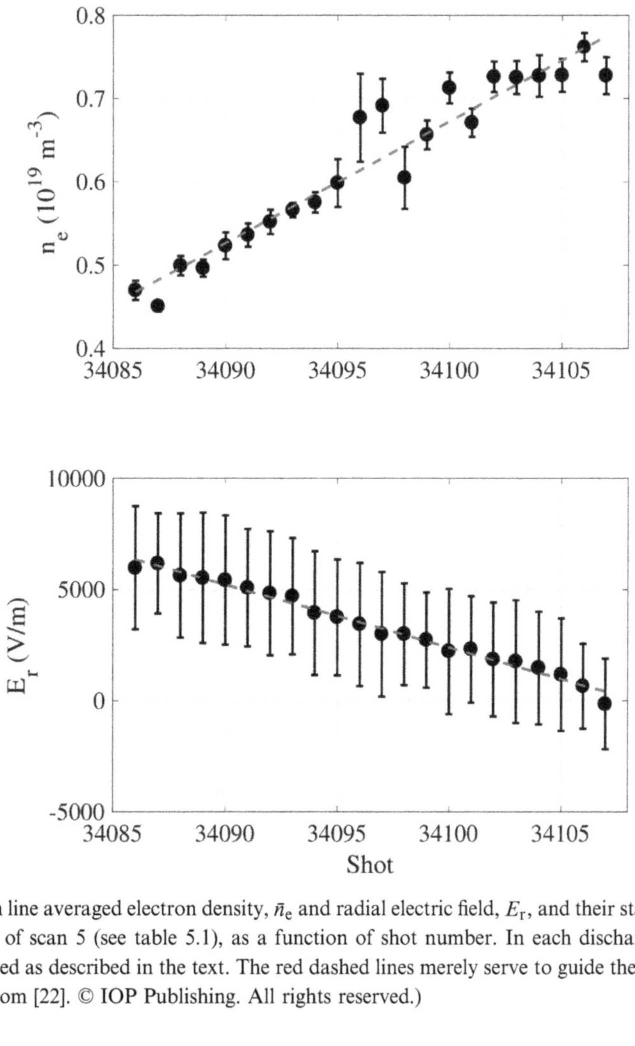

Figure 5.24. Mean line averaged electron density, $\bar{n}_e$ and radial electric field, E_r, and their standard deviations, for the discharges of scan 5 (see table 5.1), as a function of shot number. In each discharge, the rotational transform is scanned as described in the text. The red dashed lines merely serve to guide the eye. (Reproduced with permission from [22].)

plasma was close to the confinement transition (electron–ion root), leading to larger values of the standard deviation.

Figure 5.25 shows the variation of the intermittence parameter and the radial electric field with line average density. Here, the correction of ι discussed above has been applied, so that the values of ι shown correspond, to good approximation, to the local values of ι at the measurement location.

As the density increases, so does the density gradient and the turbulence drive. Nevertheless, the turbulence does not necessarily increase everywhere, as the turbulence may locally generate zonal flows that break up the turbulent eddies [27].

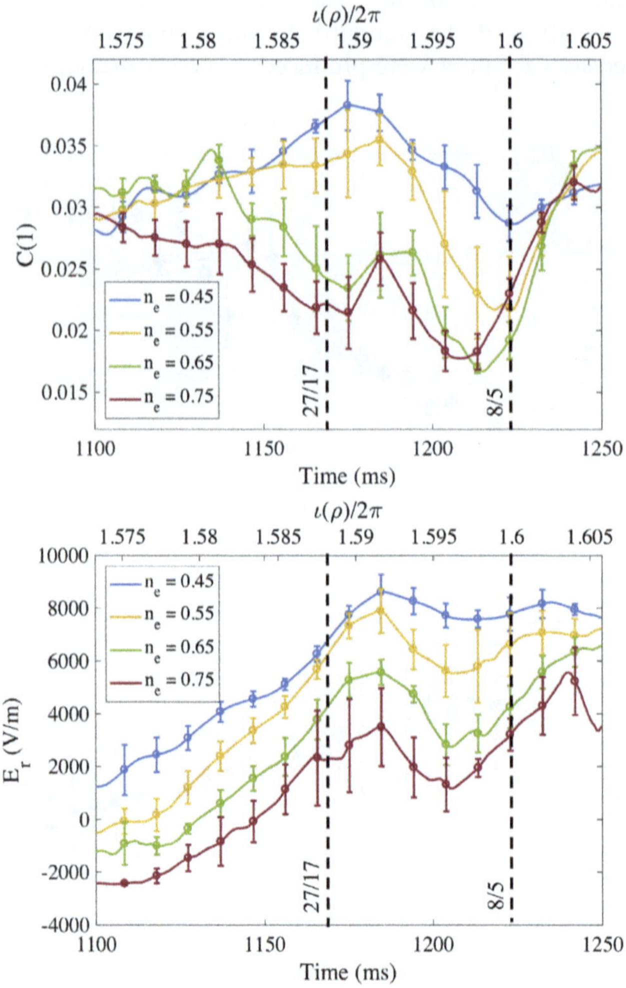

Figure 5.25. Profiles of (top) the intermittence parameter, $C(1)$, and (bottom) the radial electric field, E_r, as a function of $\bar{n}_e$ (indicated in the legends, in units of 10^{19} m^{-3}). Vertical dashed lines indicate the location of the 8/5 and 27/17 rational surfaces. Error bars indicate the experimental standard deviation in the corresponding density range (bins of width 0.1×10^{19} m^{-3}, centred around the density values indicated in the legends). (Reproduced with permission from [22]. © IOP Publishing. All rights reserved.)

We observe that the intermittence parameter gradually develops two clear local minima as the density rises. The first minimum appears close to the location of the most important low-order rational surface in this range, $\iota\!\!\!-= 8/5$. It is already present when $\bar{n}_e \simeq 0.45 \times 10^{19}$ m^{-3} and deepens as the density is raised. The second minimum, near the low-order rational surface of $\iota\!\!\!- = 27/17$, is absent at low density but becomes visible gradually for $\bar{n}_e \gtrsim 0.65 \times 10^{19}$ m^{-3}. We note also that this second minimum at 27/17 is apparent as a small dip in the theoretical curve shown in figure 5.23.

Referring to figure 5.24, one observes that the last few discharges of the density scan have nearly constant density, whereas the radial electric field, E_r, continues to vary. This offers the opportunity to study the effect of E_r on the intermittence parameter at constant $\bar{n}_e$. It should be noted that the error bars shown indicate the standard deviation of these quantities over the rotational transform scan ($1100 \leqslant t \leqslant 1250$), not the measurement error, which is much lower.

In figure 5.26 we compare the profiles of E_r and $C(1)$ between shots 34100 and 34107, having a similar value of $\bar{n}_e$, but a rather different value of E_r. At low E_r (red dashed curves), a minimum can only be seen near the rational surface 8/5, and as E_r is increased (black continuous curves), the minimum at 8/5 deepens, while a new minimum near 27/17 appears. From these figures, we deduce that an increase of E_r leads to a decrease of $C(1)$ around the corresponding low-order rational surfaces, but not elsewhere.

To study the effect of E_r in the simulation, the steady state turbulence calculations that underlie the results for the intermittence parameter shown in figure 5.21 were continued for an additional period of time in which a positive radial electric field was added. We then calculated the intermittence parameter for the potential fluctuations during this additional period. The effect of the radial electric field can be seen in figure 5.27 which shows the ratio of the curves of $C(1)$ with and without applied E_r. The figure shows that the application of positive poloidal $E \times B$ rotation reduces the intermittence parameter at the low-order rational surfaces 8/5 and 25/13, making the minima deeper, so that the system becomes more monofractal at these positions. This can be explained from the fact that the intrinsic poloidal flow generated by the turbulence in the region of these resonances is negative, so that the added positive flow reduces the net poloidal rotation. Contrariwise, the pre-existing intrinsic flow at the rational surface 13/8 is positive, so that adding a positive $E \times B$ rotation increases the flow at this location, resulting in the observed increase of $C(1)$. Thus, an imposed external flow may either reduce or increase the value of $C(1)$ near rational surfaces.

Summarizing, in a first set of experiments, performed at low electron density in 'electron root' conditions, we observed variations of $C(1)$ with $\iota\!\!\!-$, revealing minima at the locations of important low-order rational surfaces. When these results were compared to the intermittence produced by the resistive MHD turbulence calculation in similar conditions, very good agreement was obtained. The locations of five low-order rational surfaces could be identified inside the plasma, an unparalleled feat in plasma physics, to the best of our knowledge.

In a second set of experiments, we studied the variation of $C(1)$ as the line average density was raised on a shot-by-shot basis. In doing so, the TJ-II plasma performed a gradual confinement transition from 'electron' to 'ion root' and the radial electric field E_r changed sign from predominantly positive to slightly negative (see

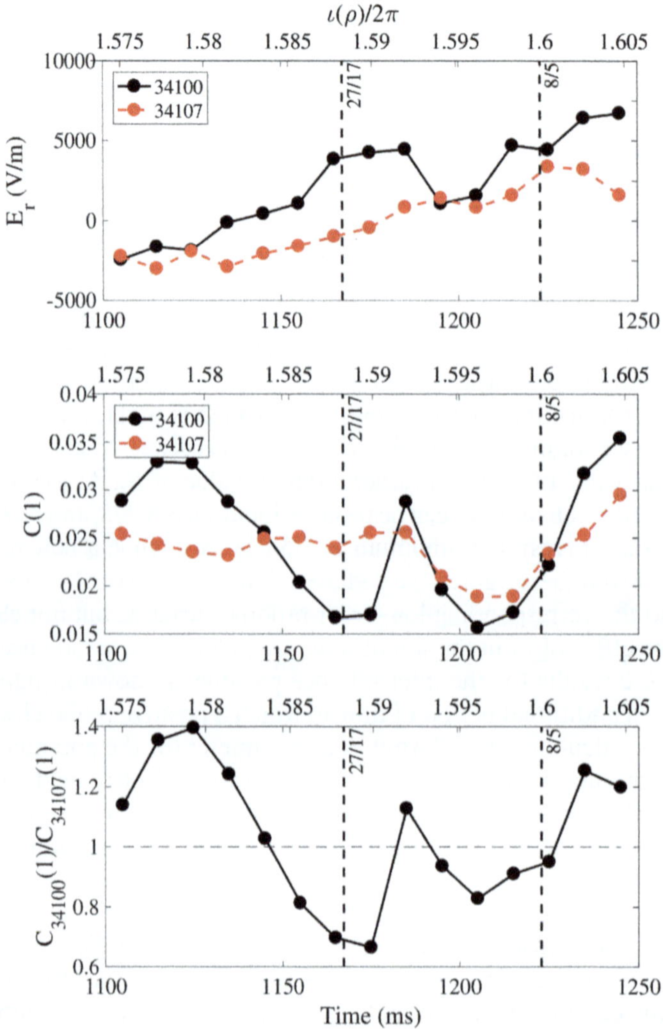

Figure 5.26. Profiles of (top) the radial electric field, E_r and (middle) the intermittence parameter, $C(1)$, for two selected discharges with a similar value of $\bar{n}_e$. Bottom: the ratio of the two curves of $C(1)$ shown, highlighting the change of intermittence due to a change in E_r only. (Reproduced with permission from [22].

figure 5.24). As the density was raised, we found that the intermittence at the rational surfaces decreased, while it increased at other locations. A possible explanation is the increased drive (increased gradients) for instabilities, which would emphasize the monofractal properties at the rational surfaces, while increasing the generation of zonal flows and the associated mixing (leading to multifractality) away from the rational surfaces. The reduction of intermittence at rational surfaces led to the detection of a new low-order rational (27/17) that had not been identified in the first set of experiments, performed at low density.

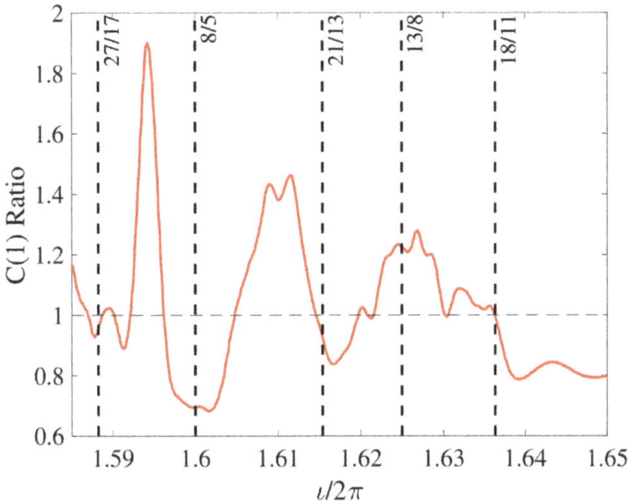

Figure 5.27. Ratio of intermittence parameter profiles after/before applying the radial electric field E_r in the resistive MHD model. (Reproduced with permission from [22]. © IOP Publishing. All rights reserved.)

A subset of discharges from the second set of experiments, at nearly constant line averaged electron density, $\bar{n}_e$, allowed determining the specific effect of the radial electric field, E_r on the intermittence parameter. We found that an increased radial electric field decreases the intermittence at the considered rational surfaces. Again, the response of the intermittence to an added electric field is fully consistent with the results of numerical calculations using the resistive MHD model.

When the turbulence drive is increased, the intermittence at a given rational surface may decrease (as increased turbulence would enhance monofractality at such locations). This is exemplified by the results shown in figure 5.25, where a new minimum corresponding to the rational 27/17 appeared when the driving gradient was raised. On the other hand, away from the low-order rationals, the increased drive can lead to an enhancement of intermittence, presumably due to an increase in the generation of zonal flows and the associated mixing (leading to multifractality).

5.6 Intermittence in the TJ-II plasma core

To study the intermittence in the plasma core region of the TJ-II stellarator, we make use of data from the heavy ion beam probe (HIBP) system located at an equivalent toroidal angle of 219° [28]. The system injects accelerated Cs^+ ions into the plasma and captures secondary Cs^{2+} ions in an energy analyzer. The energy of the secondary ions provides information about the plasma potential Φ in the sample volume of the ion beam, the beam current I_{tot} provides information about the local electron density n_e, and the toroidal displacement of the beam Z_c provides information about the poloidal current density in the plasma. Travelling through the plasma, the beam experiences attenuation, affecting I_{tot} [29].

The beam can be actively steered, so that the beam sample volume can be scanned through the plasma, from the high field side (HFS) to the low field side (LFS) and back in the corresponding poloidal cross section. The swept sample volume does not necessarily cross the magnetic axis. By performing beam tracing calculations, the position of the sampling volume can be recovered, expressed in terms of the normalized radius of the known magnetic configuration.

The beam tracing calculations also allow estimating the size of the sampling volume, which is typically around 1–2 cm poloidally and somewhat less radially. The bandwidth of the data acquisition preamplifiers is 1 MHz, sufficient to study plasma turbulence. We analyze a set of discharges in which the HIBP was scanned fast (5 ms per scan). These discharges have been analyzed in previous work, using profile data from the HIBP and fluctuations from Langmuir probes in the edge [30]. A further study was made based on information on MHD activity from Mirnov pick-up coils and poloidal flow measurements from Doppler reflectometry [31]. Thus, this set of discharges, corresponding to various magnetic configurations, has been thoroughly analyzed in the past, facilitating the interpretation of the new information presented below. Here, we focus on the spatiotemporal structure of turbulent intermittence from the HIBP data, revealing the existence of persistent structures in the core plasma. Figure 5.28 shows a typical example of HIBP data. The figure is meant to clarify the scanning operation: the bottom panel shows the radial location of the scanned sampling volume in time, while the top panel shows the measured plasma potential Φ and the middle panel the beam current $I_{\rm tot}$ (proportional to the local electron density $n_{\rm e}$).

5.6.1 Spatiotemporal mapping of turbulent structures

The bottom panel of figure 5.29 shows the time evolution of the line average electron density, $\bar{n}_{\rm e}$, in units of 10^{19} m^{-3} (as measured by the interferometer) in a particular discharge corresponding to the magnetic configuration 100_42_63. The density is initially low and is raised by means of the gas puffing system, crossing the critical density value of $\bar{n}_{\rm e} \simeq 0.6 \times 10^{19}$ m^{-3}, resulting in a transition from electron root to ion root conditions. Later, the density is lowered again, resulting in a back transition.

We calculated the intermittence parameter $C(1)$ versus time by subdividing the time trace $\Phi(t)$ in intervals of length $\Delta t = 0.1$ ms (2000 points), and estimating $C(1)$ in each interval. Each value $C(1)$ obtained is then associated with the mean time and the mean normalized radius ρ corresponding to the interval. Due to the scanning speed (see figure 5.28), the corresponding radial resolution is $\Delta\rho \simeq 0.05$.

The top panel of figure 5.29 shows the reconstructed spatiotemporal map of $C(1)$ for Φ in a discharge corresponding to the magnetic configuration labelled 100_42_63. Black slanted lines indicate the spatiotemporal trajectory of the location of the sampling volume during scanning. The 2D colourmap represents $C(1)$ after interpolation in time and space, using natural neighbour interpolation [32]. Due to the fact that the plasma is rotating while the sampling volume is scanned, the graph reveals a mixture between the poloidal and temporal structure of the turbulence, which explains part of the variability of this and the next figure.

Minima of $C(1)$ tend to occur at locations and times where relatively strong (dominant) monofractal turbulent structures are formed. The line average electron

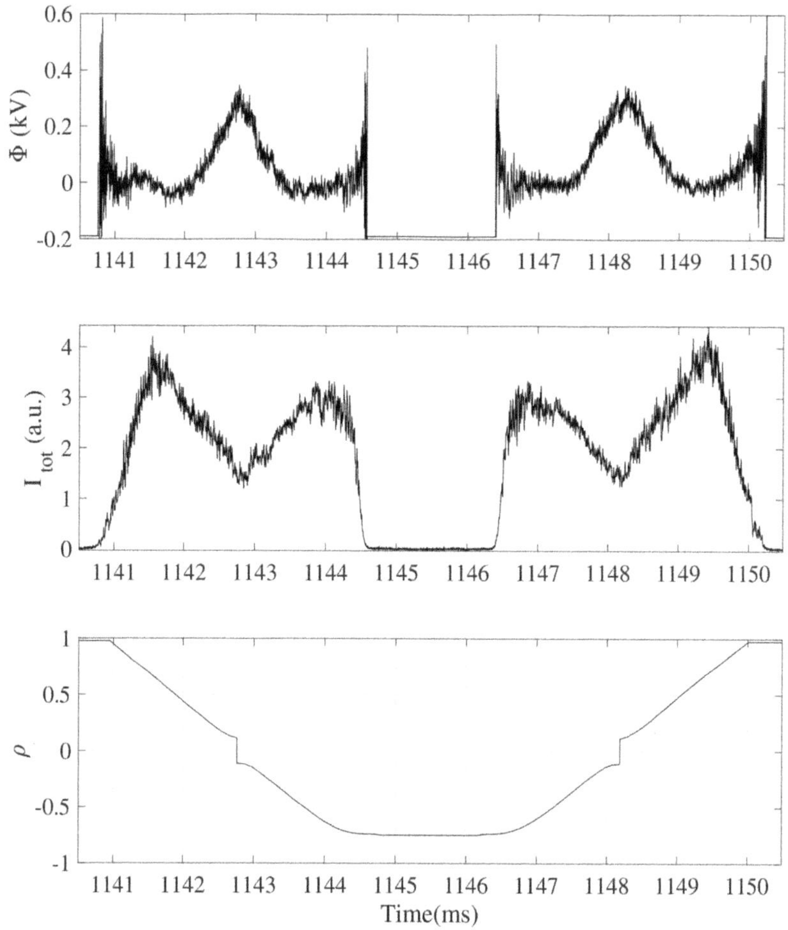

Figure 5.28. Example of data from the HIBP system during scanning operation. Top: Φ. Centre: I_{tot}. Bottom: reconstructed normalized radius, ρ. Discharge 18907. (Reproduced with permission from [12]. © IOP Publishing. All rights reserved.)

density trace in the bottom panel shows a clear increase of fluctuations at the root transition times. This has been studied to some extent in previous work [33]. The top panel shows minima (blue) at these times. Thus, this figure allows a clear identification of the radial location of the turbulence causing the fluctuations in the time trace of $\bar{n}_e$, namely, $\rho \simeq 0.6$. In previous work, the fluctuation bursts occurring at the root transitions were identified as being due to acoustic modes in the gradient region, possibly coupled to MHD modes [33]. Indeed, in view of the known vacuum iota profile, it is likely that modes associated with the 11/7 and possibly also the 14/9 rational surfaces are involved.

The radial positions of the minima of $C(1)$ do not coincide exactly with the calculated position of the main rational surfaces. Apart from the fact that there may be small reconstruction errors in the positions of the various relevant quantities, the mode structures associated with a rational surface (and its helicity) do not need to

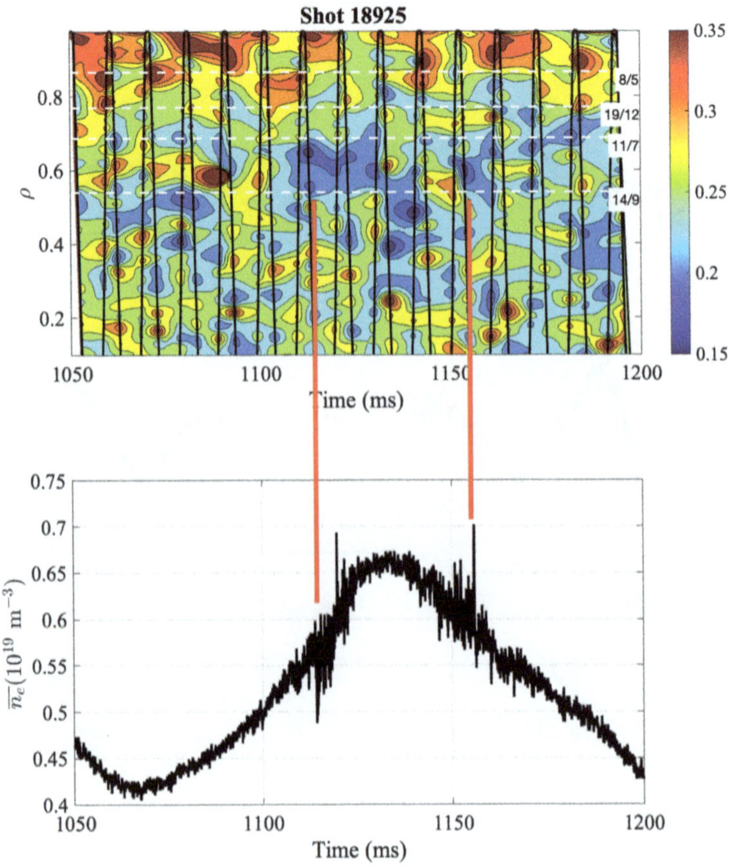

Figure 5.29. Top: spatiotemporal reconstruction of *C*(1) for Φ from an ECRH heated discharge using fast scanning. Black slanted lines indicate the spatiotemporal trajectory of the location of the sampling volume during scanning. The 2D colourmap represents *C*(1) after interpolation in time and space. Bottom: evolution of the line average density, $\bar{n}_e$. Red vertical lines are meant to guide the eye. Configuration 100_42_63. (Reproduced with permission from [12].)

coincide with the location of the rational surface itself. In addition, interactions between modes associated with different rational surfaces may lead to deformations of the mode shape.

The top panel of figure 5.30 shows the reconstructed spatiotemporal pattern of *C*(1) for Φ in a discharge corresponding to the magnetic configuration labelled 100_46_65. The description of this figure is essentially identical to that of the previous figure. Note, however, that here one observes a long-living 'trench' in the reconstructed *C*(1) map at $\rho \simeq 0.65$, for the whole duration of the ion root (high density) phase. This trench appears to be associated with the 8/5 rational surface. Interestingly, the temporal variation of mode frequency (and hence rotation speed), shown in figure 5.31, appears to match the temporal variation of the radial position of the 'trench' seen here. Thus, the variation of the magnetic frequency (first up and then down) and the radial position of the intermittence trench (first outward and

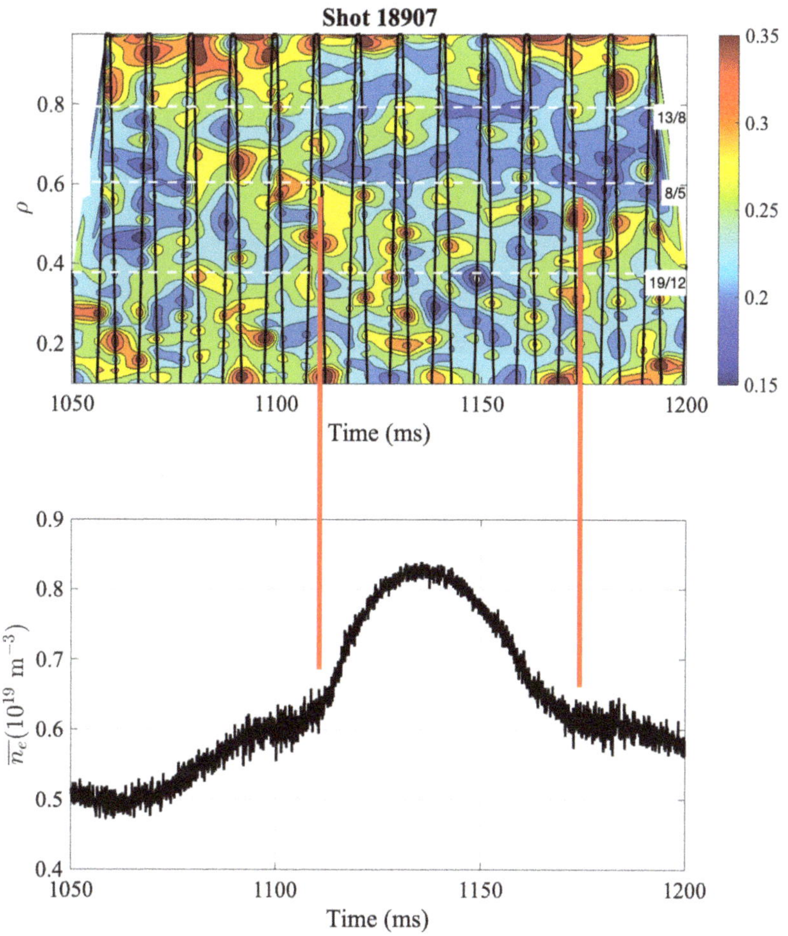

Figure 5.30. Top: spatiotemporal reconstruction of $C(1)$ for Φ from an ECRH heated discharge using fast scanning. Black slanted lines indicate the spatiotemporal trajectory of the location of the sampling volume during scanning. The 2D colourmap represents $C(1)$ after interpolation in time and space. Bottom: evolution of the line average density, $\bar{n}_e$. Red vertical lines are meant to guide the eye. Configuration 100_46_65. (Reproduced with permission from [12].)

then inward) are clearly correlated, so that the frequency change of the rotating magnetic mode can likely be explained as a change of radial position of the mode in a more or less fixed background poloidal velocity profile. In the time interval 1140–1160 ms, a second trench is visible in figure 5.30, associated with the rational 13/8. The corresponding magnetic activity has indeed been identified as corresponding to $m = 8$ in previous work [31].

In [33] we reconstructed the temporal evolution of the electron density n_e for these discharges, based on a Bayesian technique, combining diagnostic information from interferometry and reflectometry. The evolution of the electron temperature T_e could be recovered from ECE measurements. By combining this information, the evolution of the electron pressure p_e is obtained. The radial resolution offered by

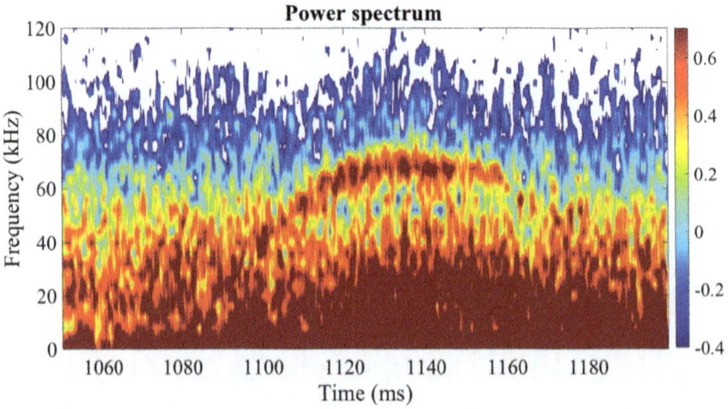

Figure 5.31. Spectrum of a Mirnov coil signal for the same discharge as the one shown in figure 5.30.

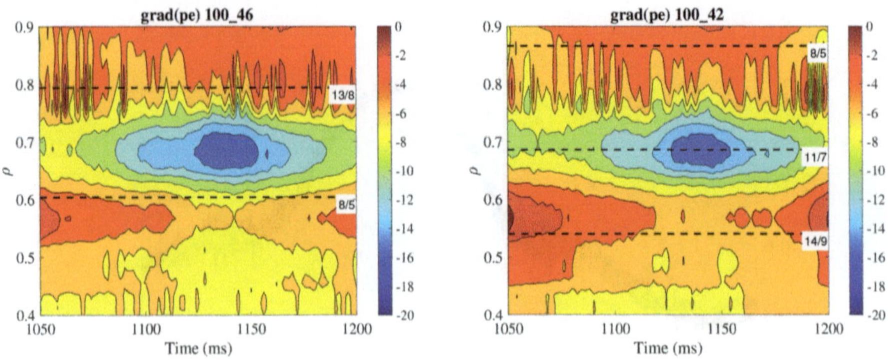

Figure 5.32. Radial pressure gradient $\nabla_r p_e$ (Pa/m), obtained by combining information on the n_e profile from interferometry and reflectometry [33] and information on the T_e profile from ECE. (Reproduced with permission from [12].)

the ECE measurements is limited (12 points along the radial direction), so that only large-scale profile structures are detectable. The pressure evolves little over the time window of interest, as any increase of n_e is largely compensated by a corresponding drop in T_e, so that the central pressure is increased only mildly in the ion root phase (by 10%–20%). Figure 5.32 shows the mean evolution of the radial pressure gradient, $\nabla_r p_e$, in the two configurations studied. For $\rho \gtrsim 0.8$, T_e is low and experimental noise is significant, so the gradient is unreliable in this region. It is rather interesting to note that the pressure gradient steepens considerably (blue colours) during the times and in the radial region where the intermittence shows persistent minima (see figures 5.29 and 5.30).

5.6.2 Topology of the turbulent structures

The rotation velocity of the plasma has been estimated in previous work, based on the frequency of magnetic oscillations and Doppler reflectometry measurements [31]. Its order of magnitude is 2000 m s^{-1} in the region of interest (see also figure 5.33 below).

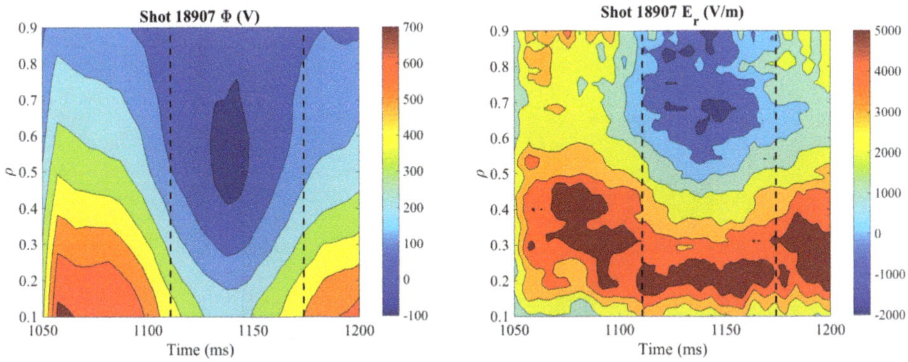

Figure 5.33. Spatiotemporal reconstruction of Φ and E_r for an ECRH heated discharge using fast scanning. Vertical dashed lines indicate the same times as the vertical lines included in figure 5.30. (Reproduced with permission from [12].)

If the poloidal plasma circumference at $\rho = 0.6$ equals $L = 2\pi\rho a \simeq 0.75$ m, the rotation frequency is typically $\omega \simeq 3$ kHz, and a poloidal rotation is completed in about 0.3 ms. Thus, given that scans are spaced by 5 ms in the fast scan experiments, the plasma typically completes various rotations between scans, and individual turbulent structures cannot be followed in time. However, if the nature of turbulence at a given location is persistent due to local characteristics (namely, the magnetic structure, i.e. MHD activity associated with a low-order rational surface), this is detectable, as shown above.

To convert the spatiotemporal map obtained in figure 5.30 of section 5.6.1 into a topological spatial map, plasma rotation has to be taken into account. As $\omega(r, t) = \dot{\theta}(r, t)$, one has

$$\theta = \int_{t_0}^{t} \omega(r, \tau) d\tau. \tag{5.8}$$

If the rotation frequency ω were constant in time, one could simplify this as $\theta = \omega(r)(t - t_0)$. However, ω is neither constant in time nor space.

Thanks to the scanning operation, the plasma potential $\Phi(r, t)$ is known [30]. One has $E_r \simeq -\nabla\Phi$ (ignoring second-order geometric effects), and, assuming rotation is dominated by the $E \times B$ term, $v_\theta = E_r/B$; finally, $\omega = v_\theta/L$, where L is the poloidal circumference, $L \simeq 2\pi\rho a$. Using this, one can therefore make a good estimate of $\omega(r, t)$, needed to perform a reconstruction of the spatial map of $C(1)$. In doing so, we will use the approximations $B = 1$ T and $a = 0.2$ m. Figure 5.33 shows the reconstructed Φ and E_r. Some spatiotemporal smoothing was applied to Φ, mainly to reduce noise in E_r.

The poloidal angle θ, resulting from equation (5.8), is shown in figure 5.34. We have selected a reference time $t_0 = 1150$ ms, whose value is not critical but should lie in the region of interest. Generally speaking, plasma rotation is extremely fast, so that θ rapidly acquires very large positive or negative values. However, due to the inversion of the radial electric field E_r, associated with the root transition, there are specific radial and temporal moments when θ remains relatively small, i.e. plasma

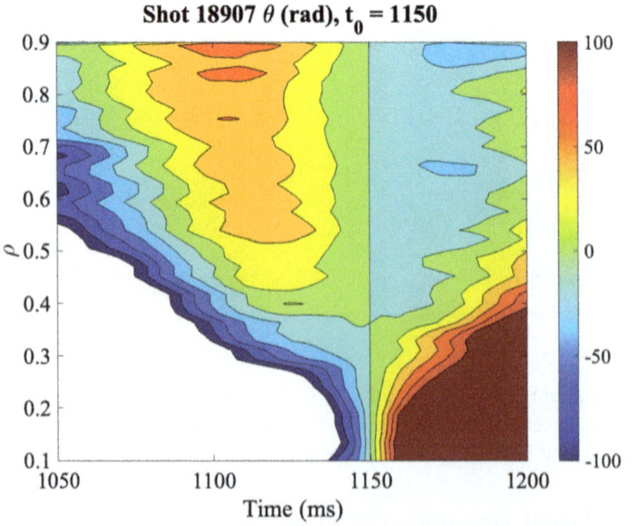

Figure 5.34. Spatiotemporal reconstruction of θ for an ECRH heated discharge using fast scanning. (Reproduced with permission from [12]. © IOP Publishing. All rights reserved.)

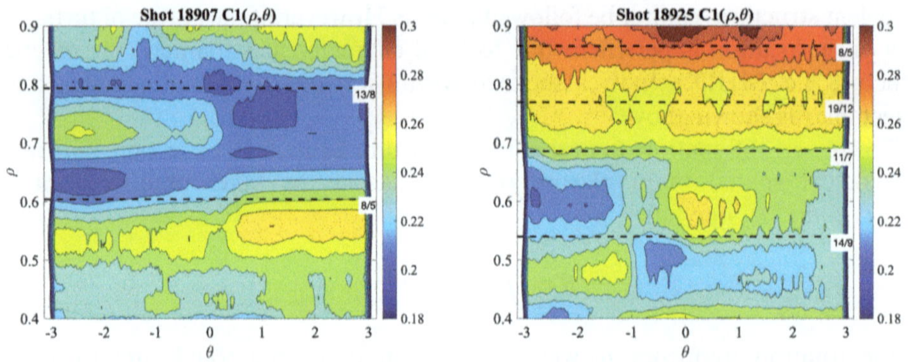

Figure 5.35. Two-dimensional spatial reconstruction of $C1(\rho, \theta)$ for an ECRH heated discharge using fast scanning. Left: configuration '100_46_65'; $t_0 = 1150$ ms (discharge 18907). Right: configuration '100_42_63'; $t_0 = 1145$ ms (discharge 18925). (Reproduced with permission from [12]. © IOP Publishing. All rights reserved.)

rotation and E_r are small. By selecting only data from regions where $-\pi \leqslant \theta \leqslant \pi$, we can thus recover the radial–poloidal mapping of $C(1)$, using the mapping $(\rho, t) \to (\rho, \theta)$, to convert the spatiotemporal map $C1(\rho, t)$ (see figure 5.30) to a topological map $C1(\rho, \theta)$.

The result is shown in figure 5.35 (some smoothing has been applied). Due to the fact that only small values of θ have been used for this reconstruction, the poloidal resolution is low. In the case of discharge 18907 (left), the modes associated with the 8/5 and 13/8 rational surfaces resonate, as the minima of $C(1)$ tend to line up in terms of poloidal angle. This is consistent with the bicoherence analysis of [31]. Also in the case of discharge 18925 (right), obtained in a similar manner, the relative importance of the radial minima of $C(1)$ near important low-order rational surfaces (here: 14/9) is clear. In this case, the mode appears to be radially asymmetric (odd) with respect to the rational surface. Something similar could be said, although less clearly, for the modes associated with the 8/5 and 13/8 modes in discharge 18907; this could be due to a certain degree of mixing between odd and even mode components.

5.6.3 Comparison with model calculations

We have run the R-MHD model, briefly described in appendix B, for the two TJ-II magnetic configurations considered here, 100_42_64 and 100_46_64. The model provides simulated fluctuations of the potential Φ on a very fast time scale, of the order of a few tens of microseconds. Since the poloidal rotation of the plasma is slow compared to this time scale, it is possible to sample the simulated fluctuations at various small intervals of the poloidal angle θ, thus obtaining simulated measurements of Φ at several poloidal positions and radii. We then apply the procedure to estimate the intermittence to those simulated measurements $\Phi(\rho, \theta, t)$ to obtain $C(1)$ at different radial and poloidal locations.

Next, it is possible to compare the resulting radial profiles of $C(1)$ at several poloidal locations to the radial profiles obtained in the experiment. This procedure is necessary as the correspondence between the poloidal angle of the cylindrical model and the experimental poloidal angle is unclear.

Examples of such comparisons are shown in figure 5.36. In a broad range of ρ (from about 0.4 to about 0.8), we find a high degree of coincidence between (a) the positions of low-order rational surfaces, (b) minima in the experimental profiles of $C(1)$, and (c) minima in the profiles of $C(1)$ calculated from the numerical model.

For $\rho > 0.85$, $C(1)$ has a tendency to rise sharply. The causes for this observation are not fully known, but could be related to the fact that the rotational transform profile steepens in the edge region, implying a larger density of (interacting) rational surfaces, leading to an increase of intermittence. Edge radial electric fields may also play a part [22].

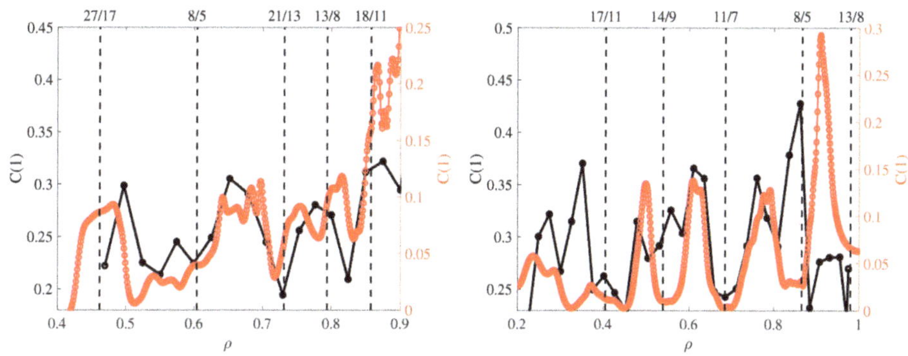

Figure 5.36. Experimental profiles of intermittence (black dots), compared to profiles from the numerical model (red dots). Left: configuration 100_46_65; right: configuration 100_42_63. (Reproduced with permission from [12].)

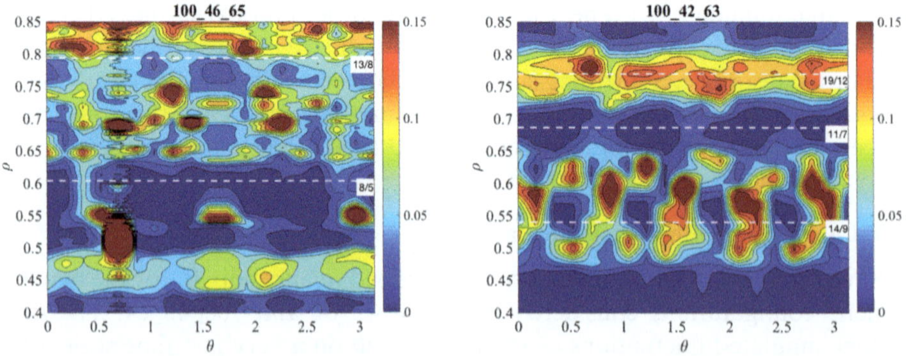

Figure 5.37. Radial–poloidal intermittence map obtained from the R-MHD model. Left: configuration 100_46_65. Right: configuration 100_42_63. (Reproduced with permission from [12].)

As the model yields $C(1)$ as a function of ρ and θ, it is possible to plot the 2D turbulent structures for these two configurations, similar to the reconstructions made in section 5.6.2. Figure 5.37 shows the 2D plots of $C(1)$ as a function of the radial and poloidal coordinates for the two discharges, corresponding to the configurations 100_46_65 and 100_42_63. The vortex structures correspond to minima of $C(1)$, the dark blue regions. One observes that these structures are either at the rational surfaces or nearby, depending on the symmetry of the components of the main helicities with respect of the rational radial position. At a given rational surface, the number of minima of $C(1)$ is consistent with the m value associated with the surface.

As mentioned above, these numerical results are not a detailed simulation of the plasma, so no direct comparison is possible with experiment, which in any case would be difficult due to the temporal variability of turbulence. As noted, the experimental reconstructions shown in section 5.6.2 do not offer sufficient poloidal resolution to properly resolve the poloidal structure.

5.6.4 Summary

This section presents the first estimations ever of the degree of intermittence of the plasma potential Φ, calculated from experimental data obtained over most of the plasma radius by a HIBP system in the TJ-II stellarator [12].

It is shown that core potential fluctuations possess a property (intermittence) that varies in a significant way with plasma conditions (electron or ion root, and the transition between the two). The formation of radial minima of $C(1)$ in relation to the electron–ion root transition is observed. The persistence of the minima of $C(1)$ over various successive scans of the HIBP system attests to their physical significance and reproducibility.

The minima are clearly associated with low-order rational surfaces. In figure 5.30, a 'trench' of low values of $C(1)$ is observed, close to the expected location of the important 8/5 rational surface in the ion root phase. Comparing this with the analysis of magnetic activity and mode numbers m performed in previous work using Mirnov coils [31], we conclude that this trench is without any doubt related to the 8/5 mode. Moreover, the temporal variation of the mode frequency detected using the Mirnov coils matches the temporal variation of the radial location of the intermittence 'trench'. This suggests that during the ion root phase, with continuously varying electron density, the mode location moves radially in accordance with the varying density, and therefore experiences a varying poloidal rotation velocity (see the velocity profiles deduced in [31]), which provides an explanation for the varying mode frequency. This sequence of events provides definitive experimental proof of the association between intermittence minima and mode location.

Further evidence is obtained by estimating the electron pressure gradient, $\nabla_r p_e$, from experimental profile measurements. Figure 5.32 shows how the pressure gradient steepens considerably at the times and radial locations where the intermittence displays minimum values. It is therefore reasonable to assume that these minima are associated with pressure gradient driven modes. In the future, this type of study might even allow the experimental determination of the critical pressure gradient to trigger these modes, if any.

From the measured low frequency (background) evolution of the potential, $\Phi(\rho, t)$, we could recover the radial electric field $E_r(\rho, t)$ and hence the poloidal rotation velocity $v_\theta(\rho, t)$ (assuming that the $E \times B$ velocity dominates plasma rotation). Using this knowledge, we could convert the spatiotemporal intermittence into an approximate topological (radial–poloidal) map of intermittence. This unique reconstruction further clarified the topological structure of turbulence.

Results obtained from an R-MHD model were compared to the experimental results. It was found there is good agreement of the experimental and model minima of $C(1)$ over a large part of the plasma radius, and that these minima tend to align with major low-order rationals. The topological reconstruction of $C(1)$ reflected the expected poloidal structure of MHD modes, radially associated with the corresponding rational and with the corresponding poloidal periodicity. Comparing with the experimental topological reconstructions, we note that the experimental radial structure was roughly as expected, whereas the experimental poloidal resolution was insufficient to resolve the periodicity.

In conclusion, near low-order rational surfaces, modes with the corresponding helicity may dominate fluctuations if they are driven (e.g. due to a locally steep pressure gradient), leading to a reduction of the degree of multifractality and hence the intermittence parameter [10]. Low-order modes tend to dominate over higher order modes, hence the intermittence parameter tends to decrease mainly near low-order rational surfaces. Such modes may facilitate the formation of turbulent vortices even in the absence of fully-formed magnetic islands. This may lead to 'trapping' and therefore affect radial particle and heat transport [14]. Through this mechanism, low-order rational surfaces are expected to affect transport.

5.7 Intermittence and causality during confinement transitions at TJ-II

Following the analysis of intermittence in the plasma interior described in the preceding section (section 5.6), we return to the Langmuir probe systems in the edge region in the same set of discharges. We will combine the study of intermittence with the analysis of causality presented in chapter 4 and thus obtain a rather complete picture of the complex evolution of the plasma across the electron to ion root transition.

The Langmuir probe systems deliver the usual quantities. The difference of the two V_f measurements of the triple probe, aligned poloidally, divided by their distance d_θ, provides an estimate of the poloidal electric field, E_θ, at the same location as the I_{sat} measurement [34]. Likewise, the radial electric field E_r can be estimated from the difference of two V_f signals divided by their radial distance d_r using two radially spaced probe pins.

The fluctuating particle flux Γ can be estimated from $\Gamma = \langle \tilde{n}\tilde{E}_\theta \rangle / B$. The ion saturation current I_{sat} is assumed proportional to the fluctuating density $\tilde{n}$. Thus, the fluctuating particle flux is proportional to the fluctuation amplitudes or root-mean-square values RMS (I_{sat}) and RMS (E_θ), as well as the cosine of the phase angle $\Delta\phi$ between these two quantities, $\cos(\Delta\phi)$ [35]. Thus, the fluctuation amplitudes and the phase angle play a prime role in the understanding of turbulent radial (particle) transport.

The probe data also allow estimating the phase velocity v_θ using the two-point correlation technique [36] applied to the mentioned two poloidally aligned V_f signals.

The phase angle is estimated using two alternative techniques. The complex coherence $\gamma(f)$ of two signals $x(t)$ and $y(t)$ over a time window $T = \{t_0 \leqslant t < t_1\}$ is calculated by subdividing this time window in a small number N of subwindows, calculating the complex Fourier transforms $X(f)$ and $Y(f)$ for each subwindow, and evaluating $\gamma = \langle XY^* \rangle_N / \sqrt{\langle |X|^2 \rangle_N \langle |Y|^2 \rangle_N}$. Here, $\langle \cdot \rangle_N$ denotes an average over the N subwindows. The frequency-averaged cross phase $\Delta\phi_{coh}$ of the coherence is estimated over a range of frequencies using $\exp(i\Delta\phi_{coh}) = \sum_f \gamma / \sum_f |\gamma|$.

Alternatively, the Hilbert cross phase is calculated for a time window T by calculating the fluctuating part of the signals, $\tilde{x}(t) = x(t) - \langle x \rangle_T$, where $\langle \cdot \rangle_T$ is the average over T, and similar for y. Then, the Hilbert transform is applied to $\tilde{x}$ and $\tilde{y}$, yielding complex signals $\hat{x}(t)$ and $\hat{y}(t)$. The Hilbert cross phase $\Delta\phi_H$ is defined from $\exp(i\Delta\phi_H) = \langle \hat{x}\hat{y}^* \rangle_T / \sqrt{\langle |\hat{x}|^2 \rangle_T \langle |\hat{y}|^2 \rangle_T}$. Figure 5.38 shows some typical time traces for the discharges analyzed here. Apart from the top time trace (a), all data are obtained from the D probe system, located inside the plasma at $\rho \simeq 0.92$. The transition occurs roughly when the line average electron density $\bar{n}_e$, figure 5.38(a), crosses the value $\sim 0.6 \times 10^{19}$ m^{-3}. The root-mean-square fluctuation amplitude of the density time trace increases during the transition, as reported in previous work [33]. Figure 5.38(b) shows that the smoothed floating potential (averaged over two poloidally spaced V_f probe tips) changes sign at the transition, as does the poloidal velocity v_θ shown in figure 5.38(c). Figure 5.38(c) also shows the smoothed radial electric field E_r. Assuming the poloidal velocity is dominated by the $E \times B$ flow, and considering that the TJ-II magnetic field $B \simeq 1$ T, the numerical values of E_r and v_θ should be very close, and so they are, except near the transition, where E_r appears to be affected by a strong zonal flow contribution. A zonal flow structure is poloidally and toroidally symmetric, and therefore would not affect v_θ, which is derived from two poloidally displaced probe pins. This interpretation is reinforced by the fact that

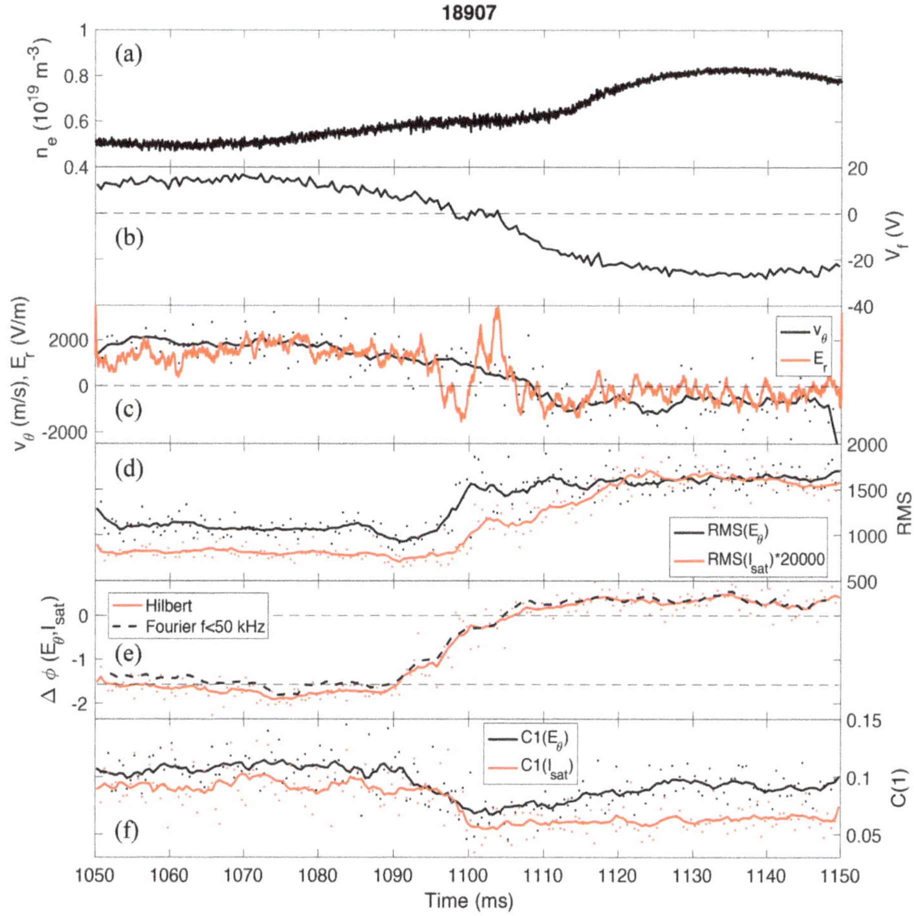

Figure 5.38. Time traces of some relevant quantities across the e–root to i-root transition (configuration '100_46_65'). The Langmuir probe position is $\rho \simeq 0.92$. Top to bottom: (a) The line integrated electron density $\bar{n}_e$ (10^{19} m^{-3}). (b) Running average of the floating potential V_f. (c) Poloidal velocity, computed from the two-point correlation technique using two poloidally spaced V_f pins and radial electric field calculated from two radially spaced V_f pins. (d) The RMS variation of E_θ (black) and I_{sat} (red) from the triple probe. (e) Cross phase between E_θ and I_{sat} calculated using two methods, as indicated in the legend. Horizontal dashed lines indicate multiples of $\pi/2$. (f) Intermittence of E_θ (black) and I_{sat} (red). In (c)-(f), dots are calculated values, the same-coloured line is a smoothed curve through the dots. (Reproduced from [38], with the permission of AIP Publishing).

long range correlations [37] between the remote B and D probes have been detected around this time [30], another hallmark of zonal flows.

Prior to the transition, the turbulence amplitude as quantified by the RMS values of E_θ and I_{sat} (shown in figure 5.38(d)) is roughly constant, but at t = 1090 ms it increases significantly. Figure 5.38(e) shows that the cross phase between E_θ and I_{sat}, calculated using the two methods described above, changes gradually from about $-\pi/2$ for t < 1090 ms to ~0 for t > 1110 ms. Figure 5.38(f) shows that the intermittence parameter C_1 for these two quantities is roughly constant for t < 1090 ms, then drops significantly at $t \simeq 1100$ ms.

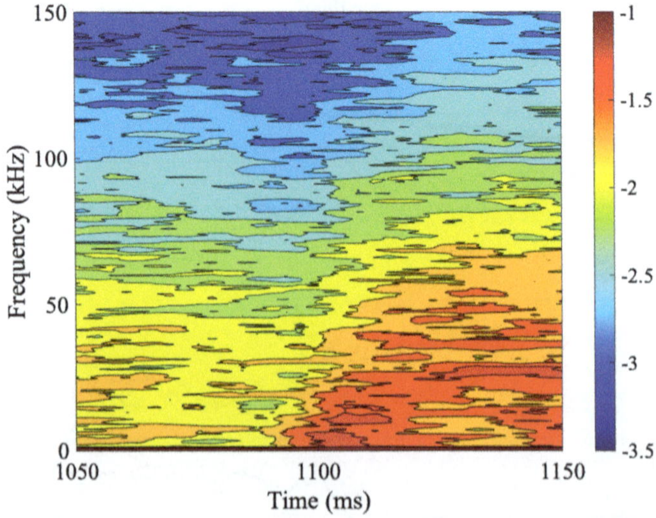

Figure 5.39. Spectrogram of E_θ for discharge 18907 (see figure 5.38). The colour bar specifies the value of $\log_{10}(S^2)$, where S^2 is the spectral power. (Reproduced from [38], with the permission of AIP Publishing.)

Figure 5.39 shows the spectrogram of E_θ for the discharge shown in figure 5.38. It shows an increase of low frequency activity following the transition, consistent with the increase of RMS (E_θ). The very low frequency activity ($f < 10$ kHz) is particularly intense around the time of the transition and is presumed to be related to the mentioned zonal flow. The later, higher frequency activity ($20 < f < 30$ kHz) is known to be related to a rotating MHD mode, as reported in [31] on the basis of Mirnov coil analysis.

In this discharge, the B probe was located at approximately the same radial position ($\rho \simeq 0.9$) as the D probe. One of the pins of the B probe was set up to measure T_e using the fast swept probe technique [39]. Results are shown in figure 5.40. It is observed that the electron temperature increases slightly following the transition, but can be discarded as the cause of the observed phase changes.

Probe B was moved radially on a shot-to-shot basis. This allows obtaining not only the temporal, but also the spatial variation of many quantities. To compare discharges with slightly varying transition times, we defined a critical time $t_{\rm crit}$ as the time at which v_θ changes sign, and then displayed the quantities as a function of $t - t_{\rm crit}$.

Results are shown in figure 5.41. The vertical axis shows the position of probe B in each shot (in terms of the normalized minor radius, ρ). Colours indicate the value of the quantity shown above each panel. Figures 5.41(a) and (b) show the evolution of the RMS fluctuation amplitude of $I_{\rm sat}$ and E_θ. Both increase significantly following the transition, more strongly so in the inner half ($\rho \lesssim 0.93$) of the examined region. Figure 5.41(c) and (d) show the intermittence of $I_{\rm sat}$ and E_θ. It decreases after the transition, mainly in the same region where the fluctuation amplitude increases. Figure 5.41(e) shows the cross phase between E_θ and $I_{\rm sat}$, calculated using the coherence technique (integrating over frequencies <50 kHz).

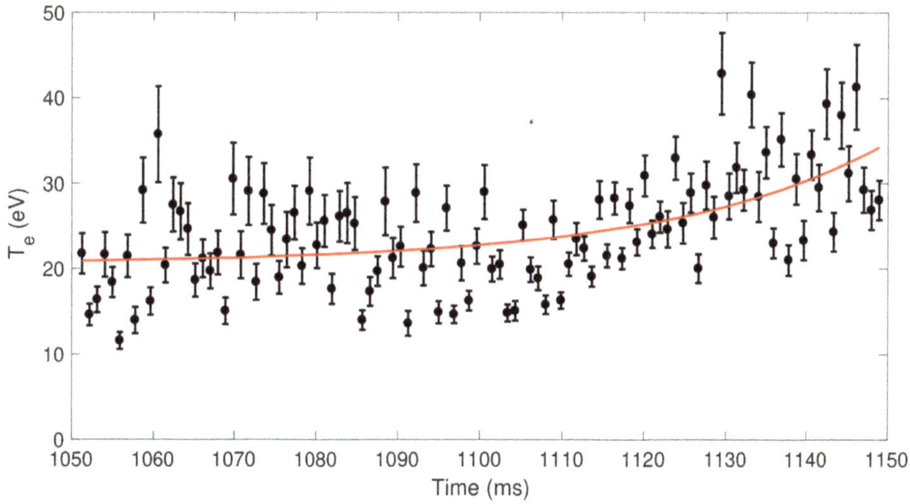

Figure 5.40. Electron temperature evolution at $\rho \simeq 0.9$ for the same discharge as figure 5.38, computed using the swept probe technique. Error bars correspond to the 1σ confidence level of the $I_{sat}(V_s)$ fit (see text). (Reproduced from [38], with the permission of AIP Publishing.)

At small values of $\rho \lesssim 0.91$, the phase increases from around $-\pi/2$ to ~ 0 or less across the transition. At larger values of ρ, the phase increases even more, similar to figure 5.38 (probe D), and reaches values $>\pi/2$ at the outermost positions. Black contours are drawn at multiples of $\pi/2$ to facilitate understanding. Before the transition, the radial phase profile is almost flat, but after the transition strong radial variation exists, such that the cross phase is maximal near $\rho \simeq 0.96$ and minimal near $\rho \simeq 0.88$. In the region of the transport barrier ($\rho \simeq 0.95$), the temporal variation of the RMS is only mild, whereas the temporal variation of the cross phase is very significant. Figure 5.41(f) shows the poloidal phase velocity, calculated using the two-point correlation technique. It evolves from positive to negative across the transition, similar to figure 5.38, although values differ according to radius. The configuration of this probe did not allow calculating E_r.

From previous work, it is known that the line average electron density $\bar{n}_e$ plays the role of a control parameter for spontaneous confinement transitions at TJ-II [40]. Figure 5.42 shows the transfer entropy (TE) between this parameter and various relevant quantities at different time lags (horizontal axis). It should be remembered that the vertical axis corresponds to the probe position, which was varied on a shot-to-shot basis.

Several things are immediately clear: (i) The TE values of the graphs $\bar{n}_e \rightarrow X$ are generally much larger than those of the graphs $X \rightarrow \bar{n}_e$, indicating that information flows from $\bar{n}_e$ to the other quantities, consistent with the idea that $\bar{n}_e$ acts as a control parameter. (ii) Radial variation is significant. The largest values of TE are achieved around $\rho \simeq 0.95 - 0.96$, roughly where the transport barrier is formed [30]. (iii) The time lags corresponding to the maxima of TE are $\tau \simeq 25$ ms, which is of the order of the transition time.

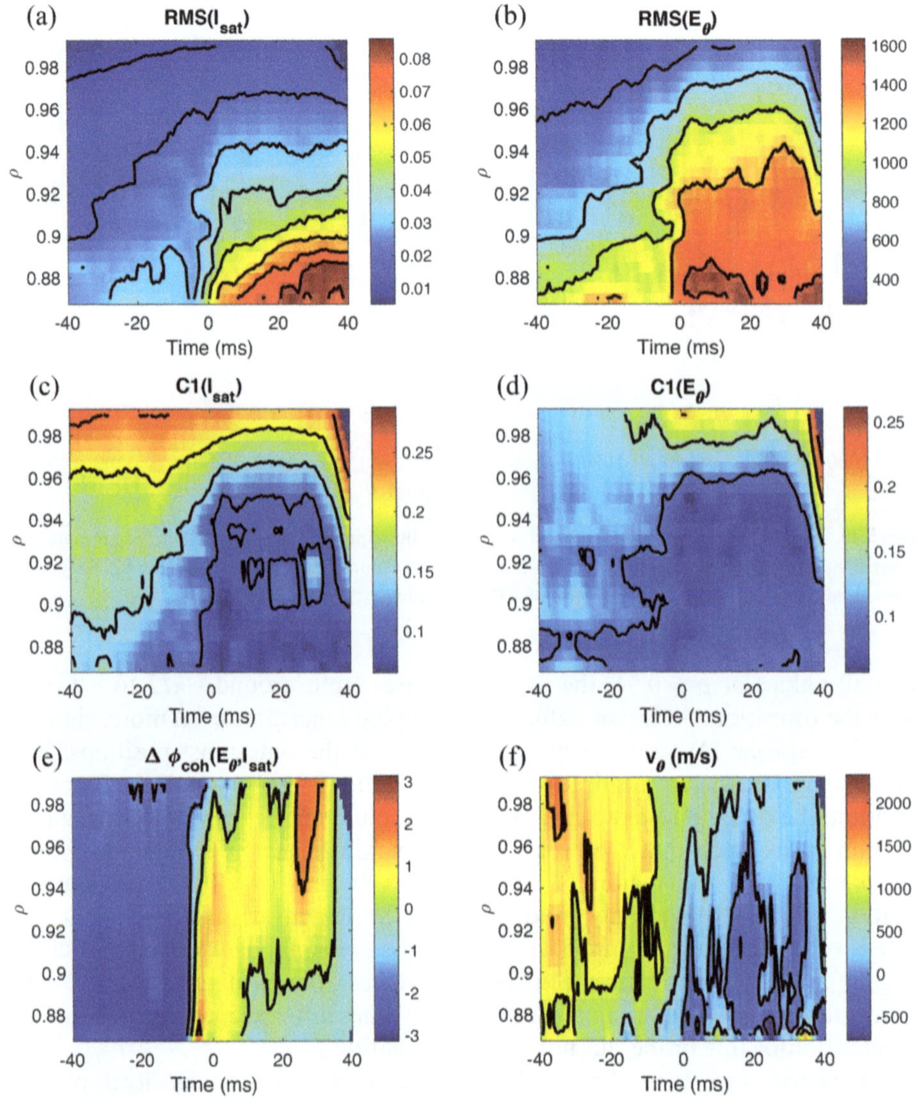

Figure 5.41. Spatiotemporal evolution of some relevant quantities across the e–root to i-root transition (configuration 100_46, probe B). The ρ coordinate corresponds to the position of probe B in each discharge. The horizontal axis shows the time minus the critical time (see the text). (Reproduced from [38], with the permission of AIP Publishing.)

We draw attention to the fact that the values of TE are very high compared to the maximum value possible ($\log_2(3) \simeq 1.6$). Interestingly, the impact of $\bar{n}_e$ on the cross phase $\Delta\phi_{coh}(E_\theta, I_{sat})$ occurs earlier for $\rho \simeq 0.91$ than for $\rho \simeq 0.96$ (see figure 5.42(h)). This is consistent with figure 5.41(e), where the phase change is seen to propagate outward following the transition time.

Figure 5.43 shows the TE between several selected quantities of interest. One particular variable is seen to have a significant causal impact: RMS (E_θ), i.e. the

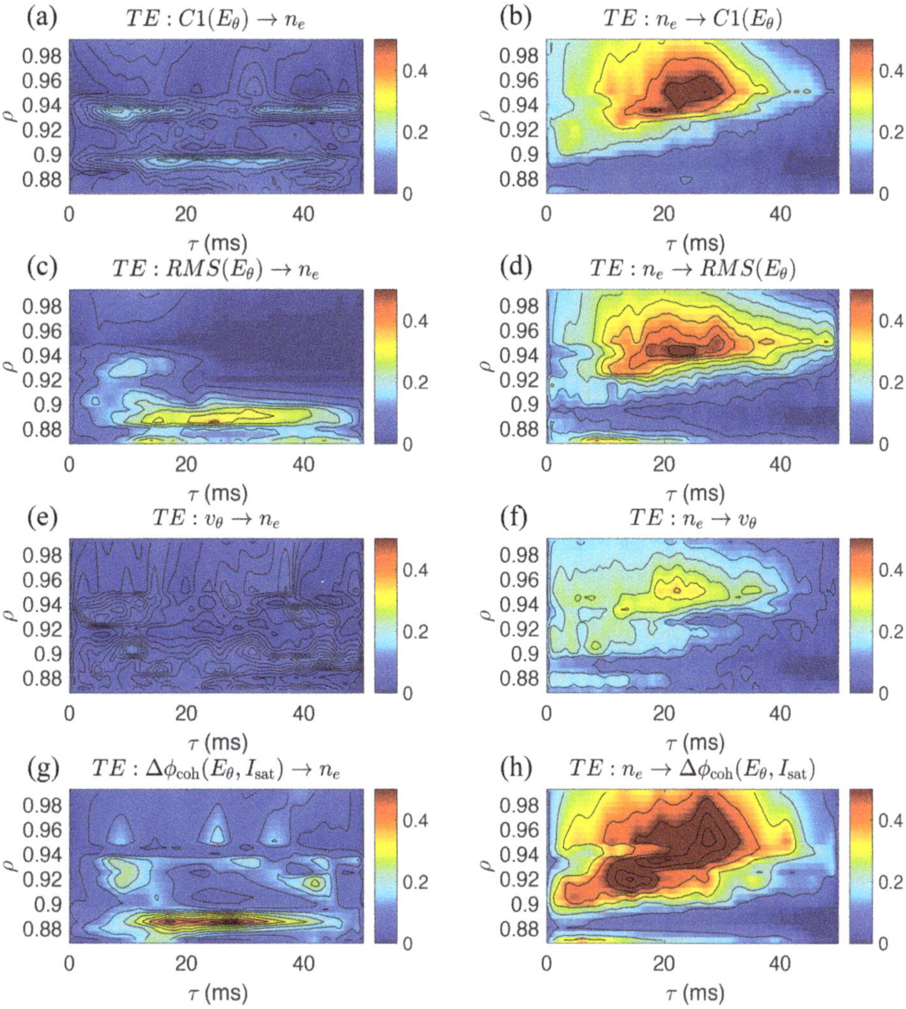

Figure 5.42. Transfer entropy between the line average electron density $\bar{n}_e$ and several variables of interest, as indicated in the plot headings. Horizontal axes: time lag τ (ms). Vertical axes: probe position in terms of normalized radius, ρ. (Reproduced from [38], with the permission of AIP Publishing.)

turbulence level; it has a large impact on the cross phase $\Delta\phi_{coh}(E_\theta, I_{sat})$ (see figure 5.43(d)), significant impact on the intermittence, $C_1(E_\theta)$ (see figure 5.43(b)), and to a lesser degree on the poloidal phase velocity v_θ (figure 5.43(f)). Here also, the influence appears to occur earlier at more inward positions of the probe.

The analysis performed here reveals the complexity of the dynamical interactions between turbulence and transport across the confinement transition. On the one hand, the plasma switches rapidly from electron to ion root conditions as the electric field is reversed [26]. And on the other, the increase of the electron density modifies the intrinsic dynamics of the turbulence by modifying the driving gradients. These two mechanisms (background transport and turbulence) interact dynamically in a complex way.

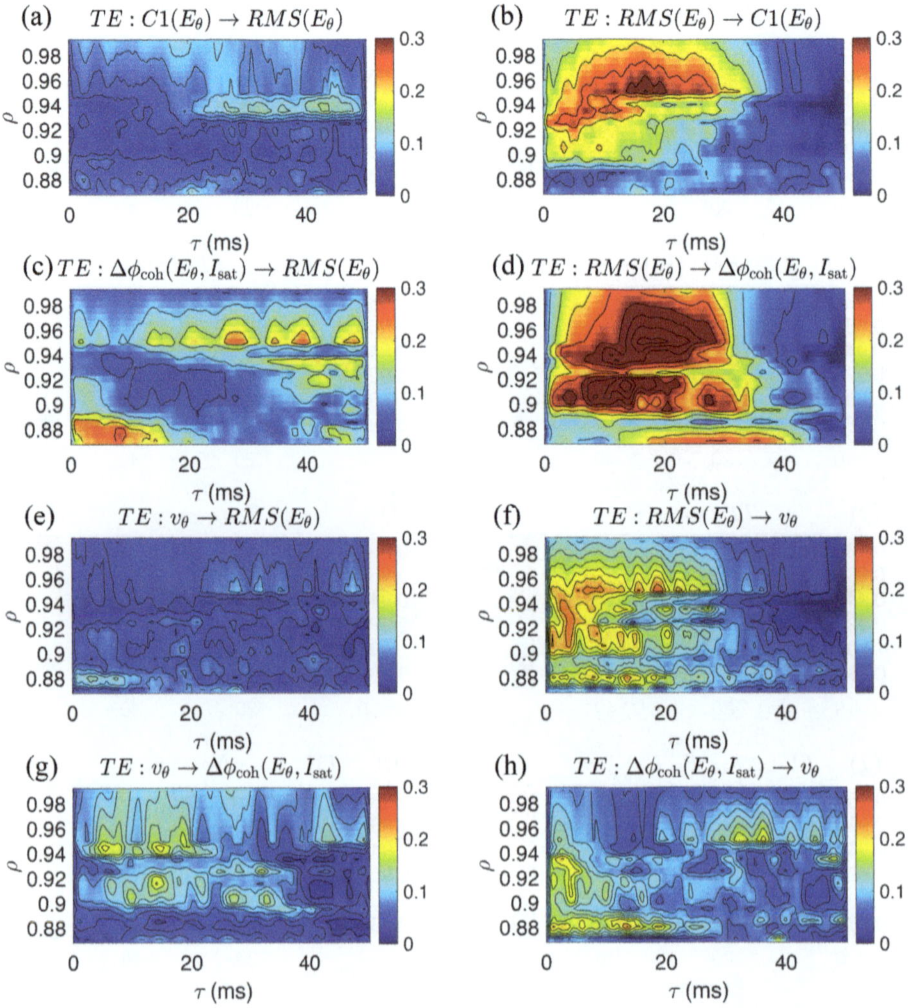

Figure 5.43. Transfer entropy between several variables of interest, as indicated in the plot headings. Horizontal axes: time lag τ (ms). Vertical axes: probe position in terms of normalized radius, ρ (Reproduced from [38], with the permission of AIP Publishing.)

We study the evolution of several key variables, in particular the cross phase between E_θ and $I_{\rm sat}$, appearing in the turbulent particle flux, $\Gamma = \langle \tilde{E}_\theta \tilde{n} \rangle / B$, providing new information regarding the transition process. First, key variables are visualized on a spatiotemporal grid in the relevant plasma edge region. Second, several additional quantities are studied to facilitate further understanding, namely the intermittence and the transfer entropy. Here, it should be noted that the e- to i-root transition itself, marked by the change of sign of v_θ and $E_{\rm r}$, occurs fast, which does not allow extracting its impact from these analyses. However, it is possible to elucidate causal relations on the slower timescale of continuous variations across the transition, as exemplified in figure 5.38.

The line integrated density, $\bar{n}_e$, is the key control parameter for the transition. Increasing $\bar{n}_e$ implies a global increase of the density gradients, while the establishment of the edge transport barrier implies a local increase of the density gradients, both of which increase the turbulence drive; see the impact on RMS(E_θ) in figure 5.42(d). It is interesting to note that this impact occurs first at inward positions, and later at positions that are further outward.

The enhanced turbulence affects the cross phase $\Delta\phi_{coh}(E_\theta, I_{sat})$ (figure 5.43(d)) and causes $C_1(E_\theta)$ to drop (figures 5.41(c) and (d) and 5.43(b)). The latter implies that a dominant low-order helicity is growing (figure 5.39), which causes changes in v_θ via Reynolds stress (figure 5.43(f)). The poloidal flow velocity v_θ is considered crucial for the actual transition.

The fact that $C_1(E_\theta)$ drops suggests that the sheared flow can at least in part be ascribed to the presence of low-order rational surfaces (e.g., $\iota\!\!\!-= 5/3$ at $\rho \simeq 1$ and $\iota\!\!\!- = 18/11$ at $\rho \simeq 0.86$) [10]. This idea might explain why the various quantities respond earlier at positions close to the inner part of the observational domain, $\rho \simeq 0.86$ ($\iota\!\!\!- = 18/11$): the transition would be initiated near that location. The interaction between turbulence and such dominant low-order modes therefore possibly plays a key role in the formation and maintenance of the sheared flow and the transport barrier.

The poloidal flow, v_θ, and cross phase, $\Delta\phi(E_\theta, I_{sat})$, visualized in figures 5.41(e) and (f), respectively, appear to evolve in parallel. On physical grounds, the sheared flow (i.e. the shearing rate) is expected to be the key driving force of the change of the cross phase, even if the corresponding graphs of figure 5.43, the TE between v_θ and $\Delta\phi(E_\theta, I_{sat})$, are not very impressive. It should be noted, though, that both v_θ and $\Delta\phi(E_\theta, I_{sat})$ are the result of a certain level of numerical processing, which may obfuscate delicate causal relationships. In any case, the zone of maximum shearing rate (roughly, the light green band in figure 5.41(f)) is seen to propagate outwards *prior* to the consequential phase change. We note that the change of cross phase is very significant ($\gtrsim\pi/2$).

At the time of the transition, when v_θ changes sign, a short-lived, large amplitude zonal flow structure is observed (figure 5.38), characterized by a significant deviation of E_r from v_θ, similar to what has been observed elsewhere [41].

After the transition, the plasma is in a new state (the ion root) and is maintained in that state by neoclassical requirements (ambipolarity) [26]. It is interesting to note that the changed cross phase is maintained at its new value (figure 5.38), associated with the new equilibrium profiles in the ion root state, in particular the v_θ or flow profile.

5.8 Intermittence and biasing at TJ-II

In the preceding section, we saw how a spontaneous confinement transition leads to a modification of intermittence. Here, we study the impact of an external radial electric field on the intermittence parameter $C(1)$. For this purpose, we make use of TJ-II discharges in which an external electric field was applied by means of a biased electrode inserted into the plasma.

The experiments were performed using the standard magnetic configuration of TJ-II, which has an edge rotational transform close to 1.65 (such that the $n/m = 8/5$

rational surface is located at $\rho \simeq 0.76$). The hydrogen plasmas were heated initially by ECRH and subsequently sustained by NBI heating. The reported results were obtained using NBI co-injection with $P_{\rm NBI} \simeq 500$ kW. The biasing and Langmuir probe system is shown in figure 2.13. The electron density was fairly constant in the interval $1165 < t < 1240$ ms, corresponding to the NBI phase, with a value of $\bar{n}_e \simeq 1 \times 10^{19}$ m^{-3}. Central electron temperatures were in the range of $300 < T_e(0) < 400$ eV. The voltage applied to the biasing probe was a 40 Hz square wave with an amplitude of −350 V and a duty cycle of 50%. In each discharge, the Langmuir probes were moved to a slightly different radial position, allowing the reconstruction of profiles over a relatively wide radial range on a shot-by-shot basis.

We quantified the measurements by detecting the 'biasing on' and 'biasing off' time intervals; each time interval lasting 12.5 ms. We rejected the first 0.5 ms and the last 0.5 ms of each time interval to avoid transient effects, so each biasing phase yielded 11.5 ms of useful data. Over the whole time window analyzed, we obtained six or seven such time intervals, depending on the discharge. Figure 5.44 shows the mean profile of the floating potential, averaged over the selected discharges and the biasing time intervals, for the two levels of biasing. The error bars indicate the standard deviation of values obtained for the various discharges and biasing time intervals.

Figure 5.45 shows the profile of the intermittence parameter of the floating potential fluctuations, $C1(V_f)$, averaged over the selected discharges and the biasing time intervals, for the two levels of biasing.

Figure 5.46 shows $-dV_f/d\rho$, calculated by fitting a fifth order polynomial to the profiles of figure 5.44 (shown as continuous lines in that figure) and taking the derivative. This procedure is particularly inaccurate near the edges of the data interval, so the apparent rapid changes seen for the smallest and largest values of ρ in each graph should be ignored. The quantity $-dV_f/d\rho$ is a proxy for the radial electric field $E_r = -dV_p/dr$ assuming electron temperature gradient effects are small, noting $d/dr \simeq (1/a)d/d\rho$ with $a \simeq 0.2$ m. Comparing biasing on and biasing off, one observes that $-dV_f/d\rho$ is generally more negative when biasing is on. In the case of the B probe, the position of the highest value of $|-dV_f/d\rho|$ (near $\rho = 0.95$) is seen to coincide approximately with the position of the highest values of $C(1)$. In the case of the D probe, resolution may not be sufficient to see this effect clearly.

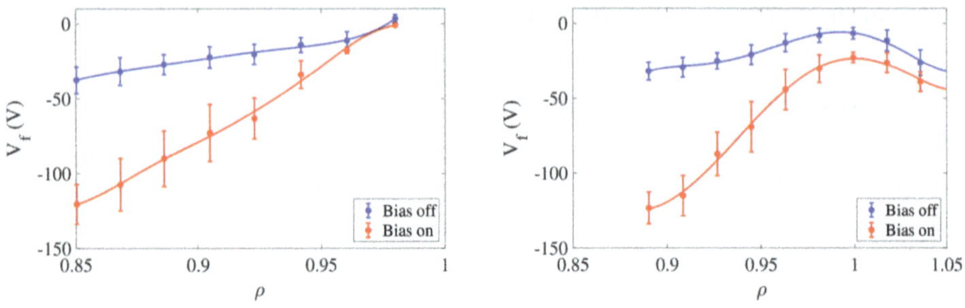

Figure 5.44. Floating potential profile, biasing on/off, NBI conditions. Left: D probe; right: B probe. (Reproduced with permission from [42]. © IOP Publishing. All rights reserved.)

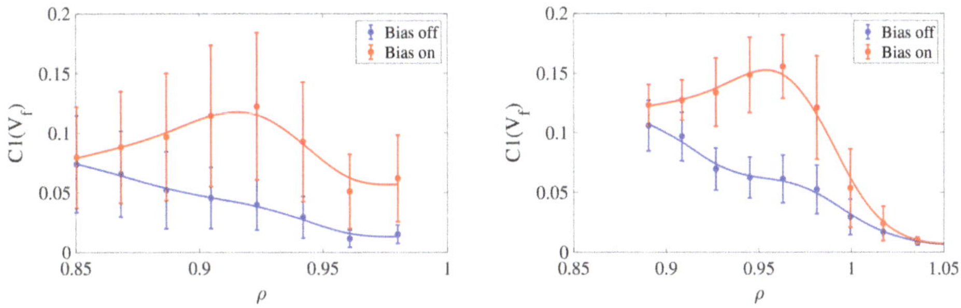

Figure 5.45. Intermittence parameter, $C(1)$, biasing on/off, NBI conditions. Left: D probe; right: B probe. (Reproduced with permission from [42]. © IOP Publishing. All rights reserved.)

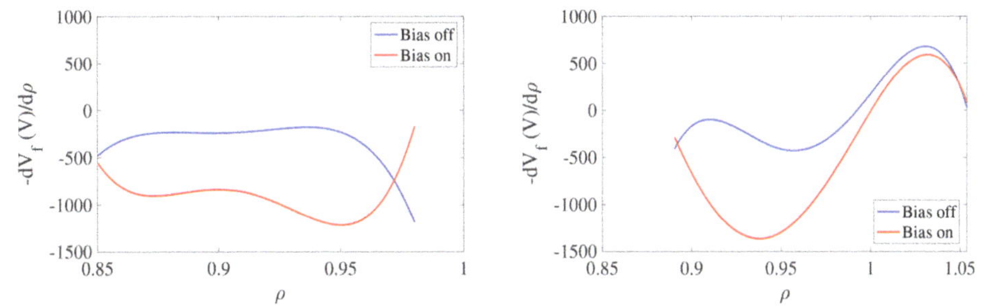

Figure 5.46. $-dV_f/d\rho$, biasing on/off, NBI conditions. Left: D probe; right: B probe. (Reproduced with permission from [42]. © IOP Publishing. All rights reserved.)

These discharges correspond to ion root conditions: negative radial electric field on average in the edge. Biasing was such as to decrease the potential and to enhance the absolute value of the negative radial electric field, E_r. The enhancement of E_r was maximum around $\rho \simeq 0.95$. Biasing led to a significant increase of the intermittence in the range $0.9 < \rho < 1$, peaking near $\rho \simeq 0.95$. We note that this position corresponds to the position of the 18/11 rational surface. This effect was similar at both the B and D probes. Thus, it is seen that the intermittence is both sensitive to the local magnetic configuration (the presence of rational surfaces [22]) and the radial electric field (or poloidal flow), as shown here, providing a window to improve our understanding of plasma turbulence.

5.9 Summary

In this chapter, we started asking ourselves how we could quantify the bursty nature of the particle flux. We compared available quantifiers of intermittence, and decided in favour of a quantifier originating in chaos theory, $C(1)$. We then calculated this quantifier for (potential) fluctuations obtained from a turbulence model, and noted that this parameter responded to the local nature of turbulence: namely, the dominance (or not) of low-order helical modes and the presence of plasma flows.

In the neighbourhood of low-order rational surfaces, and depending on local circumstances (gradients), the corresponding helical mode may dominate fluctuations. In this case, fluctuations are nearly monofractal, which results in a relatively low value of $C(1)$. On the other hand, the presence of poloidal flows, typically near but not at low-order rational surfaces, may lead to a mixing between the local fluctuations in the plasma frame of reference and fluctuations due to the poloidal structure of the turbulent structure, as observed at the fixed measurement location in the laboratory frame of reference. This would lead to an enhancement of multifractality and $C(1)$. Another reason for enhancement of $C(1)$ is the spatial overlapping of various helical modes that may occur in-between closely spaced rational surfaces.

Armed with this knowledge, we then applied the quantifier to ECE (at W7-X) and HIBP (at TJ-II) measurements in the core region. In both cases, radial minima of $C(1)$ appeared associated with the location of important low-order rational surfaces, though only in the gradient region. This suggests that a gradient is required to drive the helical modes that lead to a reduction of $C(1)$.

Careful iota scan experiments in combination with Langmuir probe measurements in the plasma edge region, reported in section 5.5, allowed determining the radial profile of the intermittence with exquisite precision, and several simultaneous radial minima could be identified and associated with low-order rational surfaces, which were found to be qualitatively consistent with numerical turbulence simulations. In this sense, the intermittence can be viewed as a 'detector' for active low-order helical modes. This study also allowed studying the impact of the radial electric field (i.e. poloidal flow) on the minima of $C(1)$.

By combining this analysis with the transfer entropy of chapter 4, the spatiotemporal interactions between various relevant turbulent variables across a confinement transition could be followed, see section 5.7, providing more insight into the complexity of the transition. In particular, it is interesting to note that while the RMS of the fluctuations increases across the confinement transition studied, the intermittence parameter drops (figure 5.38). One may speculate that the sheared flow associated with the transition selectively suppresses the larger eddies (by shearing them apart), which would reduce the intermittence, as intermittence arises through the interaction of eddies of various scales. Thus, a temporal drop of intermittence would be a clear marker of the action of sheared flows.

In view of the above, we believe the intermittence parameter opens up a new field of study, providing deeper insight into the complexity of turbulence and transport in both the plasma edge and core regions: namely, the occasional dominance of low-order helical modes and the impact of plasma flows on turbulence.

References

[1] Ott E 1993 *Chaos in Dynamical Systems* (Cambridge: Cambridge University Press)

[2] Batchelor G K 1953 *The Theory of Homogeneous Turbulence* (Cambridge: Cambridge University Press)

[3] Sandborn V A 1959 Measurements of intermittency of turbulent motion in a boundary layer *J. Fluid Mechanics* **6** 221

[4] Frisch U 1995 *Turbulence* (Cambridge: Cambridge University Press)
[5] Farge M, Schneider R and Devynck P 2006 Extraction of coherent bursts from turbulent edge plasma in magnetic fusion devices using orthogonal wavelets *Phys. Plasmas* **13** 042304
[6] Meneveau C and Sreenivasan K R 1987 Simple multifractal cascade model for fully developed turbulence *Phys. Rev. Lett.* **59** 1424
[7] Meneveau C and Sreenivasan K R 1991 The multifractal nature of turbulent energy dissipation *J. Fluid Mechanics* **224** 429
[8] Carreras B A *et al* 2000 Intermittency of plasma edge fluctuation data: multifractal analysis *Phys. Plasmas* **7** 3278
[9] Hentschel H G E and Procaccia I 1983 The infinite number of generalized dimensions of fractals and strange attractors *Physica* D **8** 435
[10] Carreras B A, García L, Nicolau J H, van Milligen B P, Hoefel U, Hirsch M and the TJ-II and W7-X Teams 2020 Intermittence and turbulence in fusion devices *Plasma Phys. Control. Fusion* **62** 025011
[11] Grenfell G, van Milligen B P, Losada U, Wu T, Liu B, Silva C, Spolaore M and Hidalgo C 2019 Measurement and control of turbulence spreading in the scrape-off layer of TJ-II *Nucl. Fusion* **59** 016018
[12] van Milligen B *et al* 2021 Topology of 2-D turbulent structures based on intermittence in the TJ-II stellarator *Nucl. Fusion* **61** 116063
[13] Wolf R C *et al* 2017 Major results from the first plasma campaign of the Wendelstein 7-X stellarator *Nucl. Fusion* **57** 102020
[14] van Milligen B P, Carreras B, García L and Nicolau J 2019 The radial propagation of heat in strongly driven non-equilibrium fusion plasmas *Entropy* **21** 148
[15] Diamond P H, Itoh S-I, Itoh K and Hahm T S 2005 Zonal flows in plasma–a review *Plasma Phys. Control. Fusion* **47** R35
[16] Sattin F, Vianello N and Valisa M 2004 On the probability distribution function of particle density at the edge of fusion devices *Phys. Plasmas* **11** 5032
[17] van Milligen B P, Nicolau J H, García L, Carreras B A, Hidalgo C, Carreras B A and the TJ-II Team 2017 The impact of rational surfaces on radial heat transport in TJ-II *Nucl. Fusion* **57** 056028
[18] van Milligen B P, Hoefel U, Nicolau J H, Hirsch M, García L, Carreras B A, Hidalgo C and the W7-X Team 2018 Study of radial heat transport in W7-X using the transfer entropy *Nucl. Fusion* **58** 076002
[19] Klinger T *et al* 2017 Performance and properties of the first plasmas of Wendelstein 7-X *Plasma Phys. Control. Fusion* **59** 014018
[20] Dinklage A *et al* 2018 Magnetic configuration effects on the Wendelstein 7-X stellarator *Nat. Phys.* **14** 855
[21] Pablant N A *et al* 2018 Core radial electric field and transport in Wendelstein 7-X plasmas *Phys. Plasmas* **25** 022508
[22] van Milligen B P, Carreras B A, García L and Hidalgo C 2020 The localization of low order rational surfaces based on the intermittence parameter in the TJ-II stellarator *Nucl. Fusion* **60** 056010
[23] Harris J H, Cantrell J L, Hender T C, Carreras B A and Morris R N 1985 A flexible heliac configuration *Nucl. Fusion* **25** 623
[24] van Milligen B P *et al* 2011 Integrated data analysis at TJ-II: the density profile *Rev. Sci. Instrum.* **82** 073503

[25] van Milligen B P, Lopez Fraguas A, Pedrosa M A, Hidalgo C, Martín de Aguilera A and Ascasíbar E 2013 Parallel and perpendicular turbulence correlation length in the TJ-II stellarator *Nucl. Fusion* **53** 093025

[26] Velasco J L, Alonso J A, Calvo I and Arévalo J 2012 Vanishing neoclassical viscosity and physics of the shear layer in stellarators *Phys. Rev. Lett.* **109** 135003

[27] Biglari H, Diamond P H and Terry P W 1990 Influence of sheared poloidal rotation on edge turbulence *Phys. Plasmas* **2** 1

[28] Melnikov A V *et al* 2017 Heavy ion beam probing-diagnostics to study potential and turbulence in toroidal plasmas *Nucl. Fusion* **57** 072004

[29] Khabanov Ph O *et al* 2019 Density profile reconstruction using HIBP in ECRH plasmas in the TJ-II stellarator *J. Instrum.* **14** C09033

[30] van Milligen B P *et al* 2011 The dynamics of the formation of the edge particle transport barrier at TJ-II *Nucl. Fusion* **51** 113002

[31] van Milligen B P, García L, Carreras B A, Pedrosa M A, Hidalgo C, Alonso J A, Estrada T and Ascasíbar E 2012 MHD mode activity and the velocity shear layer at TJ-II *Nucl. Fusion* **52** 013006

[32] Higham D J and Higham N J 2016 *MATLAB Guide* volume 150 (Philadelphia, PA: SIAM)

[33] van Milligen B P *et al* 2011 A global resonance phenomenon at the TJ-II stellarator *Nucl. Fusion* **51** 013005

[34] Pedrosa M A, Hidalgo C, Silva C, Carreras B A, Carralero D, Calvo I and the TJ-II Team 2010 Long-range correlations during plasma transitions in the TJ-II stellarator *Contrib. Plasma Phys.* **50** 507

[35] Boedo J A, Terry P W, Gray D, Ivanov R S, Conn R W, Jachmich S, van Oost G and the TJ-II Team 2000 Suppression of temperature fluctuations and energy barrier generation by velocity shear *Phys. Rev. Lett.* **84** 2630

[36] Ritz C P *et al* 1988 Advanced plasma fluctuation analysis techniques and their impact on fusion research *Rev. Sci. Instrum.* **59** 1739

[37] Pedrosa M A, Silva C, Hidalgo C, Carreras B A, Orozco R O, Carralero D and the TJ-II Team 2008 Evidence of long-distance correlation of fluctuations during edge transitions to improved-confinement regimes in the TJ-II stellarator *Phys. Rev. Lett.* **100** 215003

[38] van Milligen B, Carreras B, Voldiner I, Losada U, Hidalgo C and the TJ-II Team 2021 Causality, intermittence and crossphase evolution during confinement transitions in the TJ-II stellarator *Phys. Plasmas* **28** 092302

[39] Balbín R, Hidalgo C, Pedrosa M A, García-Cortés I and Vega J 1992 Measurement of density and temperature fluctuations using a fast-swept Langmuir probe *Rev. Sci. Instrum* **63** 4605

[40] Hidalgo C *et al* 2006 Physics of sheared flow development in the boundary of fusion plasmas *Plasma Phys. Control. Fusion* **48** S169

[41] Grenfell G, van Milligen B P, Losada U, Estrada T, Liu B, Silva C, Spolaore M and Hidalgo C 2020 The impact of edge radial electric fields on edge-SOL coupling in the TJ-II stellarator *Nucl. Fusion* **60** 014001

[42] van Milligen B P, Carreras B A, García L, Grenfell G, Voldiner I, Hidalgo C and the TJ-II Team 2022 The impact of radial electric fields and plasma rotation on intermittence in TJ-II *Plasma Phys. Control. Fusion* **64** 055006

IOP Publishing

Analysis of Turbulence in Fusion Plasmas

Boudewijn Ph van Milligen and Raul Sanchez

Chapter 6

Closing

As argued in chapter 1, turbulence is an essential ingredient of the complexity inherent in strongly driven, fusion-grade magnetically confined plasmas. Turbulence often determines a significant fraction of radial transport, spontaneously generates large-scale emergent structures, and plays an essential role in the self-regulation of the plasma state and hence the efficiency of putative future fusion reactors. In this book, we have attempted to clarify some aspects of turbulence and its interaction with the background plasma state, based on experimental data and making use of both standard analysis tools and tools that are perhaps less familiar to some readers.

6.1 Lessons learned

6.1.1 The need for data

A better understanding of turbulence in fusion plasmas requires data—a lot of it. Since turbulence is a complex phenomenon that involves spatially extended structures, point measurements—while helpful—cannot provide a complete picture. Ideally, data with sufficiently high temporal and spatial resolution, sampled at frequencies well above 100 kHz, say (see chapter 2), should be available from many points inside the plasma; preferably not only distributed along a single radial chord, but along chords at various toroidal and poloidal locations; this being particularly important in view of the fact that turbulence is generally not distributed homogeneously along flux surfaces [1]. In addition, many fields are involved: density, temperature, electric potential, and flow velocity, to name a few. Each field can reveal another aspect of the turbulent activity; and to understand the interaction of the fields, they should be measured simultaneously.

Turbulence data are not easy to obtain in the hostile environment of hot fusion plasmas. Consequently, the experimental results shown in this book overwhelmingly correspond to Langmuir probe data, obtained at a few points in the plasma edge region. Fortunately, the plasma edge region happens to be very interesting as it is a

doi:10.1088/978-0-7503-4856-0ch6

region where zonal flows and various types of transport barriers tend to form and where fluctuations are large and radial transport is intense. And since the edge is the outermost region of the confining volume, it naturally plays a very significant role in setting the overall scaling properties of the global confinement of the system. Nevertheless, information on core turbulence is necessary to complete the picture. A few types of diagnostics can deliver data with sufficient temporal and spatial resolution in the plasma core region. In this respect, we can mention the heavy ion beam probe (for potential and electron density measurements, among others) and modern high resolution ECE diagnostics (for the electron temperature). Reflectometry is another good turbulence diagnostic (for the electron density), although limited to regions with significant density gradients, while Doppler reflectometry can in addition deliver information about the perpendicular flow velocity, $v_\perp$.

6.1.2 Advanced analysis techniques

In this book, we have tried to argue that the analysis of turbulence should not be limited to determining (or predicting) its amplitude (RMS value), probability distribution, and Fourier spectrum. Nevertheless, these quantities do, of course, form the basis for further analyses. For example, the probability distribution functions (PDFs) of the particle flux, experimentally accessible using appropriate Langmuir probes and the heavy ion beam probe, contains revealing information about plasma transport. We have shown that the typical peaked shape of this PDF can largely be explained on the basis of its quadratic nature, while the m-point averaged flux PDF has self-similar scaling properties (section 2.1.2).

This self-similarity points to the fact that fusion plasmas are complex, self-regulating systems in which memory effects (long range temporal correlations as measured by, e.g. the Hurst exponent) and long range spatial correlations (zonal flows, avalanches) may play a significant role. These complex features become prominent when the plasma profiles remain close to the local critical thresholds for the onset of instabilities, which tends to happen for a range of values of the external drive. The instability thresholds should not be considered fixed, but rather dependent on the presence of, e.g. low order rational surfaces that may lower the thresholds, while spontaneously generated, strong, radially sheared poloidal flows may suppress turbulence growth.

This 'state of marginality' is not rare in fusion plasmas, but rather common, and it has a major impact on confinement properties. In part, this is due to the fact that the operators of tokamaks and stellarators generally strive to optimize the confinement properties of these systems by enhancing the external drive and carefully manipulating the profiles, such that the latter are systematically pushed towards these critical thresholds and into a marginal state. As a result of this closeness to instability thresholds and the high temperatures, the turbulence drive and resulting turbulent flux is strong, which in turn facilitates self-regulation of the profiles close to the threshold values, reduces the separation between turbulent and transport timescales, and generally enables rather complex behaviour.

The self-similarity observed over broad ranges of temporal and spatial scales indicates that a clear separation of scales between fast, small-scale turbulent fluctuations and slowly evolving, large-scale transport effects is not always feasible. It points to the spatiotemporal entwinement (or entanglement) of the plasma while in a marginal state, with significant consequences for the modelling of plasma transport; namely, it means that turbulence and profiles interact mutually and on similar timescales. The experimental verification of the idea that profiles and turbulence interact in this manner is a significant challenge (see section 5.6.1). In any case, it appears reasonable to assume that the idea that the turbulence amplitude simply responds passively to the evolving profiles, as often assumed in many theoretical and computational studies, may not generally be appropriate, and that turbulence is in fact actively involved in the regulation of overall confinement.

6.1.3 Turbulence and structures

Turbulence contains structure in and of itself (filaments), but more interestingly, some large-scale structures emerge from the complex interactions taking place in the turbulent fields. The detection of these large-scale turbulent structures is not always trivial.

Filaments in the plasma edge and scrape-off layer region can be visualized using, e.g. fast camera systems, and have indeed been detected systematically in many devices. These structures, narrow across but elongated along the magnetic field lines, can also be analyzed using two remotely positioned Langmuir probe systems (section 2.2.2) or even using ECE (section 4.6.3).

The main example of a large-scale, self-generated, emergent structure is the zonal flow, of fundamental importance for the understanding of the formation of transport barriers and improved confinement states. The corresponding structures are narrow in the radial direction but essentially homogeneous along flux surfaces ($k_\phi \simeq k_\theta \simeq 0$). We have shown how this property can be exploited to detect them, using correlations between remote multipin Langmuir probes based on the biorthogonal or singular value decomposition (section 2.5).

Thus, long range correlations between remotely placed systems (such as the dual Langmuir probe or the dual heavy ion beam probe systems at TJ-II) can be caused by both filaments and zonal flows. With some effort, it is possible to distinguish between these two phenomena, as was done using an iota scan experiment in section 2.2.2.

In addition to these structures that extend mainly along a flux surface, there exist radial structures, which are perhaps better labelled 'events', as their existence is more ephemeral and often associated with 'bursty' transport. Various types of events may give rise to radial correlation. The spatiotemporal interaction between local turbulent fluxes and profile modifications near a critical threshold may give rise to radially propagating events, sometimes called 'avalanches'. The (magnetic) resonant coupling between radially separated modes can give rise to extended radial structures. Semi-persistent radial structures, such as 'streamers' [2], may also arise from the turbulence itself. Finally, radial structures may result from the self-advection of turbulence into quieter regions with negative turbulence growth rate (known as 'turbulence spreading')—among other possibilities.

6.1.4 The interaction between turbulence and transport

The search for radially propagating, avalanche-like events has been greatly facilitated by the application of a relatively new analysis technique based on the transfer entropy. This technique gives us a statistical approach to the analysis of propagating turbulent events, far superior to what can be achieved using the standard, linear, cross-correlation technique. Thus, we have been able to show that radially propagating transport events are sometimes 'held up' at locations associated with low order rational surfaces where zonal flows tend to form; we called these regions 'trapping zones' or 'minor transport barriers'. Nevertheless, occasionally, transport events were seen to 'jump' over these barriers. This exciting novel result could perhaps explain part of the enhanced transport observed as the external drive (heating) is increased (see section 4.6.2). The physical explanation of this behaviour was suggested to involve the radial coupling between instabilities associated with different rational surfaces, based on a comparison with numerical turbulence simulations exhibiting similar effects. This would mean that the plasma is able to use different channels for the radial transport of energy; in addition to the standard diffusive and convective transport of particles and energy, some energy is transmitted ballistically, possibly via the radial coupling associated with magnetic field fluctuations. Although a fascinating hypothesis, this is still a largely open question that requires further study.

The transfer entropy is not only suited for the study of transport, but also provides a powerful method to study the interaction between fluctuating variables in complex systems—see sections 4.4 and 5.7. In principle, this allows one to improve our understanding of the mutual interaction between the various fluctuating fields by identifying causal chains [3].

As described in chapter 5, the intermittence is another interesting tool for the study of turbulence in fusion plasmas. Somewhat unexpectedly, it turns out to be sensitive to the dominance of low order helical modes (and hence allows detecting the impact on turbulence of specific rational surfaces) and plasma flows, thus providing a new window on turbulence, flows, and structure in the plasma interior. This behaviour has been verified both using a resistive MHD turbulence model (section 5.3), ECE data in the W7-X stellarator (section 5.4), and Langmuir probe data obtained in the edge of the TJ-II stellarator (section 5.5). Particularly, the latter, rather detailed study showed that the intermittence profile exhibits minima that tend to coincide with the expected location of important low order rational surfaces, in accordance with expectations from numerical turbulence calculations performed under similar conditions. This work also explored the effect of a varying radial electric field (and hence, poloidal flow) on the detected intermittence. An initial study of the intermittence in the plasma interior, based on heavy ion beam probe data obtained at the TJ-II stellarator, likewise showed how an important low order rational surface, with a corresponding helical mode driven by the pressure gradient, had a significant impact on the detected intermittence (section 5.6).

Significantly, the intermittence results confirm the image that arose from the analyses based on the transfer entropy discussed above, namely that low order rational surfaces may play a very important role in the regulation of radial transport via the associated flows.

A recent study showed, in addition, how these techniques (the transfer entropy and the intermittence) can be combined to obtain deeper insight into the various simultaneous changes and mutual field interactions occurring during a confinement transition in the TJ-II stellarator (section 5.7). Significantly, the intermittence was observed to drop when the plasma made a transition to an enhanced confinement state. We therefore expect this observation to lead to deeper insight into the evolution of turbulence across confinement transitions. Recent, as yet unpublished observations from the JET tokamak using Doppler reflectometry revealed a similar drop of intermittence across an L–H transition.

6.1.5 Remarks on the analysis and interpretation of turbulence simulations

Although the observations above are oriented towards the analysis of experimental data, similar comments can be made regarding the analysis of data from turbulence simulations, as the enormous amount of data generated in detailed turbulence simulations require analysis in order to summarize and understand the results.

Considerations regarding the importance and impact of turbulence and the assumed dominance of certain types of turbulence (ITG, ETG, TEM, etc) are often based on simulations, as sufficiently detailed experimental data are usually not available. Although obvious, we note that the conditions, limitations or assumptions that apply to a given simulation limit the results that can be expected. In particular, the simulations often do not contain a similar level of complexity as actual fusion plasmas, so that the analyses of such simulations only have a limited validity and should not be over-interpreted.

For example, a flux tube simulation cannot be expected to include long range radial events due to its limited geometric scope. In addition, flux tube simulations ignore the back-reaction of the turbulence on the plasma profiles, which is usually taken to be fixed and constant: a local gradient value. In a similar vein, simulations are often (usually, in fact) performed in a fixed magnetic geometry. In that case, no interaction with magnetic turbulence will occur—no ‘trapping zones’ or transport barriers associated with rational surfaces will be formed. Global simulations, even when they do consider the evolution of the magnetic field, may still suffer from limitations when the plasma profiles are artificially constrained to stay close to a prescribed ‘profile shape’ by dynamically adjusting sources for this purpose. This approach inherently assumes a ‘separation of scales’ between profile evolution and turbulent transport, i.e. it is assumed that profiles will not change significantly on turbulent timescales. Consequently, such simulations will not reproduce the complex dynamics resulting from the unrestrained interaction between profiles and turbulence.

But recent 'flux-driven' turbulence simulations, in which sources are kept fixed from the start and plasma profiles are allowed to evolve freely during the simulation, are starting to reveal phenomena that appear to match the observations of radial structure reported in this book. A striking example is the spontaneous appearance of an '$E \times B$ staircase' in near-marginal, low collisionality plasma states [4]. This state is characterized by a succession of radially spaced 'micro-barriers' (analogous to our 'minor barriers'), however with no apparent relation to rational surfaces. Even more recent simulations, however, did reveal a relation of such a 'staircase' with rational surfaces in long-time simulations that included kinetic electrons [5]. Although the jury is still out on this issue, certainly it appears that these recent studies provide some modelling support for the observations reported here.

6.2 Implications

6.2.1 The global picture of turbulence in fusion plasmas

When considering turbulence in fusion plasmas, one might conjure up the idea that background profiles, established through the action of transport equations, are smooth, slowly varying quantities, on top of which sits some 'noise' (turbulence) that adds some extra diffusivity to the system, but little else. We believe this idea, that may be approximately correct in other contexts, is fundamentally wrong for sufficiently hot tokamak and stellarator plasmas!

To start, profiles are not smooth. We noted that various types of transport barriers exist (labelled 'minor' if they are small or 'Internal transport barrier' (ITB) or 'H-mode pedestal' if they are intense). The transfer entropy studies presented here suggest that these barriers are not just present under extraordinary conditions (as is the case for ITBs), but rather that they are omnipresent and semipermanent, and have a significant impact on radial transport. The resulting small-scale structure in the (pressure and other) profiles is hard to detect experimentally, due to the high spatial resolution that would be required—not to mention that small 'steps' in the profile *gradient* would be hard to spot in a sea of ambient turbulence. A notable effort was made in the past using a high resolution Thomson scattering system [6, 7]. Also, there is some theoretical justification for the formation of such small-scale radial structure [8].

Next, it should hardly be necessary to insist that turbulence is not the same thing as uncorrelated noise. Far from being uncorrelated, under specific circumstances—not necessarily rare in fusion plasmas—it can exhibit strong long range correlations that can be measured directly using adequate systems. The elementary turbulent structure in a strongly magnetized fusion plasma is a filament. The parallel correlation length associated with the existence of these structures is quite amazing, reaching values of many tens to about a hundred metres, implying multiple toroidal turns (section 2.2.2). Consequently, it is hard to maintain that turbulence is somehow 'local'.

Although filaments induce long range correlations along the field lines, it might still be true that the turbulence is essentially local in the perpendicular direction to the magnetic field. While this may indeed be the case in cold (sub-marginal) plasmas, experimental evidence suggests that near-marginal conditions lead to a very different picture. Indeed, the experimental evidence suggests that turbulent fluctuations in tokamaks and stellarator can exhibit both long range radial correlations and long-time memory. The long-time memory can be detected using the methods discussed in chapter 3. Various methods, such as the rescaled range method, have been used to show that significant memory effects[1] exist for timescales varying from the turbulence correlation time all the way up to the confinement time of the plasma. We believe this evidence is convincing, and it was obtained in a wide range of devices. This means that turbulent timescales and transport timescales are not generally decoupled, at least not for sufficiently hot magnetized plasmas, but influence each other mutually via some sort of self-organization process. In this case, it is in fact reasonable to assume that the long-term memory effects exhibited by the fluctuations materialize via the 'storage of information' in the slowly evolving background profiles (turbulence → transport → profiles → turbulence).

Long range radial correlations are harder to demonstrate, due to the limited radial size of the systems studied and the strong radial variation of the background profiles. Nevertheless, chapter 4 provides some interesting examples, including *a priori* unexpected radial 'jumps' or rapid radial transport events, leading to a rather non-trivial picture of radially propagating perturbations. The physical mechanism of this type of propagation has not been fully clarified, but it was suggested that radial mode coupling might be involved [9]. The mentioned long range temporal and spatial correlations are a reflection of the fact that turbulence is essentially *spatiotemporal* in nature.

Then there are sheared flows, large-scale poloidal/toroidal structures that arise spontaneously from turbulence, associated with strong edge transport barriers in the H-mode. These structures, whose existence is beyond any doubt [10], are a plain example of the self-organization occurring in fusion plasmas and exemplify the interaction between small and large spatial scales. We maintain that these sheared flows may also exist, in a weaker form, in association with the 'minor' transport barriers associated with low order rational surfaces in the plasma interior (chapter 4).

On top of all this, the turbulence drive should also be considered: turbulence is driven by the free energy available in the steep thermodynamic gradients. This idea is reflected in all standard turbulence models (e.g. for electron temperature gradient (ETG) and ion temperature gradient (ITG) driven modes). The gradient-dependent turbulence drive is strongly non-linear: when the local (linear) turbulent growth rate becomes positive, at a specific 'critical' gradient value, this leads to rapid growth of the corresponding turbulent mode, and correspondingly a strong non-linear response of the associated transport. The enhanced transport driven by such turbulence may then in turn modify the gradient, leading to the establishment of feedback loops that control the gradient through turbulence. This feedback is not smooth and continuous, but may involve 'bursty' behaviour, sometimes referred to

[1] 'Persistence', characterized by $H > 0.5$.

as 'avalanches'. This 'bursty' behaviour may result from the strongly non-linear local flux–gradient relation and the spatiotemporal nature of the system, as in a paradigmatic sandpile: a flux event arriving at a given location from the interior of the system may modify the local gradient, thus triggering a subsequent outward transport event.

'Bursty' transport implies intermittence, as discussed in some detail in chapter 5. This intermittent behaviour can indeed be detected and quantified, as we have shown, and its relation to plasma structure (minor transport barriers associated with low order rational surfaces) has been indicated. Intermittence also implies PDFs with 'long tails', i.e. the distribution of transport events consists of many small events and a few large events that nevertheless have a significant impact on global transport.

6.2.2 Implications for our understanding of transport

The awareness that the confined hot fusion plasma often behaves as a complex system, containing a variety of instabilities and structures that affect its behaviour and its response to external sources and modulations, raises the question how this situation affects our understanding of transport processes. Which, in turn, raises the following question. *When do we actually consider that we have achieved understanding of a particular system?* Since we are dealing with very complex systems, this question is relevant—as we will never be able to manage all states and interactions mentally. Understanding would at least require the identification of all relevant (dominant) dynamical elements, and expressing these elements and their mutual relations through equations. This would amount to having a reduced model of the system. This model can then be used to predict how the system will behave when we act on it or when it evolves spontaneously. In other words, a proper understanding of complex systems such as these would require being able to formulate simplified or reduced models that highlight the essential dynamical interactions.

All equations needed to describe a plasma are known, so we could produce a first-principle, fully kinetic model of a 3D magnetized plasma with several species, including all appropriate sources. In order to elaborate such a model, one could consider improving the existing numerical models to include all fields (density, temperature, flow, the electric and magnetic field—for the main ions, electrons, and impurities), avoid the typical separation between background and fluctuating components, consider all sources and sinks, consider the whole range of timescales and length scales to include all needed instabilities and waves, and run the resulting 'complete' simulation. Such a model for the plasma interior would need to be interfaced with models for the plasma edge, scrape-off layer, and wall interaction. Although constructing an actual 'complete' model is firmly out of range of present-day computing possibilities, some progress is being made in this direction. In any case, while such a model may be helpful to understand some of the observed phenomena, it is unlikely to be practical for plasma operation, the rapid prediction of plasma behaviour, or achieving some level of understanding that will allow us to improve turbulence control, for example.

As noted, another approach would be to attempt to construct simplified models, in the best scientific tradition, that contain all the essential physics while approximating irrelevant detail. The neoclassical transport model mentioned in chapter 1 is an example of this strategy that goes a long way in the direction of modelling the plasma interior. Among other things, it models the radial transport of various fields in the involved geometry of magnetic confinement devices at the price of making significant simplifications. Although this theory has proven to be rather successful, it assumes that magnetic fields are fixed and unaffected by plasma dynamics while electric fields are not evolved dynamically but determined *a posteriori* to ensure the ambipolarity of fluxes. Usually, it is further assumed that fluxes are local, to allow the estimation of diffusivities and viscosities, although this is not a fundamental ingredient of the theory. As a result, simplified transport models based on neoclassical transport coefficients ignore, by construction, any type of complex behaviour. It is customary to include the overall effect of turbulence (resulting from the interaction between the confined charged particles and the confining electric and magnetic fields) in these simplified models by adopting the same formal structure of the local neoclassical theory, and including additional diffusive terms in the transport equation to model the turbulence-induced radial transport. It should be clear that all these choices exclude *ab initio* any possibility of displaying long range spatial and temporal effects, since the evolution of background profiles is implicitly assumed to occur on much slower timescales than the fluctuation timescales, thus ignoring any mutual interaction. As a result, many of the unusual phenomena mentioned in chapter 1 are outside the scope of such models.

To summarize, full 'first principles' models are unsatisfactory due to their complexity and the impossibility to compute 'everything' in present-day computers. On the other hand, the best simplified models mostly ignore the complexity that is the focus of this book, whereas we consider this complexity important to truly understand the behaviour of fusion-grade plasmas. Therefore, the question is: can the standard (local) approach to transport be modified or enhanced somehow to include the non-local and/or non-linear effects detected in experiment?

Several initial attempts have been made to explore generalizations of the standard diffusive transport model in order to account for at least some of the long range interactions and non-linear effects [11]. As noted in section 1.5, at this point it is useful to recall that the effective (macroscopic) transport depends on the underlying 'microscopic transport'. All diffusive models ultimately rely on the Gaussianity and Markovianity of the description of the underlying microscopic particle motion. The connection to the usual diffusive transport model is established by formulating the 'microscopic transport' in terms of either (probabilitistic) master equations or (stochastic) Langevin equations, in which fluctuations are random and Gaussian. The diffusive equations (Fick's law) are then deduced formally by taking the mathematical limit corresponding to large-distance, long-time dynamics. With this idea in mind, many authors, including ourselves, have tried to introduce long-term correlations in space and time by replacing the standard Gaussian fluctuation statistics with Lévy distributions (see chapter 3), which are stable (attracting) distributions in the framework of the central limit theorem, but with 'long tails' that produce long range correlations (these distributions being characterized by algebraic decay). Temporal correlations can be introduced in various ways (for instance, by using coloured noise).

It turns out that the resulting master or Langevin equations are capable of reproducing, at least qualitatively, some of the various unusual phenomena exhibited in fusion plasmas cited in chapter 1: power degradation, stiff profiles, rapid transport phenomena, the anomalous scaling of transport with system size, and 'uphill' transport, particularly if a critical threshold is included [12–14].

In addition, it turns out that the average behaviour in the same mathematical limit for long distances and times can be described in the framework of a family of mathematical models that share a specific feature: the replacement of the usual integer derivatives by fractional derivatives in time and space [15, 16]. Fractional derivatives are integro-differential operators with kernels that ensure that long range spatial and temporal correlations are included in the description of the system. Although some effective transport models based on these ideas have been applied to situations in fusion plasmas [17], currently these models are still excessively simple and far removed from the reality of actual plasmas. In addition, fractional derivatives, particularly *temporal* fractional derivatives, are difficult to handle computationally and it is still unclear how to extend these entities to non-Cartesian coordinate systems [11]. Therefore, it remains to be seen whether this approach will be successful in the near future.

6.3 Open questions and future perspectives

In the course of many years of research, we have learned much about turbulence in fusion plasmas, considered paradigmatic examples of strongly driven systems far from thermodynamic equilibrium. Along with many other workers in the field, we have unveiled an extraordinary richness of behaviour and complexity, highlighting the spontaneous emergence of structure and the essential spatiotemporal nature of turbulence, and its importance for and even occasional dominance in net radial transport, and so forth.

In the course of this book, we have suggested that transport in these complex systems can only be understood if some non-standard mechanisms are incorporated into the mental framework within which we understand transport, namely: non-linearity of the flux–gradient relation (critical gradient mechanisms), long range correlations (zonal flows), long-term memory effects (particle trapping in turbulent eddies, the mutual interaction between turbulence and profiles), minor transport barriers, non-locality, and ballistic transport mechanisms (mode coupling, avalanches, and other effects), etc.

When attempting to clarify these issues, much care should be taken to avoid the so-called 'confirmation bias', which consists of selecting only experimental results that confirm prior theoretical expectations or reporting only modelling results that match specific experimental observations. Avoiding this (usually unconscious) bias is notoriously difficult, particularly when addressing non-standard mechanisms: thus, one sees that observations (from experiments or simulations) are sometimes forced to match descriptions according to the standard transport paradigm, even when the evidence really would suggest otherwise[2]. This may be caused by a lack of awareness that alternative mechanisms exist, or a lack of familiarity with such alternative descriptions. In any case, the main tool to avoid this confirmation bias is simply to be aware that it may exist and to try to keep an open mind.

6.3.1 Expansion of studies

Nevertheless, may issues are still very much open to further study and improved understanding. To mention a few points:

- *Measure more fields.* The fields we have been able to analyze are mostly floating potential and ion saturation current (Langmuir probes), at several points, distributed radially, poloidally, and even toroidally. ECE measurements have provided access to electron temperature fluctuations, at several radial points. Heavy ion beam probes yielded electron density, plasma potential, and other data at several points in the plasma interior. Doppler reflectometry has provided occasional access to electron density fluctuations and perpendicular plasma flow.
- *Provide more measurements in the plasma interior.* Due to the predominance of Langmuir probe data in the presented analyses, much of this book focuses on the plasma edge. More core turbulence data are needed.
- *Measure fluctuating fields simultaneously.* To understand the (causal) interaction between the fluctuating fields, they must be measured simultaneously. For example, particle transport produced by fluctuations does not only depend on the amplitude of the density and electric field fluctuations, but also on their relative phase—as reported in section 5.7. In general though, coordination of the operation of various diagnostic systems is required to ensure that the analysis of the various measurements can be combined to improve our understanding of the turbulent interactions.
- *Measure at several locations.* Since, as we have shown, turbulence contains radially, poloidally, and toroidally extended structures, it is important to measure at several locations simultaneously. Of course, this is a big ask, as it may require the duplication of (very expensive) diagnostic systems. If this is not possible, perhaps the single diagnostic system can be deployed at different (poloidal or toroidal) locations to better understand the spatial variations of turbulent properties (sometimes referred to as the 'ballooning character' of turbulence: its amplitude tends to be larger on the low field side).
- *Expand the use of advanced data analysis techniques.* Much data are already available but are not being analyzed from the perspective of complexity, non-linearity, and causality, which may be facilitated by applying some of the methods described in this book. This is perhaps the easiest point to address: it does not require any investment, apart from time and dedication.

[2] An example: results from turbulence simulations are often summarized in terms of an effective diffusion coefficient, thus rendering any non-diffusive effects that may be present in the results invisible.

- *Analyze simulation data using the same methods.* Turbulence simulations have undoubtedly been improving very significantly, but the analysis of these data is often still based on standard methods (spectra, RMS, linear correlations) that will not reveal the inherent complexity contained in the data. We would suggest applying the methods described in this book, and others that are available or being developed, to highlight complexity and improve understanding.

6.3.2 Turbulence control

Finally, we should mention the opportunities for turbulence control. A deeper understanding of turbulence in fusion plasmas evidently should offer such opportunities, in spite of the inherent complexity. Generally speaking, this is a daunting task in view of the fact that plasmas are (often/usually) self-organized. As we have seen, this leads to the widely observed phenomenon of 'profile consistency' by which the plasma state is relatively insensitive to external variations in, e.g. the heating source. This situation will hold in any self-organized state: namely, the plasma self-regulates towards a preferred end state (close to marginal with regard to the triggering of turbulent transport). Therefore, *a priori*, external manipulation is expected to be hard—unless the intervention is sufficiently large to move the whole system significantly far away from the marginal state. As a consequence, direct feedback on turbulent fluctuations may not be the most effective way to approach this issue [18].

In spite of this, many experiments show that we *are* capable of manipulating plasma confinement, e.g. by modifying the edge radial electric field using biasing induced by probes [19]. Static or slowly modulated biasing has been applied in many devices, including TJ-II, often resulting in a significant modification of confinement [20, 21]. The analyses reported in this book reveal that such biasing has an impact on the turbulent structures, reflected by the modification of the wave–wave coupling or bicoherence (section 2.6.2) and the intermittence (section 5.8). This impact is however limited to a restricted radial zone, possibly related to the underlying magnetic configuration (the presence of low order rational surfaces). Therefore, although further work is needed to clarify this interpretation, it seems that static or slowly modulated edge biasing impacts confinement *indirectly* through the modification of edge zonal flow patterns which regulate turbulence and hence turbulent transport.

Inspired by these results, we would suggest to attempt to feed back the applied biasing voltage on relatively low frequency zonal flow oscillations, e.g. to specifically induce or enhance a zonal flow and its accompanying transport barrier. The zonal flow oscillations could be detected using, e.g. Langmuir probes or a heavy ion beam probe measuring plasma potential, or a Doppler reflectometer measuring the perpendicular plasma rotation velocity. In view of the reported results, it may only be possible to drive or enhance zonal flow oscillations at specific radial locations that are naturally susceptible to this phenomenon. The described approach is of course only feasible in devices with moderate to low performance, as it requires the insertion of a biasing probe. However, if successful, it may inspire an even more ambitious effort to act on turbulence in the hot plasma interior.

Indeed, at several locations in this book, we have argued that minor transport barriers also exist in the interior of the plasma, with their associated zonal flows. It may be possible to act on these minor transport barriers by combining a suitable detection system (e.g. Doppler reflectometry or heavy ion beam probes) with an appropriate drive (e.g. using electron cyclotron resonance current drive or applying torque to the plasma using neutral beam injection) and attempt to enhance specific barriers through feedback. The deliberate establishment and control of internal transport barriers may facilitate the path to fusion as an effective power source.

We hope that this book may contribute to enhancing awareness of the many interesting and complex aspects of turbulence in fusion plasmas and, more generally, strongly driven systems far from thermodynamic equilibrium—an intellectually stimulating field of research, if nothing else. But additionally, from a more practical viewpoint, we believe that a better understanding of turbulence could potentially lead to new ideas for turbulence control and optimization, with benefits for a wide variety of systems in which turbulence is a limiting factor.

References

[1] Birkenmeier G, Ramisch M, Manz P, Nold B and Stroth U 2011 Experimental investigation of the magnetic configuration dependence of turbulent transport *Phys. Rev. Lett.* **107** 025001

[2] Dorland W, Jenko F, Kotschenreuther M and Rogers B N 2000 Electron temperature gradient turbulence *Phys. Rev. Lett.* **85** 5579

[3] Dif-Pradalier G *et al* 2022 Transport barrier onset and edge turbulence shortfall in fusion plasmas *Commun. Phys.* **5** 229

[4] Dif-Pradalier G, Hornung G, Garbet X, Gendrih P, Grandgirard V, Latu G and Sarazin Y 2017 The $E \times B$ staircase of magnetised plasmas *Nucl. Fusion* **57** 066026

[5] Rath F, Peeters A G and Weikl A 2021 Analysis of zonal flow pattern formation and the modification of staircase states by electron dynamics in gyrokinetic near marginal turbulence *Phys. Plasmas* **28** 072305

[6] Lopes Cardozo N J *et al* 1997 Electron thermal transport in RTP: filaments, barriers and bifurcations *Plasma Phys. Control. Fusion* **39** B303

[7] Schilham A M R, Hogeweij G M D and Lopes Cardozo N J 2001 Electron thermal transport barriers in RTP: experiment and modelling *Plasma Phys. Control. Fusion* **43** 1699

[8] Ichiguchi K and Carreras B A 2011 Multi-scale MHD analysis incorporating pressure transport equation for beta-increasing LHD plasma *Nucl. Fusion* **51** 053021

[9] van Milligen B P, Carreras B A, García L, Martin de Aguilera A, Hidalgo C, Nicolau J H and the TJ-II Team 2016 The causal relation between turbulent particle flux and density gradient *Phys. Plasmas* **23** 072307

[10] Tynan G R *et al* 2013 Turbulent-driven low-frequency sheared $E \times B$ flows as the trigger for the H-mode transition *Nucl. Fusion* **53** 073053

[11] Sánchez R and Newman D E 2015 Self-organized criticality and the dynamics of near-marginal turbulent transport in magnetically confined fusion plasmas *Plasma Phys. Control. Fusion* **57** 123002

[12] van Milligen B P, Sánchez R and Newman D E 2004 Probabilistic finite-size transport models for fusion: anomalous transport and scaling laws *Phys. Plasmas* **11** 2272

[13] van Milligen B P, Sánchez R and Newman D E 2004 Uphill transport and the probabilistic transport model *Phys. Plasmas* **11** 3787

[14] Balescu R 2005 *Aspects of Anomalous Transport in Plasmas* (Bristol: Institute of Physics)

[15] del Castillo-Negrete D, Carreras B A and Lynch V E 2004 Fractional diffusion in plasma turbulence *Phys. Plasmas* **11** 3854

[16] Sánchez R, Carreras B A and van Milligen B P 2005 Fluid limit of nonintegrable continuous-time random walks in terms of fractional differential equations *Phys. Rev.* E **71** 011111

[17] del Castillo-Negrete D 2006 Fractional diffusion models of nonlocal transport *Phys. Plasmas* **13** 082308

[18] Uckan T, Richards B, Bengtson R D, Carreras B A, Li L, Hurwitz P D, Rowan W L, Tsui H Y W and Wootton A J 1995 Plasma edge turbulence probing and feedback control and stabilization experiments *Nucl. Fusion* **35** 487–94

[19] Taylor R J, Brown M L, Fried B D, Grote H, Liberati J R, Morales G J, Pribyl P, Darrow D and Ono M 1989 H-mode behavior induced by cross-field currents in a tokamak *Phys. Rev. Lett.* **63** 2365

[20] Pedrosa M A *et al* 2003 Edge turbulence during limiter biasing experiments in the TJ-II stellarator *Czech. J. Phys.* **53** 877

[21] Hidalgo C, Pedrosa M A, García L and Ware A 2004 Experimental evidence of coupling between sheared-flow development and an increase in the level of turbulence in the TJ-II stellarator *Phys. Rev.* E **70** 067402

IOP Publishing

Analysis of Turbulence in Fusion Plasmas

Boudewijn Ph van Milligen and Raul Sanchez

Appendix A

Devices

In this appendix, we briefly summarize the main characteristics of some of the confinement devices mentioned in the book, ordered according to increasing plasma volume.

A.1 TJ-K

TJ-K is a small stellarator of the torsatron type [1], with a major radius $R = 0.6$ m and a minor radius $a = 0.10$ m, corresponding to a total plasma volume of about 0.5 m^3. The magnetic field is produced by a set of coils, among which a helical winding with symmetry $l = 1$, $m = 6$, leading to a rotational transform in the range 0.20–0.30. Plasmas are heated by microwaves at 2.45 and 8 GHz, with a power of $1 - 6$ kW.

Typical plasma parameters are electron density $\sim 5 \times 10^{17}$ m^{-3}, electron temperature $\sim$10 eV, and ion temperature $\sim$1 eV (figure A.1).

TJ-K was constructed at CIEMAT in Madrid, Spain; its original denomination was TJ-IU. In 1999, TJ-K was moved to Kiel, Germany—hence the 'K'—and in 2005 it moved again to the IPF in Stuttgart, Germany. TJ-K disposes of a unique poloidal Langmuir probe array, covering a full poloidal circumference, shown in figure A.2.

A.2 TJ-II

TJ-II is a stellarator of the heliac type with a major radius of $R_0 = 1.5$ m, a minor radius of $a \simeq 0.2$ m, and four field periods [3]. The total plasma volume is about 1 m^3. A photo is shown in figure A.3. Plasmas are fuelled by means of gas puff and optionally pellet injection. They can be heated by electron cyclotron resonant heating (ECRH) and neutral beam injection (NBI).

The ECRH system consists of two gyrotrons with a frequency of 53.2 GHz, allowing the injection of up to 2×300 kW of heating power [4]. The half width of the power deposition profile is $w_{\rm ECRH} \simeq 3$ cm [5].

doi:10.1088/978-0-7503-4856-0ch7

Figure A.1. TJ-K. Image kindly provided by Alf Köhn-Seemann, University of Stuttgart CC BY 4.0.

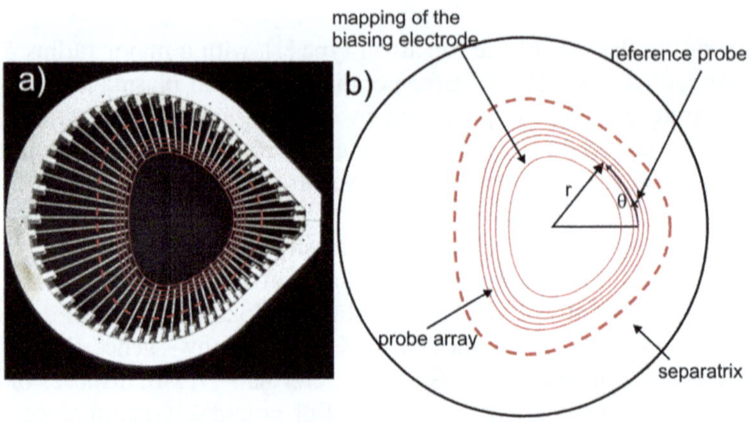

Figure A.2. (a) Photograph of the poloidal probe array. (b) Alignment of the probe array with respect to the flux surfaces of TJ-K. The dashed line depicts the separatrix. (Reproduced from [2], with the permission of AIP Publishing.)

The NBI system consists of two injectors (co and counter with respect to the magnetic field, B_ϕ), with an injected power of up to about 500 kW port-through each.

The TJ-II vacuum magnetic geometry is completely determined by the currents flowing in the external coil sets. The magnetic field is typically adjusted to yield 0.95 T on the magnetic axis at the ECRH injection point, in order to guarantee central absorption of the ECR heating power [6]. In ECRH heated discharges at TJ-II, the normalized pressure $\langle\beta\rangle$ remains rather low and currents flowing inside the

Figure A.3. TJ-II in 2012; front left: Thomson scattering; left and bottom right: the two neutral beam injection systems; top: the inclined structures are two HIBP systems. Also visible on top, up front, is the reciprocating Langmuir probe D. Image credit: © CIEMAT http://www.ciemat.es.

plasma are generally quite small (unless explicitly driven), so that the actual magnetic configuration is typically rather close to the vacuum magnetic configuration [7]. Thus, the external coil currents provide exquisite control of the magnetic configuration in ECRH plasmas.

TJ-II is uniquely prepared to study long range correlations. First, it fitted with a double set of Langmuir probes. There are two reciprocating probe drives, located at toroidal angles $\phi = 38.2°$ (the D probe, entering the plasma from above) and at $\phi = 195°$ (the B probe, entering the plasma from below) [8]. Each drive can be fitted with different probe heads.

Second, it disposes of two heavy ion beam probe (HIBP) systems, located at equivalent toroidal angles of 219° (system I) and 129° (system II) [9]. Each system has various slits and can be operated at a fixed position or scanned through the plasma to obtain plasma profiles (figure A.4).

Finally, TJ-II also disposes of a Doppler reflectometer [10].

A.3 W7-AS

The Wendelstein 7 Advanced Stellarator has been operative from 1988 to 2002, and is the precursor device to the W7-X stellarator (appendix A.4). It was called ‘advanced’ because of the use of modular (shaped, non-planar) field coils. It had a

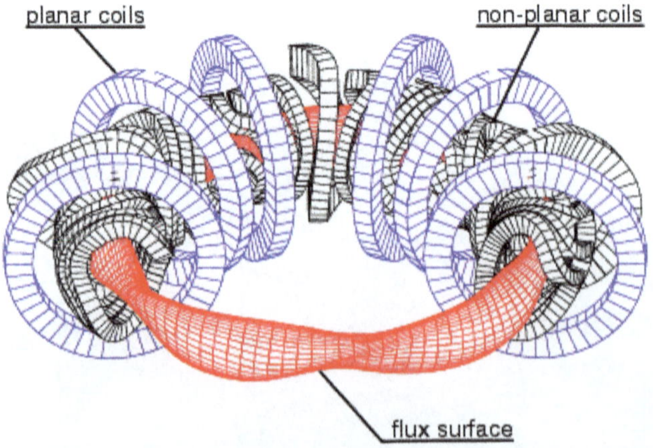

Figure A.4. Sketch of the W7-AS device, showing the field coils and the plasma. Reproduced from [12], with the permission of AIP Publishing.

major radius of 2 m, a minor radius of about 0.18 m, yielding a plasma volume of about 1 m^3 [11].

A.4 W7-X

Wendelstein 7-X is a helical advanced stellarator (HELIAS) with major radius $R = 5.5$ m, minor radius $a \simeq 0.5$ m, 5 field periods, and a toroidal magnetic field of up to ~3 T. The total plasma volume is about 30 m^3, making it the world's largest stellarator. The design of the device is based, among others, on the optimization of neoclassical particle transport and MHD stability [13, 14]. Plasmas are heated using electron cyclotron resonance heating [15], neutral beam injection, and ion cyclotron resonance heating. A repetitive pellet injection system is also available (figure A.5). It disposes of a wide range of diagnostics; in this book, we make use of electron temperature measurements obtained using an electron cyclotron emission diagnostic, covering a large part of the minor radius on both sides of the magnetic axis [16, 17].

A.5 JET

JET is a large tokamak with major radius $R_0 = 2.96$ m and minor radius $a \simeq 1$ m, yielding a total plasma volume of about 100 m^3. Plasmas can be heated and fuelled using various methods, including NBI, ion cyclotron resonance heating, and pellet injection.

In 1997 JET set the a world record for fusion power production, reaching $Q = 0.67$ and producing 16 MW of fusion power while injecting 24 MW of thermal power to heat the fuel [18]. Recently (2022), new D–T experiments have further increased that record [19] (figure A.6).

In this book, we make use of electron temperature measurements from a high resolution electron cyclotron emission radiometer, having 96 radially distributed channels with a radial separation of less than 1 cm [20].

Figure A.5. Wendelstein 7-X in 2011, during construction. Image credit: Max-Planck Institut für Plasmaphysik https://commons.wikimedia.org/w/index.php?curid=25105258, Tino Schulz—Public Relations Department, Max-Planck-Institut, CC BY-SA 3.0.

Figure A.6. The Joint European Torus. Image credit: © the EUROfusion consortium members.

References

[1] Krause N, Lechte C, Stöber J, Stroth U, Ascasibar E, Alonso J and Niedner S 2002 The torsatron TJ-K, an experiment for the investigation of turbulence in a toroidal low-temperature plasma *Rev. Sci. Instrum.* **73** 3474–81

[2] Manz P, Ramisch M and Stroth U 2009 Poloidal mode structure of long-distance correlation of fluctuations under strong $E \times B$ shear in the torsatron TJ-K *Phys. Plasmas* **16** 042309

[3] Alejaldre C *et al* 1999 First plasmas in the TJ-II flexible heliac *Plasma Phys. Control. Fusion* **41** A539

[4] Fernández A *et al* 2009 Performance of the TJ-II ECRH system with the new −80 kV 50 A high voltage power supply *Fusion Eng. Des.* **84** 772

[5] Eguilior S, Castejón F, de la Luna E, Cappa A, Likin K, Fernández A and the TJ-II Team 2003 Heat wave experiments on TJ-II flexible heliac *Plasma Phys. Control. Fusion* **45** 105

[6] van Milligen B P, Pedrosa M A and Hidalgo C 2011 The dynamics of the formation of the edge particle transport barrier at TJ-II *Nucl. Fusion* **51** 113002

[7] van Milligen B P *et al* 2011 Integrated data analysis at TJ-II: the density profile *Rev. Sci. Instrum.* **82** 073503

[8] Grenfell G, van Milligen B P, Losada U, Estrada T, Liu B, Silva C, Spolaore M and Hidalgo C 2020 The impact of edge radial electric fields on edge-SOL coupling in the TJ-II stellarator *Nucl. Fusion* **60** 014001

[9] Melnikov A V *et al* 2017 Heavy ion beam probing-diagnostics to study potential and turbulence in toroidal plasmas *Nucl. Fusion* **57** 072004

[10] Happel T, Estrada T, Blanco E, Tribaldos V, Cappa A and Bustos A 2009 Doppler reflectometer system in the stellarator TJ-II *Rev. Sci. Instrum.* **80** 073502

[11] Hirsch M *et al* 2008 Major results from the stellarator Wendelstein 7-AS *Plasma Phys. Control. Fusion* **50** 053001

[12] Wagner F *et al* 2005 W7-AS: one step of the Wendelstein stellarator line *Phys. Plasmas* **12** 072509

[13] Wanner M, Feist J-H, Renner H, Sapper J, Schauer F, Schneider H, Erckmann V and Niedermeyer H 2001 Design and construction of Wendelstein 7-X *Fusion Eng. Des.* **56-57** 155–62

[14] Wolf R C and The Wendelstein 7-X Team 2008 A stellarator reactor based on the optimization criteria of Wendelstein 7-X *Fusion Eng. Des.* **83** 990

[15] Klinger T *et al* 2017 Performance and properties of the first plasmas of Wendelstein 7-X *Plasma Phys. Control. Fusion* **59** 014018

[16] Marushchenko N B, Dinklage A, Hartfuss H J, Hirsch M, Maassberg H and Turkin Y 2006 Optimization of ECE diagnostics for the W7-X stellarator *Fusion Sci. Technol.* **50** 395

[17] König R *et al* 2015 The set of diagnostics for the first operation campaign of the Wendelstein 7-X stellarator *J. Instrum.* **10** P10002

[18] Keilhacker M, Gibson A, Gormezano C and Rebut P H 2001 The scientific success of JET *Nucl. Fusion* **41** 1925–66

[19] Gibney E 2022 Nuclear-fusion reactor smashes energy record *Nature* **602** 371

[20] de la Luna E *et al* 2004 Electron cyclotron emission radiometer upgrade on the Joint European Torus (JET) tokamak *Rev. Sci. Instrum.* **75** 3831

IOP Publishing

Analysis of Turbulence in Fusion Plasmas

Boudewijn Ph van Milligen and Raul Sanchez

Appendix B

Resistive MHD turbulence model

In this appendix, we briefly summarise the details of the two-fluid resistive magnetohydrodynamic (MHD) model that is used throughout the book to help in understanding and interpreting some experimental results [1–3]. It is based on the reduced MHD equations [4], the dominant instability being pressure gradient modes. The geometry is that of a periodic cylinder, with minor radius a and length $L = 2\pi R_0$. We use a coordinate system (r, θ, ζ), in which r is the radius of the cylindrical surface, θ is the poloidal angle, and $\zeta = z/R_0$, where z is the coordinate along the axis of the cylinder, so ζ is an effective toroidal angle when the cylinder is bent in a torus.

The magnetic field line curvature κ includes the averaged effect of the toroidal and helical components of the magnetic field and is given by

$$\kappa \equiv \frac{r}{R_0} B_0^2 V''.$$

Here, the prime denotes the derivative with respect to toroidal flux, and $V' = \int dl/B$ is the specific volume enclosed by a flux surface.

The four equations of the model are summarized here in their dimensionless form,

$$\frac{\partial \widetilde{\psi}}{\partial t} = \nabla_{\|}\Phi - S\varpi_{*e}\left(\frac{T_{eq}}{n_{eq}}\nabla_{\|}n + \nabla_{\|}T_e\right) + \eta\widetilde{J}_{\|}, \tag{B.1}$$

$$\frac{\partial \widetilde{U}}{\partial t} = -\mathbf{v}_{\perp}\cdot\nabla U + S^2\nabla_{\|}J_{\|} - S^2\frac{\beta_0}{2\varepsilon^2}\kappa\left(\frac{T_{eq}}{n_{eq}}\frac{1}{r}\frac{\partial\widetilde{n}}{\partial\theta} + \frac{1}{r}\frac{\partial\widetilde{T}_e}{\partial\theta}\right) + \mu\nabla_{\perp}^2\widetilde{U}, \tag{B.2}$$

$$\frac{\partial \widetilde{n}}{\partial t} = -\mathbf{v}_{\perp}\cdot\nabla n + \frac{S}{\varpi_{ci}}\nabla_{\|}J_{\|} + D_{\perp}\nabla_{\perp}^2\widetilde{n}, \tag{B.3}$$

$$\frac{\partial \widetilde{T}_e}{\partial t} = -\mathbf{v}_{\perp}\cdot\nabla T_e + \frac{S}{\varpi_{ci}}\frac{T_{eq}}{n_{eq}}\nabla_{\|}J_{\|} + \chi_{\perp}\nabla_{\perp}^2\widetilde{T}_e + \nabla_{\|}\left(\chi_{\|}\nabla_{\|}T_e\right). \tag{B.4}$$

The first equation, describing the evolution of the poloidal magnetic flux ψ, is derived from Faraday's and Ohm's laws. The second one is the momentum balance equation where U is the z component of the vorticity. The other two equations give the evolution of density n and electron temperature T_e. All the quantities appearing in equations (B.1)–(B.4) are decomposed in an equilibrium component f_{eq} and a fluctuating component $\tilde{f}$, so $f = f_{\text{eq}} + \tilde{f}$.

The perpendicular velocity is

$$\mathbf{v}_\perp = -\nabla\Phi \times \mathbf{z}, \tag{B.5}$$

where $\mathbf{z}$ is the unit vector in the toroidal direction. The gradient parallel to the magnetic field $\nabla_\parallel$ is

$$\nabla_\parallel = \frac{\partial}{\partial\zeta} - \frac{1}{q}\frac{\partial}{\partial\theta} + \frac{\partial\tilde{\psi}}{\partial r}\frac{1}{r}\frac{\partial}{\partial\theta} - \frac{1}{r}\frac{\partial\tilde{\psi}}{\partial\theta}\frac{\partial}{\partial r}. \tag{B.6}$$

Here, Φ is the electrostatic potential, the parallel current density is $J_\parallel = \nabla_\perp^2\psi$, and $U = \nabla_\perp^2\Phi$. The resistivity is η, the viscosity is μ, and the perpendicular diffusivity is $D_\perp$. $\chi_\parallel$ and $\chi_\perp$ are the parallel and perpendicular thermal diffusivity, respectively. β_0 is the ratio of the plasma pressure, p, and the magnetic pressure, $B_z^2/2\mu_0$, at the plasma axis: $\beta_0 = 2\mu_0 p(0)/B_z^2$, where B_z is the toroidal magnetic field and μ_0 is the vacuum permeability; τ_R is the resistive time at the magnetic axis, $\tau_R = \mu_0 a^2/\eta(0)$, where $\eta(0)$ is the resistivity at the magnetic axis; and τ_A is the Alfvén time, $\tau_A = R_0\sqrt{\mu_0 m_i n_i}/B_z$, where m_i and n_i are the ion mass and density, respectively. R_0 and a are the major and minor radii, respectively, and the inverse aspect ratio is $\varepsilon = a/R_0$. The Lundquist number $S = \tau_R/\tau_A$. $\overline{\omega}_{*e} = \tau_A\omega_{*e}$, where $\omega_{*e} = T_e/ea^2B_z$ is the electron diamagnetic frequency; $\overline{\omega}_{ci} = \tau_A\omega_{ci}$, where $\omega_{ci} = eB_z/m_i$ is the ion cyclotron frequency.

In the set of equations (B.1)–(B.4) density is normalized to $n_i(0)$, magnetic fields are normalized to B_z, lengths are normalized to minor radius a, and time to τ_R. The electron temperature and resistivity are normalized to their respective values at the magnetic axis.

In the numerical calculations, we represent the different physical variables f by a Fourier series

$$f(r, \theta, \zeta, t) = \sum_{m,n} f_{mn}^{s}(r, t)\sin(m\theta + n\zeta) + \sum_{m,n} f_{mn}^{c}(r, t)\cos(m\theta + n\zeta). \tag{B.7}$$

Nonlinear calculations can be performed in two ways: imposing single or multiple helicity. In single helicity calculations, the only (n, m) values that are included are multiples of the dominant helicity n/m of interest [5]. Figure B.1 shows a snapshot of a typical modelling result obtained with this model in stellarator-like (low shear) cylindrical geometry. The area of the graph corresponds to a region of the poloidal–radial (θ, r) plane at constant toroidal angle (ϕ = constant). The graph shows vortices (trapping regions where the flow roughly moves in circles), such as the poloidally periodic structures seen near $r/a = 0.7$, related to a corresponding

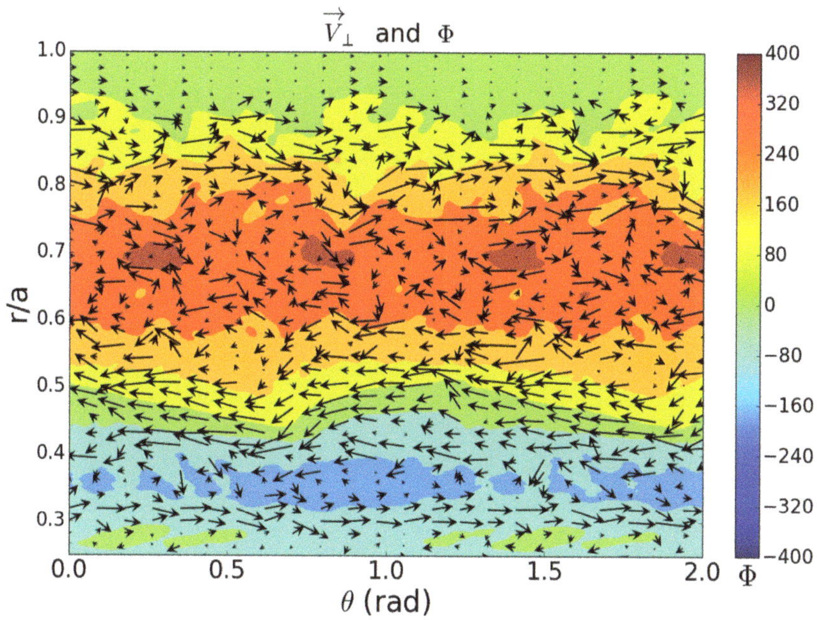

Figure B.1. Modeling results, showing zonal flow regions and vortices.

rational surface. Also visible are zonal flow regions (horizontally elongated structures with predominantly horizontal flow velocities in both directions), on both sides of the vortex sequence.

References

[1] García L, Carreras B A, Lynch V E, Pedrosa M A and Hidalgo C 2001 Sheared flow amplification by vacuum magnetic islands in stellarator plasmas *Phys. Plasmas* **8** 4111

[2] van Milligen B P, Nicolau J H, García L, Carreras B A, Hidalgo C and the TJ-II Team 2017 The impact of rational surfaces on radial heat transport in TJ-II *Nucl. Fusion* **57** 056028

[3] van Milligen B P, Hoefel U, Nicolau J H, Hirsch M, García L, Carreras B A, Hidalgo C and the W7-X Team 2018 Study of radial heat transport in W7-X using the transfer entropy *Nucl. Fusion* **58** 076002

[4] Strauss H R 1976 Nonlinear, three-dimensional magnetohydrodynamics of noncircular tokamaks *Phys. Fluids* **19** 134

[5] Carreras B A, García L, Nicolau J H, van Milligen B P, Hoefel U, Hirsch M and the TJ-II and W7-X Teams 2020 Intermittence and turbulence in fusion devices *Plasma Phys. Control. Fusion* **62** 025011

www.ingramcontent.com/pod-product-compliance
Ingram Content Group UK Ltd.
Pitfield, Milton Keynes, MK11 3LW, UK
UKHW051042120926
8556IPUK00030B/710

9 780750 348546